MATERIALS BEHAVIOR

Research Methodology and
Mathematical Models

MATERIALS BEHAVIOR

Research Methodology and Mathematical Models

Edited by
Mihai Ciocoiu, PhD

A. K. Haghi, PhD, and Gennady E. Zaikov, DSc
Reviewers and Advisory Board Members

Apple Academic Press Inc. | Apple Academic Press Inc.
3333 Mistwell Crescent | 9 Spinnaker Way
Oakville, ON L6L 0A2 | Waretown, NJ 08758
Canada | USA

First issued in paperback 2021

Exclusive worldwide distribution by CRC Press, a member of Taylor & Francis Group

No claim to original U.S. Government works

ISBN 13: 978-1-77463-425-7 (pbk)
ISBN 13: 978-1-77188-737-3 (hbk)

Library and Archives Canada Cataloguing in Publication

Materials behavior : research methodology and mathematical models/edited by Mihai Ciocoiu, PhD; A.K. Haghi, PhD, and Gennady E. Zaikov, DSc, reviewers and advisory board members.

Includes bibliographical references and index.
ISBN 978-1-77188-737-3 (hardcover).--ISBN 978-0-429-45431-8 (PDF)
1. Statistical mechanics. 2. Materials--Testing. 3. Polymers--Testing. 4. Surface chemistry.
5. Molecular dynamics. I. Ciocoiu, Mihai, editor

QC174.8.M38 2015   620.1'10721   C2015-902886-8

Library of Congress Cataloging-in-Publication Data

Materials behavior : research methodology and mathematical models / Mihai Ciocoiu, PhD [editor] ; A.K. Haghi, PhD, and Gennady E. Zaikov, DSc, reviewers and advisory board members.

pages cm
Includes bibliographical references and index.
ISBN 978-1-77188-737-3 (alk. paper)--ISBN 978-0-429-45431-8 (ebook)
1. Statistical mechanics. 2. Materials--Testing. 3. Polymers--Testing. 4. Surface chemis-try. 5. Molecular dynamics. I. Ciocoiu, Mihai. II. Haghi, A. K. III. Zaikov, G. E. (Gennadii Efremovich), 1935-

QC174.8.M355 2015   620.1'10721--dc23   2015015266

Apple Academic Press also publishes its books in a variety of electronic formats. Some content that appears in print may not be available in electronic format. For information about Apple Academic Press products, visit our website at **www.appleacademicpress.com** and the CRC Press website at **www.crcpress.com**

ABOUT THE EDITOR

Mihai Ciocoiu, PhD

Mihai Ciocoiu, PhD, is a Professor of Textiles-Leather and Industrial Management at Gheorghe Asachi Technical University of Iasi, Romania. He is the founder and Editor-In-Chief of the *Romanian Textile and Leather Journal*. He is currently a senior consultant, editor, and member of the academic board of the *Polymers Research Journal* and the *International Journal of Chemoinformatics and Chemical Engineering*.

REVIEWERS AND ADVISORY BOARD MEMBERS

A. K. Haghi, PhD

A. K. Haghi, PhD, holds a BSc in urban and environmental engineering from University of North Carolina (USA); a MSc in mechanical engineering from North Carolina A&T State University (USA); a DEA in applied mechanics, acoustics and materials from Université de Technologie de Compiègne (France); and a PhD in engineering sciences from Université de Franche-Comté (France). He is the author and editor of 65 books as well as 1000 published papers in various journals and conference proceedings. Dr Haghi has received several grants, consulted for a number of major corporations, and is a frequent speaker to national and international audiences. Since 1983, he served as a professor at several universities. He is currently Editor-in-Chief of the *International Journal of Chemoinformatics and Chemical Engineering* and *Polymers Research Journal* and on the editorial boards of many international journals. He is a member of the Canadian Research and Development Center of Sciences and Cultures (CRDCSC), Montreal, Quebec, Canada.

Gennady E. Zaikov, DSc

Gennady E. Zaikov, DSc, is Head of the Polymer Division at the N. M. Emanuel Institute of Biochemical Physics, Russian Academy of Sciences, Moscow, Russia, and Professor at Moscow State Academy of Fine Chemical Technology, Russia, as well as Professor at Kazan National Research Technological University, Kazan, Russia. He is also a prolific author, researcher, and lecturer. He has received several awards for his work, including the Russian Federation Scholarship for Outstanding Scientists. He has been a member of many professional organizations and on the editorial boards of many international science journals.

CONTENTS

List of Contributors ....*xi*

List of Abbreviations ....*xiii*

List of Symbols ....*xv*

Preface ....*xix*

1. **Understanding Modeling and Simulation of Aerogels Behavior: From Theory to Application** 1

   M. Dilamian

2. **Biodegradable Polymer Films on Low Density Polyethylene and Chitosan Basis: A Research Note** 83

   M. V. Bazunova and R. M. Akhmetkhanov

3. **A Detailed Review on Behavior of Ethylene-Vinyl Acetate Copolymer Nanocomposite Materials** 89

   Dhorali Gnanasekaran, Pedro H. Massinga Jr., and Walter W. Focke

4. **The Influence of the Electron Density Distribution in the Molecules of (N)-Aza-Tetrabenzoporphyrins on the Photocatalytic Properties of Their Films** 125

   V. A. Ilatovsky, G. V. Sinko, G. A. Ptitsyn, and G. G. Komissarov

5. **On Fractal Analysis and Polymeric Cluster Medium Model** 149

   G. V. Kozlov, I. V. Dolbin, Jozef Richert, O. V. Stoyanov, and G. E. Zaikov

6. **Polymers as Natural Composites: An Engineering Insight** 161

   G. V. Kozlov, I. V. Dolbin, Jozef Richert, O. V. Stoyanov, and G. E. Zaikov

7. **A Cluster Model of Polymers Amorphous: An Engineering Insight** 209

   G. V. Kozlov, I. V. Dolbin, Jozef Richert, O. V. Stoyanov, and G. E. Zaikov

8. **A Note On Modification of Epoxy Resins by Polyisocyanates** 255

   N. R. Prokopchuk, E. T. Kruts'ko, and F. V. Morev

9. **Trends in Application of Hyperbranched Polymers (HBPs) in the Textile Industry** 263

   Mahdi Hasanzadeh

10. **A Comprehensive Review on Characterization and Modeling of
     Nonwoven Structures**..**271**

     M. Kanafchian

     Index...**329**

LIST OF CONTRIBUTORS

Dhorali Gnanasekaran
Institute of Applied Materials, Department of Chemical Engineering, University of Pretoria, Pretoria 0002, South Africa, Tel.: (+27) 12 420 3728, Fax: (+27) 12 420 2516

E. T. Kruts'ko
Doctor of Technical Sciences, Professor (BSTU), Sverdlova Str.13a, Minsk, Republic of Belarus

F. V. Morev
Postgraduate, Belarusian State Technological University, Sverdlova Str.13a, Minsk, Republic of Belarus

G. A. Ptitsyn
N.N. Semenov Institute of Chemical Physics, Russian Academy of Sciences, 4 Kosygin str., 119991 Moscow, Russia

G. E. Zaikov
N.M. Emanuel Institute of Biochemical Physics of Russian Academy of Sciences, Moscow 119334, Kosygin st., 4, Russian Federation, E-mail: Chembio@sky.chph.ras.ru

G. G. Komissarov
N.N. Semenov Institute of Chemical Physics, Russian Academy of Sciences, 4 Kosygin str., 119991 Moscow, Russia; E-mail: gkomiss@yandex.ru; komiss@chph.ras.ru

G. V. Kozlov
Kabardino-Balkarian State University, Nal'chik – 360004, Chernyshevsky st., 173, Russian Federation

G. V. Sinko
N.N. Semenov Institute of Chemical Physics, Russian Academy of Sciences, 4 Kosygin str., 119991 Moscow, Russia

I. V. Dolbin
Kabardino-Balkarian State University, Nal'chik – 360004, Chernyshevsky st., 173, Russian Federation, E-mail: I_dolbin@mail.ru

Jozef Richert
Institut Inzynierii Materialow Polimerowych I Barwnikow, 55 M. Sklodowskiej-Curie str., 87-100 Torun, Poland, E-mail: j.richert@impib.pl

M. Dilamian
University of Guilan, Rasht, Iran

M. Kanafchian
University of Guilan, Rasht, Iran

M. V. Bazunova
Bashkir State University, 32 ZakiValidi Street, 450076 Ufa, Republic of Bashkortostan, Russia, Tel.: (347) 2299686; Fax: (347) 2299707; E-mail: mbazunova@mail.ru

Mahdi Hasanzadeh
Department of Textile Engineering, University of Guilan, Rasht, Iran; E-mail: hasanzadeh_mahdi@yahoo.com

N. R. Prokopchuk

Corresponding Member of National Academy of Sciences of Belarus, Doctor of Chemical Sciences, Professor, Head of Department (BSTU), Sverdlova Str.13a, Minsk, Republic of Belarus, E-mail: prok_nr@mail.by

O. V. Stoyanov

Kazan National Research Technological University, Kazan, Tatarstan, Russia, E-mail: OV_Stoyanov@mail.ru

Pedro H. Massinga Jr.

Universidade Eduardo Mondlane, Faculdade de Ciências, Campus Universitário Principal, Av. Julius Nyerere, PO Box 257, Maputo, Moçambique

R. M. Akhmetkhanov

Bashkir State University, 32 ZakiValidi Street, 450076 Ufa, Republic of Bashkortostan, Russia, Tel.: (347) 2299686; Fax: (347) 2299707

V. A. Ilatovsky

N.N. Semenov Institute of Chemical Physics, Russian Academy of Sciences, 4 Kosygin str., 119991 Moscow, Russia

Walter W. Focke

Institute of Applied Materials, Department of Chemical Engineering, University of Pretoria, Pretoria 0002, South Africa, Tel.: (+27) 12 420 3728, Fax: (+27) 12 420 2516

LIST OF ABBREVIATIONS

ABS	Acrylonitrile–Butadiene–Styrene
ANOVA	Analysis of Variance
BPE	Branched Polyethylenes
CCD	Central Composite Design
CD	Cross-Direction
CNT	Classical Nucleation Theory
CV	Coefficient of Variation
CSC	Crystallites with Stretched Chains
DSC	Differential Scanning Calorimetry
EDANA	European Disposables and Nonwovens Association
EP	Epoxy Polymer
EVA	Ethylene-co-Vinyl Acetate
FH	Fluorohectorite
FOD	Fiber Orientation Distribution
FT	Fourier Transform
HBP	Hyper Branched Polymer
HRR	Heat Release Rate
HRTEM	High Resolution Transmission Electron Microscopy
HT	Hectorite
HT	Hough Transform
I(e)	Informational Entropy
IP	Inclined Plates
IRDP	Institutional Research Development Programme
LDHs	Layered Double Hydroxides
LDPE	Low Density Polyethylene
LOI	Loss on Ignition
MC	Monte Carlo
MD	Machine Direction
MD	Molecular Dynamics
MFI	Melt Flow Index
MMT	Montmorillonite
NRF	National Research Foundation
NSMs	Nano Structured Materials
PA	Polyurethane
PAr	Polyarylate
Pc	Phthalo Cyanines

PC	Polycarbonate
PET	Poly(ethylene terephthalate)
PGD	Pores Geometry Distribution
PHRR	Peak of Heat Release
PMMA	Poly(methyl methacrylate)
POSS	Polyhedral Oligomeric Silse Squioxaneo
PP	Polypropylene
PVD	Pore Volume Distributions
REP	Rarely Cross-Linked Epoxy Polymer
RSM	Response Surface Methodology
SEM	Scanning Electron Microscope
SR	Smoke Release
TBP	Tetrabenzoporphyrin
TEM	Transmission Electron Microscopy
TGA	Thermogravimetric Analysis
THR	Total Heat Release
TPC	Tetra Pyrrole Compounds
TPP	Tetraphenyl Porphyrin
TTI	Time to Ignition
VA	Vinyl Acetate
WL	Weight Loss
0DNSM	Zero-Dimensional Nanostructured Materials
1DNSM	One-Dimensional Nanostructured Materials
2DNSM	Two-Dimensional Nanostructured Materials

LIST OF SYMBOLS

a	the acceleration
a and b	integers
a_i	the acceleration of particle i
b	Burgers vector
c	speed of light in m/s
C_∞	characteristic ratio
d	dimension of Euclidean space
D_p	nanofiller particles diameter in nm
d_{surf}	nanocluster surface fractal dimension
d_u	fractal dimension of accessible for contact ("nonscreened") indicated particle surface
d_w	dimension of random walk
E	the potential energy of the system
E_a	the distance from the surface acceptor level to the E_v
E_n and E_m	elasticity moduli of nanocomposites and matrix polymer, respectively
F	the force exerted on the particle
F_i	the force exerted on particle i
F_s	the distance from the Fermi level at the surface to E_v
G	shear modulus
G_∞	equilibrium shear modulus
G_c, G_m and G_f	shear moduli of composite, polymer matrix and filler, respectively
G_{cl}	the shear modulus
h	Planck constant
I	the scattering intensity
I_0	a reference value of intensity
I_{ph}	photocurrent in μA
k	Boltzmann constant
K_s	stress concentration coefficient
K_T	bulk modulus
L	filler particle size
l_0	main chain skeletal length
l_k	specific spatial scale of structural changes
l_{st}	statistical segment length
m	the mass
M	the total sampling number

m and n	exponents in the Mie equation
$m_{absorbed\ water}$	weight of the saturated condensed vapors of volatile liquid, g
M_{cl}	molecular weight of the chain part between cluster
M_e	molecular weight of chain part between entanglements
m_i	the mass of particle i
m_{sample}	weight of dry sample, g
N	the number of atoms in the system
N_A	Avogadro number
n_{cl}	statistical segments number per one nanocluster
N_α and N_β	the numbers of particles of the entities of type α and β, respectively
p	solid-state component volume fraction
P_c	percolation threshold
q	the parameter
q	the wave number
Q_1 and Q_2	the charges
R	a hydrogen atom or an organic group
r	the position
R	universal gas constant
r_{ij}	the distance between a pair of atoms i and j
r^N	the complete set of 3N atomic coordinates
S	macromolecule cross-sectional area
T, T_g and T_m	testing, glass transition and melting temperatures, respectively
$u(r)$	an externally applied potential field
v	the velocity
V	the volume of the system
W	absorbed light power W
w	activation energy of the transition to the charged form
W_n	nanofiller mass contents in mas.%,
Z_i	the effective charge of the i-th ion

Greek Symbols

$f_\alpha^{(0)}$	the equilibrium distribution
$\langle\rangle$	ensemble average
σ_f^n	nominal (engineering) fracture stress
σ_f^c and σ_f^m	fracture stress of composite and polymer matrix, respectively
a	the efficiency constant
α_i	the electric polarizability of the i-th ion
β	coefficient
β_p and ν_p	critical exponents (indices) in percolation theory
ΔS	entropy change in this process course
ε	misfit strain arising from the difference in lattice parameters

ε_0	the permittivity of free space
ε_f	strain at fracture
ε_Y	the yield strain
η	exponent
J	total concentration of adsorbed molecules
l	wavelength m
λ_b	the smallest length of acoustic irradiation sequence
λ_k	length of irradiation sequence
n	Poisson's ratio
ν_{cl}	cluster network density
ν_p	correlation length index in percolation theory
r	nanofiller (nanoclusters) density
ρ	polymer density
ρ_{cl}	the nanocluster density
ρ_d	the density if linear defects
ρ_α and ρ_β	the corresponding densities of α and β subsystems
τ	the relaxation time (dimensionless)
τ_{in}	the initial internal stress
t_{IP}	the shear stress in IP (cluster)
φ_n	nanofiller volume contents
c	the relative fraction of elastically deformed polymer
Γ	Eiler gamma-function

PREFACE

This book covers a wide variety of recent research on advanced materials and their applications. It provides valuable engineering insights into the developments that have lead to many technological and commercial developments.

This book also covers many important aspects of applied research and evaluation methods in chemical engineering and materials science that are important in chemical technology and in the design of chemical and polymeric products. This book gives readers a deeper understanding of physical and chemical phenomena that occur at surfaces and interfaces. Important is the link between interfacial behavior and the performance of products and chemical processes. Helping to fill the gap between theory and practice, this book explains the major concepts of new advances in high performance materials and their applications.

This book has an important role in advanced materials in macro and nanoscale. Its aim is to provide original, theoretical, and important experimental results that use nonroutine methodologies often unfamiliar to the usual readers. It also includes chapters on novel applications of more familiar experimental techniques and analyzes of composite problems that indicate the need for new experimental approaches.

CHAPTER 1

UNDERSTANDING MODELING AND SIMULATION OF AEROGELS BEHAVIOR: FROM THEORY TO APPLICATION

M. DILAMIAN

University of Guilan, Rasht, Iran

CONTENTS

Abstract 2
1.1   Theory 2
1.2   Applications 35
1.3   Conclusion 75
Keywords 77
References 77

ABSTRACT

A deeper understanding of phenomena on the microscopic scale may lead to completely new fields of application. As a tool for microscopic analysis, molecular simulation methods such as the molecular dynamics (MD), Monte Carlo (MC) methods have currently been playing an extremely important role in numerous fields, ranging from pure science and engineering to the medical, pharmaceutical, and agricultural sciences. MC methods exhibit a powerful ability to analyze thermodynamic equilibrium, but are unsuitable for investigating dynamic phenomena. MD methods are useful for thermodynamic equilibrium but are more advantageous for investigating the dynamic properties of a system in a nonequilibrium situation. The importance of these methods is expected to increase significantly with the advance of science and technology. The purpose of this study is to consider the most suitable method for modeling and characterization of aerogels. Initially, giving an introduction to the Molecular Simulations and its methods help us to have a clear vision of simulating a molecular structure and to understand and predict properties of the systems even at extreme conditions. Considerably, molecular modeling is concerned with the description of the atomic and molecular interactions that govern microscopic and macroscopic behaviors of physical systems. The connection between the macroscopic world and the microscopic world provided by the theory of statistical mechanics, which is a basic of molecular simulations. There are numerous studies mentioned the structure and properties of aerogels and xerogels via experiments and computer simulations. Computational methods can be used to address a number of the outstanding questions concerning aerogel structure, preparation, and properties. In a computational model, the material structure is known exactly and completely, and so structure/property relationships can be determined and understood directly. Techniques applied in the case of aerogels include both "mimetic" simulations, in which the experimental preparation of an aerogel is imitated using dynamical simulations, and reconstructions, in which available experimental data is used to generate a statistically representative structure. In this section, different simulation methods for modeling the porous structure of silica aerogels and evaluating its structure and properties have been mentioned. Many works in the area of simulation have been done on silica aerogels to better understand these materials. Results from different studies show that choosing a suitable potential leads to a more accurate aerogel model in the other words if the interatomic potential does not accurately describe the interatomic interactions, the simulation results will not be representative of the actual material.

1.1   THEORY

1.1.1   INTRODUCTION

The idea of using molecular dynamics (MD) for understanding physical phenomena goes back centuries. Computer simulations are hopefully used to understand the properties of assemblies of molecules in terms of their structure and the microscopic interactions between them. This serves as a complement to conventional experiments, enabling us to learn something new, something that cannot be found out in other ways. The main concept of molecular simulations for a given intermolecular "exactly" predict the thermodynamic (pressure, heat capacity, heat of adsorption, structure) and transport (diffusion coefficient, viscosity) properties of the system. In some cases, experiment is impossible (inside of stars weather forecast), too dangerous (flight simulation explosion simulation), expensive (high pressure simulation wind channel simulation), and blind (Some properties cannot be observed on very short time-scales and very small space-scales). The two main families of simulation technique are MD and Monte Carlo (MC); additionally, there is a whole range of hybrid techniques, which combine features from both. In this lecture we shall concentrate on MD. The obvious advantage of MD over MC is that it gives a route to dynamical properties of the system: transport coefficients, time-dependent responses to perturbations, rheological properties and spectra. Computer simulations act as a bridge Fig. 1.1) between microscopic length and time scales and the macroscopic world of the laboratory: we provide a guess at the interactions between molecules, and obtain 'exact' predictions of bulk properties. The predictions are 'exact' in the sense that they can be made as accurate as we like, subject to the limitations imposed by our computer budget. At the same time, the hidden detail behind bulk measure ments can be revealed. An example is the link between the diffusion coefficient and velocity autocorrelation function (the former easy to measure experimentally, the latter much harder). Simulations act as a bridge in another sense: between theory and experiment. We may test a theory by conducting a simulation using the same model. We may test the model by comparing with experimental results. We may also carry out simulations on the computer that are difficult or impossible in the laboratory (e.g., working at extremes of temperature or pressure) [1].

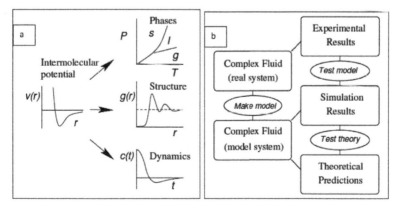

FIGURE 1.1   Simulations as a bridge between (a) microscopic and macroscopic; (b) theory and experiments.

The purpose of Molecular Simulations is described as below:

a.   Mimic the real world:
   • predicting properties of (new) materials;
   • computer 'experiments' at extreme conditions (Carbon phase behavior at very high pressure and temperature);
   • understanding phenomena on a molecular scale (protein conformational change with MD, empirical potential, including bonds, angles dihedrals).

b.   Model systems:
   • test theory using same simple model;
   • explore consequences of model;
   • explain poorly understood phenomena in terms of essential physics.

Molecular scale simulations are usually accomplished in three stages by developing a molecular model, calculating the molecular positions, velocities and trajectories, and finally collecting the desired properties from the molecular trajectories. It is the second stage of this process that characterizes the simulation method. For MD, the molecular positions are deterministically generated from the Newtonian equations of motion. In other methods, for instance the MC method, the molecular positions are generated randomly by stochastic methods. Some methods have a combination of deterministic and stochastic features. It is the degree of this determinism that distinguishes between different simulation methods [43].

In other words, MD simulations are in many respects very similar to real experiments. When we perform a real experiment, we proceed as follows. We prepare a sample of the material that we wish to study. We connect this sample to a measuring instrument (e.g., a thermometer, nometer, or viscosimeter), and we measure the property of interest during a certain time interval. If our measurements are subject to statistical noise (like most of the measurements), then the longer we average, the more accurate our measurement becomes. In a MD simulation, we follow exactly the same approach. First, we prepare a sample: we select a model system consisting of N particles and we solve Newton's equations of motion for this system until the properties of the system no longer change with time (we equilibrate the system). After equilibration, we perform the actual measurement. In fact, some of the most common mistakes that can be made when performing a computer experiment are very similar to the mistakes that can be made in real experiments (e.g., the sample is not prepared correctly, the measurement is too short, the system undergoes an irreversible change during the experiment, or we do not measure what we think) [5].

1.1.2   HISTORICAL BACKGROUND

Before computer simulation appeared on the scene, there was only one-way to predict the properties of a molecular substance, namely by making use of a theory that provided an approximate description of that material. Such approximations are inevitable precisely because there are very few systems for which the equilibrium properties can be computed exactly (e.g., the ideal gas, the harmonic crystal, and a number of lattice models, such as the two-dimensional Ising model for ferromagnets). As a result, most properties of real materials were predicted on the basis of approximate theories (e.g., the van der Waals equation for dense gases, the Debye-Huckel theory for electrolytes, and the Boltzmann equation to describe the transport properties of dilute gases).

Given sufficient information about the intermolecular interactions, these theories will provide us with an estimate of the properties of interest. Unfortunately, our knowledge of the intermolecular interactions of all but the simplest molecules is also quite limited. This leads to a problem if we wish to test the validity of a particular theory by comparing directly to experiment. If we find that theory and experiment disagree, it may mean that our theory is wrong, or that we have an incorrect estimate of the intermolecular interactions, or both. Clearly, it would be very nice if we could obtain essentially exact results for a given model system without having to rely on approximate theories. Computer simulations allow us to do precisely that. On the one hand, we can now compare the calculated properties of a model system with those of an experimental system: if the two disagree, our model is inadequate; that is, we have to improve on our estimate of the intermolecular interactions. On the other hand, we can compare the result of a simulation of a given model system with the predictions of an approximate analytical theory applied to the same model. If we now find that theory and simulation disagree, we know that the *theory* is flawed. So, in this case, the computer simulation plays the role of the experiment designed to test the theory. This method of screening theories before we apply them to the real world is called a *computer experiment.* This application of computer simulation is of tremendous importance. It has led to the revision of some very respectable theories, some of them dating back to Boltzmann. And it has changed the way in which we construct new theories. Nowadays it is becoming increasingly rare that a theory is applied to the real world before being tested by computer simulation. But note that the computer as such offers us no understanding, only numbers. And, as in a real experiment, these numbers have statistical errors. So what we get out of a simulation is never directly a theoretical relation. As in a real experiment, we still have to extract the useful information [29].

The early history of computer simulation illustrates this role of computer simulation. Some areas of physics appeared to have little need for simulation because very good analytical theories were available (e.g., to predict the properties of dilute gases or of nearly harmonic crystalline solids). However, in other areas, few if any exact theoretical results were known, and progress was much hindered by the lack of unambiguous tests to assess the quality of approximate theories. A case in point was the theory of dense liquids. Before the advent of computer simulations, the only way to model liquid was by mechanical simulation [3–5] of large assemblies of macroscopic spheres (e.g., ball bearings). Then the main problem becomes show to arrange these balls in the same way as atoms in a liquid. Much work on this topic was done by the famous British scientist J. D. Bernal, who built and analyzed such mechanical models for liquids.

The first proper MD simulations were reported in 1956 by Alder and Wainwright at Livermore, who studied the dynamics of an assembly of hard spheres. The first MD simulation of a model for a "real" material was reported in 1959 (and published in 1960) by the group led by Vineyard at Brookhaven, who simulated radiation damage in crystalline Cu. The first MD simulation of a real liquid (argon) was reported in 1964 by Rahman at Argonne. After that, computers were increasingly becoming available to scientists outside the US government labs, and the practice of simulation started spreading to other continents. Much of the methodology of computer simulations has been developed since then, although it is fair to say that the basic algorithms for MC and MD have hardly changed since the 1950s. The most common application of computer simulations is to predict the properties of materials. The need for such simulations may not be immediately obvious [29].

1.1.3   MOLECULAR DYNAMIC: INTERACTIONS AND POTENTIALS

The concept of the MD method is rather straightforward and logical. The motion of molecules is generally governed by Newton's equations of motion in classical theory. In MD simulations, particle motion is simulated on a computer according to the equations of motion. If one molecule moves solely on a classical mechanics level, a computer is unnecessary because mathematical calculation with pencil and paper is sufficient to solve the motion of the molecule. However, since molecules in a real system are numerous and interact with each other, such mathematical analysis is impracticable. In this situation, therefore, computer simulations become a powerful tool for a microscopic analysis [76]. MD simulation consists of the numerical, step-by-step, solution of the classical equations of motion, which for a simple atomic system may be written as,

$$m_i \ddot{r_i} = f_i \qquad (1)$$

$$f_i = -\frac{\partial}{\partial_{r_i}} v \qquad (2)$$

For this purpose we need to be able to calculate the forces fi acting on the atoms, and these are usually derived from a potential energy U (rN), where rN = (r1; r2; rN) represents the complete set of 3N atomic coordinates [5].

The energy E is a function of the atomic positions, R, of all the atoms in the system, these are usually expressed in term of Cartesian coordinates. The value of the energy is calculated as a sum of internal, or bonded, terms E-bonded, which describe the bonds, angles and bond rotations in a molecule, and a sum of external or nonbonded terms, Enonbonded, These terms account for interactions between nonbonded atoms or atoms separated by 3 or more covalent bonds.

$$V(R) = E_{bonded} + E_{non-bonded} \tag{3}$$

1.1.3.1 NON-BONDED INTERACTIONS

There are two potential functions we need to be concerned about between non-bonded atoms:
- Van der Waals Potential
- Electrostatic Potential

The energy term representing the contribution of nonbonded interactions in the CHARMM potential function has two components, the Van der Waals interaction energy and the electrostatic interaction energy. Some other potential functions also include an additional term to account for hydrogen bonds. In the CHARMM potential energy function, these interactions are account for by the electrostatic and Van der Waals interactions.

$$E_{non-bonded} = E_{van-dar-waals} + E_{electrostactic} \tag{4}$$

The van der Waals interaction between two atoms arises from a balance between repulsive and attractive forces. The repulsive force arises at short distances where the electron-electron interaction is strong. The attractive force, also referred to as the dispersion force, arises from fluctuations in the charge distribution in the electron clouds. The fluctuation in the electron distribution on one atom or molecules gives rise to an instantaneous dipole, which in turn, induces a dipole in a second atom or molecule-giving rise to an attractive interaction. Each of these two effects is equal to zero at infinite atomic separation r and become significant as the distance decreases. The attractive interaction is longer range than the repulsion but as the distance become short, the repulsive interaction becomes dominant. This gives rise to a minimum in the energy. Positioning of the atoms at the optimal distances stabilizes the system. Both value of energy at the minimum E* and the optimal separation of atoms r* (which is roughly equal to the sum of Van der Waals radii of the atoms) depend on chemical type of these atoms (Fig. 1.2).

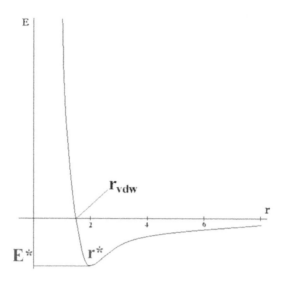

FIGURE 1.2   Potential energy vs. atomic distances.

The van der Waals interaction is most often modeled using the Lennard-Jones 6–12 potential which expresses the interaction energy using the atom-type dependent constants A and C [Eq (5), Fig. 1.3]. Values of A and C may be determined by a variety of methods, like nonbonding distances in crystals and gas-phase scattering measurements.

$$E_{van-der-waals} = \sum_{nonbonded\ pairs} \left(\frac{A_{ik}}{r_{ik}^{12}} - \frac{C_{ik}}{r_{ik}^{6}} \right) \tag{5}$$

The part of the potential energy $U_{nonbonded}$ representing nonbonded interactions between atoms is traditionally split into 1-body, 2-body, 3-body, etc., terms:

$$u_{non-bonded}\left(r^{N}\right) = \sum_{i} u(r_{i}) + \sum_{i}\sum_{j>i} v(r_{i}r_{j}) \tag{6}$$

The $u(r)$ term represents an externally applied potential field or the effects of the container walls; it is usually dropped for fully periodic simulations of bulk systems. Also, it is usual to concentrate on the pair potential $v\ (r_{i};\ r_{j}) = v\ (r_{ij})$ and neglect three-body (and higher order) interactions. There is an extensive literature on the way these potentials are determined experimentally, or modeled theoretically [1–4]. In some simulations of complex fluids, it is sufficient to use the simplest models that faithfully represent the essential physics. In this chapter we shall concentrate on continuous, differentiable pair-potentials (although discontinuous potentials such as hard spheres and spheroids have also played a role [5].

The most common model that describes matter in its different forms is a collection of spheres that we call "atoms" for brevity. These "atoms" can be a single atom such as Carbon (C) or Hydrogen (H) or they can represent a group of atoms such as CH_2 or CS_2. These spheres can be connected together to form larger molecules. The interactions between these atoms are governed by a force potential that maintains the integrity of the matter and prevents the atoms from collapsing. The most commonly used potential that was first used for liquid argon [57], is the Lennard-Jones potential [44]. This potential has the following general form:

$$\varnothing_{LJ}(r) = A\varepsilon\left[\left(\frac{\sigma}{r_{ij}}\right)^m - \left(\frac{\sigma}{r_{ij}}\right)^n\right] \tag{7}$$

where $r_{ij} = r_i - r_j$ is the distance between a pair of atoms i and j. This potential has a short-range repulsive force that prevents the atoms from collapsing into each other and also a longer-range attractive tail that prevents the disintegration of the atomic system. Parameters m and n determine the range and the strength of the attractive and repulsive forces applied by the potential where normally m is larger than n. The common values used for these parameters are $m - 12$ and $n = 6$. The constant A depends on m and n and with the mentioned values, it will be $A = 4$. The result is the well-known 6–12 Lennard-Jones potential. Two other terms in the potential, namely ε and σ, are the energy and length parameters. In the case of a molecular system several interaction sites or atoms are connected together to form a long chain, ring or a molecule in other forms. In this case, in addition to the mentioned Lennard-Jones potential, which governs the Intermolecular potential, other potentials should be employed. An example of an intermolecular potential is the torsional potential, which was first introduced by Ryckaert and Bellemans [43, 74]. The electrostatic interaction between a pair of atoms is represented by Coulomb potential; D is the effective dielectric function for the medium and r is the distance between two atoms having charges q_i and q_k.

$$E_{electrostatic} = \sum_{nonbonded\ pairs} \frac{q_i q_k}{Dr_{ik}} \tag{8}$$

For applications in which attractive interactions are of less concern than the excluded volume effects, which dictate molecular packing, the potential may be truncated at the position of its minimum, and shifted upwards to give what is usually termed the WCA model. If electrostatic charges are present, we add the appropriate Coulomb potentials:

$$v^{coulomb}(r) = \frac{Q_1 Q_2}{4\pi\varepsilon_0 r} \tag{9}$$

where Q_1, Q_2 are the charges and ε_0 is the permittivity of free space. The correct handling of long-range forces in a simulation is an essential aspect of polyelectrolyte simulations.

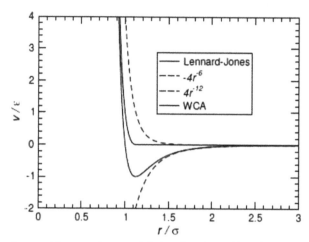

FIGURE 1.3   Lennard-Jones pair potential showing the r–12 and r–6 contributions. Also shown is the WCA shifted repulsive part of the potential [43].

1.1.3.2   BONDING POTENTIALS

For molecular systems, we simply build the molecules out of site-site potentials of the form of Eq. (7) or similar. Typically, a single-molecule quantum-chemical calculation may be used to estimate the electron density throughout the molecule, which may then be modeled by a distribution of partial charges via Eq. (9), or more accurately by a distribution of electrostatic multipoles [3, 81]. For molecules we must also consider the intermolecular bonding interactions. The simplest molecular model will include terms of the following kind:

The E-bonded term is a sum of three terms:

$$E_{bonded} = E_{bond-strech} + E_{angel-bend} + E_{rotate-along-bond} \tag{10}$$

which correspond to three types of atom movement:
- stretching along the bond;
- bending between bonds;
- rotating around bonds.

The first term in the above equation is a harmonic potential representing the interaction between atomic pairs where atoms are separated by one covalent bond, that is, 1,2-pairs. This is the approximation to the energy of a bond as a function of displacement from the ideal bond length, b_0. The force constant, K_b, determines the strength of the bond. Both ideal bond lengths b_0 and force constants K_b are specific for each pair of bound atoms, that is, depend on chemical type of atoms-constituents.

$$E_{bond-strach} = \sum_{1,2\ pairs} K_b (b - b_0)^2 \tag{11}$$

Values of force constant are often evaluated from experimental data such as infrared stretching frequencies or from quantum mechanical calculations. Values of bond length can be inferred from high-resolution crystal structures or microwave spectroscopy data. The second term in above equation is associated with alteration of bond angles theta from ideal values q_0, which is also represented by a harmonic potential. Values of q_0 and K_q depend on chemical type of atoms constituting the angle. These two terms describe the deviation from an ideal geometry; effectively, they are penalty functions and that in a perfectly optimized structure, the sum of them should be close to zero.

$$E_{bond-bend} = \sum_{angles} K_\theta (\theta - \theta_0)^2 \tag{12}$$

The third term represents the torsion angle potential function which models the presence of steric barriers between atoms separated by 3 covalent bonds (1,4 pairs). The motion associated with this term is a rotation, described by a dihedral angle and coefficient of symmetry $n = 1, 2, 3$), around the middle bond. This potential is assumed to be periodic and is often expressed as a cosine function. In addition to these terms, the CHARMM force field has two additional terms; one is the Urey-Bradley term, which is an interaction based on the distance between atoms separated by two bonds (1,3 interaction). The second additional term is the improper dihedral term (see the section on CHARMM), which is used to maintain chirality and planarity.

$$E_{rotate-along-bond} = \sum_{1,4\ pairs} K_\varnothing (1 - \cos(n\varnothing)) \tag{13}$$

The parameters for these terms, K_b, K_q, K_f, are obtained from studies of small model compounds and comparisons to the geometry and vibrational spectra in the gas phase (IR and Raman spectroscopy), supplemented with ab initio quantum calculations [5].

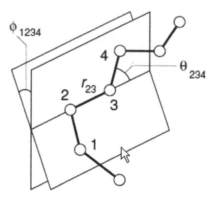

FIGURE 1.4   Geometry of a simple chain molecule, illustrating the definition of interatomic distance r, bend angle θ, and torsion angle ⌀ [5].

$$u_{intramolecular} = \frac{1}{2}\sum_{bonds} k_{ij}^r \left(r_{ij} - r_{eq} \right)^2 \tag{14a}$$

$$+\frac{1}{2}\sum k_{ijk}^\theta \left(\theta_{ijk} - \theta_{eq} \right)^2 \tag{14b}$$

$$+\frac{1}{2}\sum_{\substack{torsion \\ angles}} \sum_m k_{ijkl}^{\varnothing,m} (1+\cos\left(m\varnothing_{ijkl} - \gamma_m \right)) \tag{14c}$$

The geometry is illustrated in Fig. 1.4. The "bonds" will typically involve the separation $r_{ij} = \left| r_i - r_j \right|$ between adjacent pairs of atoms in a molecular framework, and we assume in Eq (14a) above a harmonic form with specified equilibrium separation, although this is not the only possibility. The "bend angles" θ_{ijk} are between successive bond vectors such as $r_i - r_j$ and $r_j - r_k$, and therefore involve three atom coordinates:

$$\cos\theta_{ijk} = \hat{r}_{ij}.\hat{r}_{jk} = \left(r_{ij}.r_{ij} \right)^{-\frac{1}{2}} \left(r_{jk}.r_{jk} \right)^{-\frac{1}{2}} \left(r_{ij}.r_{jk} \right) \tag{15}$$

where $\hat{r} = r/r$. Usually this bending term is taken to be quadratic in the angular displacement from the equilibrium value, as in Eq (14b), although periodic functions are also used. The "torsion angles" $\varnothing_{ijk}$ are defined in terms of three connected bonds, hence four atomic coordinates: [5]

$$\cos\varnothing_{ijkl} = -\hat{n}_{ijk} .\hat{n}_{jkl}, \; where \; n_{ijk} = r_{ij} \times r_{jk}, \; n_{jkl} = r_{jk} \times r_{kl} \tag{16}$$

and $\hat{n} = n/n$, the unit normal to the plane dened by each pair of bonds. Usually the-torsional potential involves an expansion in periodic functions of order m = 1,2,... (see, Eq (14c)). A simulation package force field will specify the precise form of Eq. (14), and the various strength parameters k and other constants therein. Actually, Eq. (14) is a considerable oversimplification. Molecular mechanics force fields, aimed at accurately predicting structures and properties, will include many cross-terms (e.g., stretch-bend): MM3 [7,53] andMM4 [7] are examples. Quantum mechanical calculations may give a guide to the "best" molecular force field; also comparison of simulation results with thermophysical properties and vibration frequencies is invaluable in force-field development and refinement. A separate family of force fields, such as AMBER [8], CHARMM [15] and OPLS [45] are geared more to larger molecules (proteins, polymers) in condensed phases; their functional form is simpler, closer to that of Eq. (14), and their parameters are typically determined by quantum chemical calculations combined with thermophysical and phase coexis-tence data. This field is too broad to be reviewed here; several molecular modeling texts (albeit targeted at biological applications) should be consulted by the interested reader. The modeling of long chain molecules will be of particular interest to us, especially as an illustration of the scope for progressively simplifying and "coarse-graining" the potential model. Various explicit-atom potentials have been devised for the n-alkanes. More approximate potentials have also been constructed [26, 28] in which the CH_2 and CH_3 units are represented by single "united atoms." These potentials are typically less accurate and less transferable than the explicit-atom po-tentials, but significantly less expensive; comparisons have been made between the two approaches [29]. For more complicated molecules this approach may need to be modified. In the liquid crystal field, for instance, a compromise has been suggested [32]: use the united-atom approach for hydrocarbon chains, but model phenyl ring hydrogens explicitly.

In polymer simulations, there is frequently a need to economize further and coarse-grain the interactions more dramatically: significant progress has been made in recent years in approaching this problem systematically [85]. Finally, the most fundamental properties, such as the entanglement length in a polymer melt [72], may be investigated using a simple chain of pseudo atoms or beads (modeled using the WCA potential of Fig. 1.3, and each representing several monomers), joined by an attractive finitely extensible nonlinear elastic (FENE) potential, which is illus-trated in Fig. 1.4.

$$v^{FENE}(r) = \begin{cases} -\frac{1}{2}kR_0^2 \ln\left(1-\left(r/R_0\right)^2\right) & r < R_0 \\ \infty & r \geq R_0 \end{cases} \qquad (17)$$

The key feature of this potential is that it cannot be extended beyond $r = R_0$, en-suring (for suitable choices of the parameters k and R_0) that polymer chains cannot move through one another. The empirical potential energy function is differentiable with respect to the atomic coordinates; this gives the value and the direction of the force acting on an atom and thus it can be used in a MD simulation. The empirical potential function has several limitations, which result in inaccuracies in the calculated potential energy. One limitation is due to the fixed set of atom types employed when determining the parameters for the force field. Atom types are used to define an atom in a particular bonding situation, for example an aliphatic carbon atom in an sp_3 bonding situation has different properties than a carbon atom found in the His ring. Instead of presenting each atom in the molecule, as a unique one described by unique set of parameters, there is a certain amount of grouping in order minimizes the number of atom types. This can lead to type-specific errors. The properties of certain atoms, like aliphatic carbon or hydrogen atoms, are less sensitive to their surroundings and a single set of parameters may work quite well, while other atoms like oxygen and nitrogen are much more influenced by their neighboring atoms. These atoms require more types and parameters to account for the different bonding environments.

Another important point to take into consideration is that the potential energy function does not include entropic effects. Thus, a minimum value of E calculated as a sum of potential functions does not necessarily correspond to the equilibrium, or the most probable state; this corresponds to the minimum of free energy. Because of the fact that experiments are generally carried out under isothermal-isobaric conditions (constant pressure, constant system size and constant temperature) the equilibrium state corresponds to the minimum of Gibb's Free Energy, G. While just an energy calculation ignores entropic effects, these are included in MD simulations.

1.1.3.3 STATISTICAL MECHANICS

Molecular simulations are based on the framework of statistical mechanics/thermodynamics. MD simulations generate information at the microscopic level, including atomic positions and velocities. The conversion of this microscopic information to macroscopic observables such as pressure, energy, heat capacities, etc., requires statistical mechanics. In a MD simulation, one often wishes to explore the macroscopic properties of a system through microscopic simulations, for example, to calculate changes in the binding free energy of a particular drug candidate, or to examine the energetics and mechanisms of conformational change. The connection between microscopic simulations and macroscopic properties is made via statistical mechanics, which provides the rigorous mathematical expressions that relate macroscopic properties to the distribution and motion of the atoms and molecules of the N-body system; MD simulations provide the means to solve the equation of motion of the particles and evaluate these mathematical formulas. With MD simulations, one can study both thermodynamic properties and/or time dependent (kinetic) phenomenon [10].

Statistical mechanics is the branch of physical sciences that studies macroscopic systems from a molecular point of view. The goal is to understand and to predict macroscopic phenomena from the properties of individual molecules making up the system. The system could range from a collection of solvent molecules to a solvated protein-DNA complex. In order to connect the macroscopic system to the microscopic system, time independent statistical averages are often introduced. The thermodynamic state of a system is usually defined by a small set of parameters, for example, the temperature, T, the pressure, P, and the number of particles, N. Other thermodynamic properties may be derived from the equations of state and other fundamental thermodynamic equations. The mechanical or microscopic state of a system is defined by the atomic positions, q, and momenta, p; these can also be considered as coordinates in a multidimensional space called phase space. For a system of N particles, this space has 6N dimensions. A single point in phase space, denoted by G, describes the state of the system. An ensemble is a collection of points in phase space satisfying the conditions of a particular thermodynamic state. A MD simulations generates a sequence of points in phase space as a function of time; these points belong to the same ensemble, and they correspond to the different conformations of the system and their respective momenta. Several different ensembles are described below [11]. An ensemble is a collection of all possible systems, which have different microscopic states but have an identical macroscopic or thermodynamic state.

1.1.3.4 NEWTON'S EQUATION OF MOTION

We view materials as a collection of discrete atoms. The atoms interact by exerting forces on each other. Force Field is defined as a mathematical expression that describes the dependence of the energy of a molecule on the coordinates of the atoms in the molecule. The MD simulation method is based on Newton's second law or the equation of motion, F=ma, where F is the force exerted on the particle, m is its mass and "a" is its acceleration. From knowledge of the force on each atom, it is possible to determine the acceleration of each atom in the system. Integration of the equations of motion then yields a trajectory that describes the positions, velocities and accelerations of the particles as they vary with time. From this trajectory, the average values of properties can be determined. The method is deterministic; once the positions and velocities of each atom are known, the state of the system can be predicted at any time in the future or the past. MD simulations can be time consuming and computationally expensive. However, computers are getting faster and cheaper. Based on the interaction model, a simulation computes the atoms' trajectories numerically. Thus a molecular simulation necessarily contains the following ingredients [5,17, 43, 76].

- The model that describes the interaction between atoms. This is usually called the interatomic potential: V ({ri}), where {ri} represent the position of all atoms.
- Numerical integrator that follows the atoms equation of motion. This is the heart of the simulation. Usually, we also need auxiliary algorithms to set up the initial and boundary conditions and to monitor and control the systems state (such as temperature) during the simulation.
- Extract useful data from the raw atomic trajectory information. Compute materials properties of interest. Visualization [17].

The dynamics of classical objects follow the three laws of Newton. Here, we review the Newton's laws (Fig. 1.5):

FIGURE 1.5   Sir Isaac Newton (1643–1727 England).

First law: Every object in a state of uniform motion tends to remain in that state of motion unless an external force is applied to it. This is also called the Law of inertia.

Second law: An object's mass m, its acceleration a, and the applied force F are related by

$$F_i = m_i a_i \tag{18}$$

where F_i is the force exerted on particle i, m_i is the mass of particle i and a_i is the acceleration of particle i.

Third law: For every action there is an equal and opposite reaction.

The *second law* gives us the equation of motion for classical particles. Consider a particle, its position is described by a vector r = (x, y, z). The velocity is how fast r changes with time and is also a vector: v = dr/dt = (v_x, v_y, v_z). In component form, v_x = dx/dt, v_y = dy/dt, v_z = dz/dt. The acceleration vector is then time derivative of velocity, that is, a = dv/dt. The force can also be expressed as the gradient of the potential energy,

$$F_i = -\nabla_i V \qquad (19)$$

Combining of equation of motion and gradient of the potential energy yields:

$$-\frac{dV}{d_{r_i}} = m_i \frac{d^2 r_i}{dt^2} \qquad (20)$$

where V is the potential energy of the system. Newton's equation of motion can then relate the derivative of the potential energy to the changes in position as a function of time. The kinetic energy is given by the velocity,

$$E_{kin} = \frac{1}{2} m |v|^2 \qquad (21)$$

The total energy is the sum of potential and kinetic energy contributions,

$$E_{tot} = E_{kin} + V \qquad (22)$$

When we express the total energy as a function of particle position r and momentum p = mv, it is called the Hamiltonian of the system,

$$H(r, p) = \frac{|P|^2}{2m} + V(r) \qquad (23)$$

The Hamiltonian (i.e., the total energy) is a conserved quantity as the particle moves. To see this, let us compute its time derivative,

$$\frac{dE_{tot}}{dt} = mv \cdot \frac{dv}{dt} + \frac{dV(r)}{dr} \cdot \frac{dr}{dt} \qquad (24)$$

$$= m \frac{dr}{dt} \cdot \left[-\frac{1}{m} \frac{dV(r)}{dr} \right] + \frac{dV(r)}{dr} \cdot \frac{dr}{dt} = 0$$

Therefore the total energy is conserved if the particle follows the Newton's equation of motion in a conservative force field (when force can be written in terms of spatial derivative of a potential field), while the kinetic energy and potential energy can interchange with each other. The Newton's equation of motion can also be written in the Hamiltonian form, [17]

$$\frac{dr}{dt} = \frac{\partial H(r, p)}{\partial p} \qquad (25)$$

$$\frac{dp}{dt} = -\frac{\partial H(r, p)}{\partial r}$$

Hence,

$$\frac{dH}{dt} = \frac{\partial H(r,p)}{\partial r} \cdot \frac{dr}{dt} + \frac{\partial H(r,p)}{\partial p} \cdot \frac{dp}{dt} = 0 \qquad (26)$$

FIGURE 1.6   Sir William Rowan Hamilton (1805–1865 Ireland).

Newton's Second Law of motion: a simple application,

$$F = ma = m\frac{dv}{dt} = m\frac{d^2x}{dt^2} \qquad (27)$$

$$a = \frac{dv}{dt}$$

Taking the simple case where the acceleration is constant. We obtain an expression for the velocity after integration:

$$v = at + v_0 \qquad (28)$$

And since

$$v = \frac{dx}{dt} \qquad (29)$$

We can once again integrate to obtain

$$x = vt + x_0 \qquad (29a)$$

Combining this equation with the expression for the velocity, we obtain the following relation which gives the value of x at time t as a function of the acceleration, a, the initial position, x_0, and the initial velocity, v_0.

$$x = vt + x_0$$

$$x = at^2 + v_0 t + x_0 \tag{30}$$

The acceleration is given as the derivative of the potential energy with respect to the position, r,

$$a = -\frac{1}{m}\frac{dE}{dr} \tag{31}$$

Therefore, to calculate a trajectory, one only needs the initial positions of the atoms, an initial distribution of velocities and the acceleration, which is determined by the gradient of the potential energy function. The equations of motion are deterministic, for example, the positions and the velocities at time zero determine the positions and velocities at all other times, t. The initial positions can be obtained from experimental structures, such as the x-ray crystal structure of the protein or the solution structure determined by NMR spectroscopy. The initial distribution of velocities are usually determined from a random distribution with the magnitudes conforming to the required temperature and corrected so there is no overall momentum, that is,

$$P = \sum_{i=1}^{N} m_i v_i = 0 \tag{32}$$

The velocities, v_i, are often chosen randomly from a Maxwell-Boltzmann or Gaussian distribution at a given temperature, which gives the probability that an atom i has a velocity v_x in the x direction at a temperature T.

$$p(v_{ix}) = (\frac{m_i}{2\pi k_B T})^{1/2} exp\left[-\frac{1}{2}\frac{m_i v_{ix}^2}{k_B T} \right] \tag{33}$$

The temperature can be calculated from the velocities using the relation

$$T = \frac{1}{(3N)}\sum_{i=1}^{N} \frac{|p_i|}{2m_i} \tag{34}$$

where N is the number of atoms in the system.

1.1.3.5   INTEGRATION ALGORITHMS

Solving Newton's equations of motion does not immediately suggest activity at the cutting edge of research. The MD algorithm in most common use today may even have been known to Newton [41]. Nonetheless, the last decade has seen a rapid development in our understanding of numerical algorithms; a forthcoming review [21] and a book summarize the present state of the field.

Continuing to discuss, for simplicity, a system composed of atoms with coordinates $r^N = (r_1, r_2, \ldots r_N)$ and potential energy $u(r^N)$, we introduce the atomic momenta $p^N = (p_1, p_2 \cdots p_N)$, in terms of which the kinetic energy may be written $\kappa(p^N) = \sum_{i=1}^{N} |p_i|^2 / 2m_i$. Then the energy, or Hamiltonian, may be written as a sum of kinetic and potential terms $H = K + U$. Write the classical equations of motion as

$$\dot{r}_i = \frac{p_i}{m_i} \qquad (35)$$

$$\dot{p}_i = f_i$$

This is a system of coupled ordinary differential equations. Many methods exist to perform step-by-step numerical integration of them. Characteristics of these equations are: (a) they are 'stiff', that is, there may be short and long timescales, and the algorithm must cope with both; (b) calculating the forces is expensive, typically involving a sum over pairs of atoms, and should be performed as infrequently as possible. Also we must bear in mind that the advancement of the coordinates fulfills two functions: (i) accurate calculation of dynamical properties, especially over times as long as typical correlation times τ_a of properties a of interest (we shall define this later); (ii) accurately staying on the constant-energy hyper surface, for much longer times $\tau_{run} \gg \tau_a$, in order to sample the correct ensemble [5].

To ensure rapid sampling of phase space, we wish to make the time step as large as possible consistent with these requirements. For these reasons, simulation algorithms have tended to be of *low order* (i.e., they do not involve storing high derivatives of positions, velocities etc.): this allows the time step to be increased as much as possible without jeopardizing energy conservation. It is unrealistic to expect the numerical method to accurately follow the true trajectory for very long times τ_{run}. The 'ergodic' and 'mixing' properties of classical trajectories, that is, the fact that nearby trajectories diverge from each other exponentially quickly, make this impossible to achieve. All these observations tend to favor the Verlet algorithm in one form or another, and we look closely at this in the following section. For historical reasons only, we mention them are general class of predictor-corrector methods which have been optimized for classical mechanical equations [35, 52].

The potential energy is a function of the atomic positions (3N) of all the atoms in the system. Due to the complicated nature of this function, there is no analytical solution to the equations of motion; they must be solved numerically. Numerous numerical algorithms have been developed for integrating the equations of motion. Different algorithms have been listed below:

- Verlet algorithm
- Leap-frog algorithm
- Velocity Verlet
- Beeman's algorithm

It is important to attention which algorithm to use, one should consider the following criteria:

- The algorithm should conserve energy and momentum.
- It should be computationally efficient
- It should permit a long time step for integration.

All the integration algorithms assume the positions, velocities and accelerations can be approximated by a Taylor series expansion:

$$y(t_{i+1}) = y(t_i) + hy'(t_i) + \frac{h^2}{2}y''(t_i) + \ldots + \frac{h^n}{n!}y^{(n)}(t_i) + \frac{h^{n+1}}{(n+1)!}y^{(n+1)}(\xi_i) \qquad (36)$$

$$r(t + \delta t) = r(t) + v(t)\delta t + \frac{1}{2}a(t)\delta t^2 + \ldots$$

$$v(t + \delta t) = v(t) + a(t)\delta t + \frac{1}{2}b(t)\delta t^2 + \ldots$$

$$u(t + \delta t) = a(t) + b(t)\delta t + \ldots$$

where r is the position, v is the velocity (the first derivative with respect to time), a is the acceleration (the second derivative with respect to time), etc. To derive the Verlet algorithm one can write

$$r(t + \delta t) = r(t) + v(t)\delta t + \frac{1}{2}a(t)\delta t^2 \qquad (37)$$

$$r(t - \delta t) = r(t) - v(t)\delta t + \frac{1}{2}a(t)\delta t^2$$

Summing these two equations, one obtains

$$r(t + \delta t) = 2r(t) - r(t - \delta t) + a(t)\delta t^2 \qquad (38)$$

1.1.3.5.1 VERLET ALGORITHMS

The Verlet algorithm uses positions and accelerations at time t and the positions from time t-dt to calculate new positions at time t+dt. The Verlet algorithm uses no explicit velocities. The advantages of the Verlet algorithm are: (i) it is straightforward, and (ii) the storage requirements are modest. The disadvantage is that the algorithm is of moderate precision. The advantages and disadvantages of this algorithm have been summarized in Table 1.1 [5, 76].

TABLE 1.1   Verlet's Algorithm Characteristics

Pros	Cons
• Simple and Effective	• Not as accurate as RK
• Low Memory and CPU Requirements (don't need to store velocities or perform multiple force calculations)	• We never calculate velocities
• Time Reversible	
• Very stable even with large numbers of interacting particles	

We can estimate the velocities using a finite difference:

$$v(t) = \frac{1}{2\Delta t}\left[r(t+\Delta t) - r(t-\Delta t)\right] + O\left(\Delta t^2\right) \tag{39}$$

There are variations of the Verlet algorithm, such as the leapfrog algorithm, which seek to improve velocity estimations.

1.1.3.5.2   THE LEAP-FROG ALGORITHM

In this algorithm, the velocities are first calculated at time t+1/2dt; these are used to calculate the positions, r, at time t+dt. In this way, the velocities leap over the positions, then the positions leap over the velocities. The advantage of this algorithm is that the velocities are explicitly calculated, however, the disadvantage is that they are not calculated at the same time as the positions. The velocities at time t can be approximated by the relationship [76]:

$$v(t) = \frac{1}{2}\left[v\left(t-\frac{1}{2}\delta t\right) + v\left(t+\frac{1}{2}\delta t\right)\right] \tag{40}$$

1.1.3.5.3   THE VELOCITY VERLET ALGORITHM

This algorithm yields positions, velocities and accelerations at time t. There is no compromise on precision.

$$r(t+\delta t) = r(t) + v(t)\delta t + \frac{1}{2}a(t)\delta t^2 \tag{41}$$

$$r\ (t+\delta t)= v\ (t)+\frac{1}{2}\left[a(t)+a(t+\delta t)\right]\delta t$$

The MD method is applicable to both equilibrium and nonequilibrium physical phenomena, which makes it a powerful computational tool that can be used to simulate many physical phenomena (if computing power is sufficient). The main procedure for conducting the MD simulation using the velocity Verlet method is shown in the following steps:

1. Specify the initial position and velocity of all molecules.
2. Calculate the forces acting on molecules.
3. Evaluate the positions of all molecules at the next time.
4. Evaluate the velocities of all molecules at the next time step.
5. Repeat the procedures from Step 2.

In the above procedure, the positions and velocities will be evaluated at every time interval h in the MD simulation. The method of evaluating the system averages is necessary to make a comparison with experimental or theoretical values. Since microscopic quantities such as positions and velocities are evaluated at every time interval in MD simulations, a quantity evaluated from such microscopic values, for example, the pressure will differ from that measured experimentally. In order to compare with experimental data, instant pressure is sampled at each time step, and these values are averaged during a short sampling time to yield a macroscopic pressure. This average can be expressed as,

$$\bar{A} = \sum_{n=1}^{N} A_n\ /\ N \tag{42}$$

in which A_n is the n-th sampled value of an arbitrary physical quantity A, and $\bar{A}$, called the "time average," is the mathematical average of N sampling data [76].

1.1.3.5.4   THE BEEMAN'S ALGORITHM

This algorithm is closely related to the Verlet algorithm:

$$r(t+\delta t) = r(t)+v(t)\delta t+\frac{2}{3}a(t)\delta t^2 -\frac{1}{6}a(t-\delta t)\delta t^2 \tag{43}$$

$$v(t+\delta t) = v(t)+v(t)\delta t+\frac{1}{3}a(t)\delta t+\frac{5}{6}a(t)\delta t -\frac{1}{6}a\ (t-\delta t)\delta t$$

The advantages and disadvantages of this algorithm have been summarized in Table 1.2.

TABLE 1.2   Beeman's Algorithm Characteristic

Pros	Cons
• Provides a more accurate expression for the velocities and better energy conservation.	• More complex expressions make the calculation more expensive.

TABLE 1.3   The Comparison Between Classical MD and Quantum MD

Classical MD		
Simulate/predict processes		**Characteristics**
1. Polypeptide folding 2. Biomolecular association 3. Partitioning between solvents 4. Membrane/micelle formation	Thermodynamic equilibrium Governed by weak (Nonbonded) forces	• Degrees of freedom: atomic (solute + solvent) • Equations of motion: classical dynamics • Governing theory: statistical mechanics
Quantum MD		
5. Chemical reactions, enzyme catalysis 6. Enzyme catalysis 7. Photochemical reactions, electron transfer	Chemical transformations Governed by strong forces	• Degrees of freedom: electronic, nuclear • Equations of motion: quantum dynamics • Governing theory: quantum statistical mechanics

1.1.3.6   PERIODIC BOUNDARY CONDITIONS

Small sample size means that, unless surface effects are of particular interest, periodic boundary conditions need to be used. Consider 1000 atoms arranged in a $10 \times 10 \times 10$ cube. Nearly half the atoms are on the outer faces, and these will have a large effect on the measured properties. Even for $10^6 = 1003$ atoms, the surface atoms amount to 6% of the total, which is still nontrivial. Surrounding the cube with replicas of itself takes care of this problem. Provided the potential range is not too long, we can adopt the minimum image convention that each atom interacts with the nearest atom or image in the periodic array. In the course of the simulation, if an atom leaves the basic simulation box, attention can be switched to the incoming image (Fig. 1.7). Of course, it is important to bear in mind the imposed artificial periodicity when considering properties, which are influenced by long-range correlations. Special attention must be paid to the case where the potential range is not short: for example for charged and dipolar systems [5, 76].

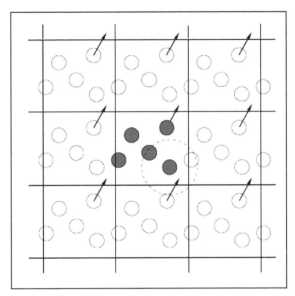

FIGURE 1.7 Periodic boundary conditions. As a particle moves out of the simulation box, an image particle moves in to replace it. In calculating particle interactions within the cutoff range, both real and image neighbors are included [5].

1.1.3.7 NEIGHBOR LIST

Computing the nonbonded contribution to the interatomic forces in an MD simulation involves, in principle, a large number of pairwise calculations: we consider each atom I and loop over all other atoms j to calculate the minimum image separations r_{ij}. Let us assume that the interaction potentials are of short range, $v(r_{ij}) = 0$ if $r_{ij} > r_{cut}$, the potential cutoff. In this case, the program skips the force calculation, avoiding expensive calculations, and considers the next candidate j. Nonetheless, the time to examine all pair separations is proportional to the number of distinct pairs, $\frac{1}{2}N(N-1)$ in an N-atomsystem, and for every pair one must compute at least r_{ij}^2; this still consumes a lot of time. Some economies result from the use of lists of nearby pairs of atoms. Verlet [7] suggested such a technique for improving the speed of a program. The potential cutoff sphere, of radius r_{cut}, around a particular atom is surrounded by a 'skin,' to give a larger sphere of radius r_{list}. At the first step in a simulation, a list is constructed of all the neighbors of each atom, for which the pair separation is within r_{list}. Over the next few MD time steps, only pairs appearing in the list are checked in the force routine. From time to time the list is reconstructed: it is important to do this before any unlisted pairs have crossed the safety zone and come within interaction range. It is possible to trigger the list reconstruction automatically, if a record is kept of the distance traveled by each atom since the last update. The choice of list cutoff distance r_{list} is a compromise: larger lists will need to be reconstructed less frequently, but will not give as much of a saving on cpu time as smaller lists. This choice can easily be made by experimentation.

For larger systems ($N \geq 1000$ or so, depending on the potential range) another technique becomes preferable. The cubic simulation box (extension to noncubic cases is possible) is divided into a regular lattice of $n_{cell} \times n_{cell} \times n_{cell}$ cells. These cells are chosen so that the side of the cell $l_{cell} = L / n_{cell}$ is greater than the potential cutoff distance cut. If there is a separate list of atoms in each of those cells, then searching through the neighbors is a rapid process: it is only necessary to look at atoms in the same cell as the atom of interest, and nearest neighbor cells. The cell structure may be set up and used by the method of linked lists [43, 55]. The first part of the method involves sorting all the atoms into their appropriate cells. This sorting is rapid, and may be performed every step. Then, within the force routine, pointers are used to scan through the contents of cells, and calculate pair forces. This approach is very efficient for large systems with short-range forces. A certain amount of unnecessary work is done because the search region is cubic, not (as for the Verlet list) spherical [5].

1.1.3.8   TIME DEPENDENCE

Knowledge of time-dependent statistical mechanics is important in three general areas of simulation. Firstly, in recent years there have been significant advances in the understanding of MD algorithms, which have arisen out of an appreciation of the formal operator approach to classical mechanics. Second, an understanding of equilibrium time correlation functions, their link with dynamical properties, and especially their connection with transport coefficients, is essential in making contact with experiment. Third, the last decade has seen a rapid development of the use of non-equilibrium MD, with a better understanding of the formal aspects, particularly the link between the dynamical algorithm, dissipation, chaos, and fractal geometry [5].

1.1.3.9   APPLICATION AND ACHIEVEMENTS

Given the modeling capability of MD and the variety of techniques that have emerged, what kinds of problem can be studied? Certain applications can be eliminated owing to the classical nature of MD. There are also hardware imposed limitations on the amount of computation that can be performed over a given period of time – be it an hour or a month – thus restricting the number of molecules of a given complexity that can be handled, as well as storage limitations having similar consequences (to some extent, the passage of time helps alleviate hardware restrictions). The phenomena that can be explored must occur on length and time scales that are encompassed by the computation. Some classes of phenomena may require repeated runs based on different sets of initial conditions to sample adequately the kinds of behavior that can develop, adding to the computational demands. Small system size enhances the fluctuations and sets a limit on the measurement accuracy; finite-size effects – even the shape of the simulation region – can also influence certain results. Rare events present additional problems of observation and measurement. Liquids represent the state of matter most frequently studied by MD methods. This is due to historical reasons, since both solids and gases have well-developed theoretical foundation, but there is no general theory of liquids. For solids, theory begins by assuming that the atomic constituents undergo small oscillations about fixed lattice positions; for gases, independent atoms are assumed and interactions are introduced as weak perturbations. In the case of liquids, however, the interactions are as important as in the solid state, but there is no underlying ordered structure to begin with. The following list includes a somewhat random and far from complete assortment of ways in which MD simulation is used:

- Fundamental studies: equilibration, tests of molecular chaos, kinetic theory, diffusion, transport properties, size dependence, tests of models and potential functions.
- Phase transitions: first- and second-order, phase coexistence, order parameters, critical phenomena.
- Collective behavior: decay of space and time correlation functions, coupling of translational and rotational motion, vibration, spectroscopic measurements, orientational order, dielectric properties.
- Complex fluids: structure and dynamics of glasses, molecular liquids, pure water and aqueous solutions, liquid crystals, ionic liquids, fluid interfaces, films and monolayers.
- Polymers: chains, rings and branched molecules, equilibrium conformation, relaxation and transport processes.
- Solids: defect formation and migration, fracture, grain boundaries, structural transformations, radiation damage, elastic and plastic mechanical properties, friction, shock waves, molecular crystals, epitaxial growth.
- Biomolecules: structure and dynamics of proteins, protein folding, micelles, membranes, docking of molecules.
- Fluid dynamics: laminar flow, boundary layers, rheology of non-Newtonian fluids, unstable flow. And there is much more.

The elements involved in an MD study, the way the problem is formulated, and the relation to the real world can be used to classify MD problems into various categories. Examples of this classification include whether the interactions are short- or long-ranged; whether the system is thermally and mechanically isolated or open to outside influence; whether, if in equilibrium, normal dynamical laws are used or the equations of motion are modified to produce a particular statistical mechanical ensemble; whether the constituent particles are simple structure less atoms or more complex molecules and, if the latter, whether the molecules are rigid or flexible; whether simple interactions are represented by continuous potential functions or by step potentials; whether interactions involve just pairs of particles or multiparticle contributions as well; and so on and so on [70].

Despite the successes, many challenges remain. Multiple phases introduce the issue of interfaces that often have a thickness comparable to the typical simulated region size. In homogeneities such as density or temperature gradients can be difficult to maintain in small systems, given the magnitude of the inherent fluctuations. Slow relaxation processes, such as those typical of the glassy state, diffusion that is hindered by structure as in polymer melts, and the very gradual appearance of spontaneously forming spatial organization, are all examples of problems involving temporal scales many orders of magnitude larger than those associated with the underlying molecular motion [70].

1.1.4 MONTE CARLO METHOD

In the MD method, the motion of molecules (particles) is simulated according to the equations of motion and therefore it is applicable to both thermodynamic equilibrium and non-equilibrium phenomena. In contrast, the MC method generates a series of microscopic states under a certain stochastic law, irrespective of the equations of motion of particles. Since the MC method does not use the equations of motion, it cannot include the concept of explicit time, and thus is only a simulation technique for phenomena in thermodynamic equilibrium. Hence, it is unsuitable for the MC method to deal with the dynamic properties of a system, which are dependent on time. In the following paragraphs, we explain important points of the concept of the MC method.

How do microscopic states arise for thermodynamic equilibrium in a practical situation? We discuss this problem by considering a two-particle attractive system using Fig. 1.8. As shown in Fig. 1.8A, if the two particles overlap, then a repulsive force or significant interaction energy arises. As shown in Fig. 1.2B, for the case of close proximity, the interaction energy becomes low and an attractive force acts on the particles. If the two particles are sufficiently distant, as shown in Fig. 1.8C, the interactive force is negligible and the interaction energy can be regarded as zero.

In actual phenomena, microscopic states which induce a significantly high energy, as shown in Fig. 1.8A, seldom appear, but microscopic states which give rise to a low-energy system, as shown in Fig. 1.8B, frequently arise. However, this does not mean that only microscopic states that induce a minimum energy system appear. Consider the fact that oxygen and nitrogen molecules do not gather in a limited area, but distribute uniformly in a room. It is seen from this discussion that, for thermodynamic equilibrium, microscopic states do not give rise to a minimum of the total system energy, but to a minimum free energy of a system. For example, in the case of a system specified by the number of particles N, temperature T, and volume of the system V, microscopic states arise such that the following Helmholtz free energy F becomes a minimum:

$$F = E - TS \tag{44}$$

where E is the potential energy of the system, and S is the entropy. In the preceding example, the reason why oxygen or nitrogen molecules do not gather in a limited area can be explained by taking into account the entropy term on the right-hand side in Eq. (44). That is, the situation in which molecules do not gather together and form flocks but expand to fill a room gives rise to a large value of the entropy. Hence, according to the counterbalance relationship of the energy and the entropy, real microscopic states arise such that the free energy of a system is at minimum [76].

(A) Overlapping   (B) Close proximity   (C) Sufficiently distant

FIGURE 1.8   Typical energy situations for a two-particle system [76].

Next, we consider how microscopic states arise stochastically. We here treat a system composed of N interacting spherical particles with temperature T and volume V of the system; these quantities are given values and assumed to be constant. If the position vector of an arbitrary particle i (i=1, 2, ..., N) is denoted by r_i, then the total interaction energy U of the system can be expressed as a function of the particle positions; that is, it can be expressed as $U=U(r_1, r_2, ..., r_N)$. For the present system specified by given values of N, T, and V, the appearance of a microscopic state that the particle i (i=1, 2, ..., N) exits within the small range of $r_i \sim (r_i+\Delta r_i)$ is governed by the probability density function $\rho(r_1, r_2, ..., r_N)$. This can be expressed from statistical mechanics [57] as

$$\rho(r_1, r_2, ..., r_N) = \frac{exp\{-U(r_1, r_2, ...r_N)/kT\}}{\int_V ... \int_V exp\{-U(r_1, r_2, ...r_N)/kT\} dr_1 dr_2 .. dr_N} \tag{45}$$

If a series of microscopic states is generated with an occurrence according to this probability, a simulation may have physical meaning. However, this approach is impracticable, as it is extraordinarily difficult and almost impossible to evaluate analytically the definite integral of the denominator in Eq. (45). In fact, if we were able to evaluate this integral term analytically, we would not need a computer simulation because it would be possible to evaluate almost all physical quantities analytically. The "Metropolis method" [58] overcomes this difficulty for MC simulations.

In the Metropolis method, the transition probability from microscopic states i to j, p_{ij}, is expressed as

$$p_{ij} = \begin{cases} 1 \ (for \ \rho_j \ / \ \rho_i \geq 1) \\ \dfrac{\rho_j}{\rho_i} \ (for \ \dfrac{\rho_j}{\rho_i} < 1) \end{cases} \tag{46}$$

in which ρ_j and ρ_i are the probability density functions for microscopic states j and i appearing, respectively. The ratio of ρ_j/ρ_i is obtained from Eq. (45) as

$$\frac{\rho_j}{\rho_i} = exp\left\{ -\frac{1}{kT}(U_j - U_I) \right\} \tag{47}$$

$$= exp\left[-\frac{1}{kT}\left\{ U\left(r_1^j, r_2^j, ..., r_N^j\right) - U\left(r_1^i, r_2^i, .., r_N^i\right) \right\} \right]$$

In the above equations, U_i and U_j are the interaction energies of microscopic states i and j, respectively. The superscripts attached to the position vectors denote the same meanings concerning microscopic states. Equation (46) implies that, in the transition from microscopic states i to j, new microscopic state j is adopted if the system energy decreases, with the probability ρ_j/ρ_i (1) if the energy increases. As clearly demonstrated by Eq. (47), for ρ_j/ρ_i the denominator in Eq. (45) is not required in Eq. (47), because ρ_j is divided by ρ_i and the term is canceled through this operation. This is the main reason for the great success of the Metropolis method for MC simulations. That a new microscopic state is adopted with the probability ρ_j/ρ_i, even in the case of the increase in the interaction energy, verifies the accomplishment of the minimum free-energy condition for the system. In other words, the adoption of microscopic states, yielding an increase in the system energy, corresponds to an increase in the entropy [76].

The above discussion is directly applicable to a system composed of nonspherical particles. The situation of nonspherical particles in thermodynamic equilibrium can be specified by the particle position of the mass center, r_i (i=1, 2, ..., N), and the unit vector e_i (i=1, 2, ..., N) denoting the particle direction. The transition probability from microscopic states i to j, p_{ij} can be written in similar form to Eq. (46). The exact expression of ρ_j/ρ_i becomes

$$\frac{\rho_j}{\rho_i} = exp\left\{ -\frac{1}{kT}(U_j - U_i) \right\} = exp\left[-\frac{1}{kT}\left\{ U\ (r_1^j, r_2^j, r_N^j, e_1^j, e_2^j, ..., e_N^j) - U\ (r_1^i, r_2^i, r_N^i, e_1^i, e_2^i, ..., e_N^i) \right\} \right] \tag{48}$$

The main procedure for the MC simulation of a nonspherical particle system is as follows:

1. Specify the initial position and direction of all particles.
2. Regard this state as microscopic state i, and calculate the interaction energy U_i.

3. Choose an arbitrary particle in order or randomly and call this particle "particle α."
4. Make particle α move transnationally using random numbers and calculate the interaction energy U_j for this new configuration.
5. Adopt this new microscopic state for the case of $Uj \leq Ui$ and go to Step 7.
6. Calculate ρ_j/ρ_i in Eq. (48) for the case of $U_j > U_i$ and take a random number R_1 from a uniform random number sequence distributed from zero to unity.
   6.1. If $R_1 \leq \rho_j/\rho_i$, adopt this microscopic state j and go to Step 7.
   6.2. If $R_1 > \rho_j/\rho_i$, reject this microscopic state, regard previous state i as new microscopic state j, and go to Step 7.
7. Change the direction of particle α using random numbers and calculate the interaction energy U_k for this new state.
8. If $U_k \leq U_j$, adopt this new microscopic state and repeat from Step 2.
9. If $U_k > U_j$, calculate ρk/ρj in Eq. (48) and take a random number R2 from the uniform random number sequence.
   9.1. If $R_2 \leq \rho_k/\rho_j$, adopt this new microscopic state k and repeat from Step 2.
10. If $R_2 > \rho_k/\rho_j$, reject this new state, regard previous state j as new microscopic state k, and repeat from Step 2.

Although the treatment of the translational and rotational changes is carried out separately in the above algorithm, a simultaneous procedure is also possible in such a way that the position and direction of an arbitrary particle are simultaneously changed, and the new microscopic state is adopted according to the condition in Eq. (46). However, for a strongly interacting system, the separate treatment may be found to be more effective in many cases. We will now briefly explain how the translational move is made using random numbers during a simulation. If the position vector of an arbitrary particleα in microscopic state i is denoted by $r_a = (x_a, y_a, z_a)$, this particle is moved to a new position $r'_\alpha = (x'_\alpha, y'_\alpha, z'_\alpha)$ by the following equations using randomnumbers R_1, R_2, and R_3, taken from a random number sequence ranged from zero to unity:

$$\begin{cases} x'_\alpha = x_\alpha + R_1 \delta r_{max} \\ y'_\alpha = y_\alpha + R_2 \delta r_{max} \\ z'_\alpha = z_\alpha + R_3 \delta r_{max} \end{cases} \tag{49}$$

These equations imply that the particle is moved to an arbitrary position, determined by random numbers, within a cube centered at the particle center with side length of $2\delta r_{max}$. A series of microscopic states is generated by moving the particles according to the above-mentioned procedure. Finally, we show the method of evaluating the average of a physical quantity in MC simulations. These averages, called "ensemble averages," are different from the time averages that are obtained from MD simulations. If a physical quantity A is a function of the microscopic states of a system, and A_n is the n-th sampled value of this quantity in an MC simulation, then the ensemble average $\langle A \rangle$ can be evaluated from the equation

$$\langle A \rangle = \sum_{n=1}^{M} A_n / M \qquad (50)$$

where M is the total sampling number. In actual simulations, the sampling procedure is not conducted at each time step but at regular intervals. This may be more efficient because if the data have significant correlations they are less likely to be sampled by taking a longer interval for the sampling time. The ensemble averages obtained in this way may be compared with experimental data [76].

1.1.5   LATTICE BOLTZMANN METHOD

Whether or not the lattice Boltzmann method is classified into the category of molecular simulation methods may depend on the researcher, but this method is expected to have a sufficient feasibility as a simulation technique for polymeric liquids and particle dispersions. In the lattice Boltzmann method [82], a fluid is assumed to be composed of virtual fluid particles, and such fluid particles move and collide with other fluid particles in a simulation region. A simulation area is regarded as a lattice system, and fluid particles move from site to site; that is, they do not move freely in a region. The most significant difference of this method in relation to the MD method is that the lattice Boltzmann method treats the particle distribution function of velocities rather than the positions and the velocities of the fluid particles. Figure 1.9 illustrates the lattice Boltzmann method for a two-dimensional system. Figure 1.9A shows that a simulation region is divided into a lattice system. Figure 1.9B is a magnification of a unit square lattice cell. Virtual fluid particles, which are regarded as groups or clusters of solvent molecules, are permitted to move only to their neighboring sites, not to other, more distant sites. That is, the fluid particles at site 0 are permitted to stay there or to move to sites 1, 2, ..., 8 at the next time step. This implies that fluid particles for moving to sites 1, 2, 3, and 4 have the velocity $c = (\Delta x/\Delta t)$, and those for moving to sites 5, 6, 7, and 8 have the velocity $\sqrt{2}c$, in which Δx is the lattice separation of the nearest two sites and Δt is the time interval for simulations. Since the movement speeds of fluid particles are known as $\sqrt{2}c$ or, macroscopic velocities of a fluid can be calculated by evaluating the number of particles moving to each neighboring lattice site. In the usual lattice Boltzmann method, we treat the particle distribution function, which is defined as a quantity such that the above-mentioned number is divided by the volume and multiplied by the mass occupied by each lattice site. This is the concept of the lattice Boltzmann method. The two-dimensional lattice model shown in Fig. 1.9 is called the "D2Q9" model because fluid particles have nine possibilities of velocities, including the quiescent state (staying at the original site) [76].

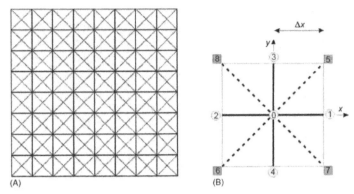

FIGURE 1.9   Two-dimensional lattice model for the lattice Boltzmann method (D2Q9 model) [76].

The velocity vector for fluid particles moving to their neighboring site is usually denoted by cα and, for the case of the D2Q9 model, there are nine possibilities, such as c_0, c_1, c_2, ..., c_8. For example, the velocity of the movement in the left direction in Fig. 1.8B is denoted byc2, and c_0 is zero vector for the quiescent state ($c_{0=}0$). We consider the particle distribution function f_α (r,t) at the position r (at point 0 in Fig. 1.8B) at time t in the α-direction. Since f_α (r,t) is equal to the number density of fluid particles moving in the α-direction, multiplied by the mass of a fluid particle, the summation of the particle distribution function concerning all the directions (α=0, 1, ..., 8) leads to the macroscopic density ρ (r,t):

$$\rho(r,t) = \sum_{\alpha=0}^{8} f_\alpha (r,t) \tag{51}$$

Similarly, the macroscopic velocity **u** (**r**,t) can be evaluated from the following relationship of the momentum per unit volume at the position **r**:

$$\rho(r,t) u(r,t) = \sum_{\alpha=0}^{8} f_\alpha (r,t) \, C_\alpha \tag{52}$$

In Eqs. (51) and (52), the macroscopic density ρ (r,t) and velocity u (r,t) can be evaluated if the particle distribution function is known. Since fluid particles collide with the other fluid particles at each site, the rate of the number of particles moving to their neighboring sites changes. In the rarefied gas dynamics, the well-known Boltzmann equation is the basic equation specifying the velocity distribution function while taking into account the collision term due to the interactions of gaseous molecules; this collision term is a complicated integral expression. The Boltzmann equation is quite difficult to solve analytically, so an attempt has been made to simplify the collision term. One such simplified model is the Bhavnagar-Gross-Krook

(BGK) collision model. It is well known that the BGK Boltzmann method gives rise to reasonably accurate solutions, although this collision model is expressed in quite simple form. We here show the lattice Boltzmann equation based on the BGK model. According to this model, the particle distribution function f_α (r+c$_\alpha$Δt,t+Δt) in the α-direction at the position (r+c$_\alpha$Δt) at time (t+Δt) can be evaluated by the following equation [76].

$$f_\alpha\left(r+c_\alpha\Delta t,t+\Delta t\right)=f_\alpha\left(r,t\right)+\frac{1}{\tau}\left\{f_\alpha^{(0)}\left(r,t\right)-f_\alpha\left(r,t\right)\right\} \tag{53}$$

This equation is sometimes expressed in separate expressions indicating explicitly the two different processes of collision and transformation:

$$f_\alpha\left(r+c_\alpha\Delta t,t+\Delta t\right)=\tilde{f}_\alpha\left(r,t\right) \tag{54}$$

$$\tilde{f}_\alpha\left(r,t\right)=f_\alpha\left(r,t\right)+\frac{1}{\tau}\left\{f_\alpha^{(0)}\left(r,t\right)-f_\alpha\left(r,t\right)\right\}$$

where τ is the relaxation time (dimensionless) and $f_\alpha^{(0)}$ is the equilibrium distribution, expressed for the D2Q9 model as

$$f_\alpha^{(0)}=\rho w_\alpha\left\{1+3\frac{c_\alpha.u}{c^2}-\frac{3u^2}{2c^2}+\frac{9}{2}.\frac{\left(c_\alpha.u\right)^2}{c^4}\right\} \tag{55}$$

$$w_\alpha=\begin{cases}\frac{4}{9} \text{ for }\alpha=0\\ \frac{1}{9}\text{ for }\alpha=1,2,3,4\\ \frac{1}{36}\text{ for }\alpha=5.6.7.8\end{cases}\left|c_\alpha\right|=\begin{cases}0\text{ for }\alpha=0\\ c\text{ for }\alpha=1,2,3,4\\ \sqrt{2}c\text{ for }\alpha=5,6,7,8\end{cases} \tag{56}$$

In these equations ρ is the local density at the position of interest, u is the fluid velocity (u=$|U|$), c=Δx/Δt, and w$_\alpha$ is the weighting constant. The important feature of the BGK model shown in Eq. (54) is that the particle distribution function in the α-direction is independent of the other directions. The particle distributions in the other directions indirectly influence f$_\alpha$ (r+c$_\alpha$Δt,t+Δt) through the fluid velocity u and the density ρ. The second expression in Eq. (55) implies that the particle distribution f$_\alpha$ (r,t) at the position r changes into f$_\alpha$ (r,t) after the collision at the site at time t, and the first expression implies that $\tilde{f}_\alpha\left(r,t\right)$ becomes the distribution f$_\alpha$ (r+c$_\alpha$Δt,t+Δt) at (r+c$_\alpha$Δt) after the time interval Δt [76].

The main procedure of the simulation is as follows:
1. Set appropriate fluid velocities and densities at each lattice site.

2. Calculate equilibrium particle densities $f_\alpha^{(0)}$ (α=0, 1, ..., 8) at each lattice site from Eq. (55) and regard these distributions as the initial distributions, $f_\alpha = f_\alpha^{(0)}$ (α=0, 1, ..., 8).
3. Calculate the collision terms $\tilde{f}_\alpha(r,t)$ (α=0, 1, ..., 8) at all sites from the second expression of Eq. (54).
4. Evaluate the distribution at the neighboring site in the α-direction $f_\alpha(r + c_\alpha \Delta t, t + \Delta t)$ from the first expression in Eq. (54).
5. Calculate the macroscopic velocities and densities from Eqs. (51) and (52), and repeat the procedures from Step 3.

In addition to the above-mentioned procedures, we need to handle the treatment at the boundaries of the simulation region. These procedures are relatively complex and are explained in detail in Chapter 8. For example, the periodic boundary condition, which is usually used in MD simulations, may be applicable. For the D3Q19 model shown in figure 9, which is applicable for three-dimensional simulations, the equilibrium distribution function is written in the same expression of Eq. (55), but the weighting constants are different from Eq. (56). The basic equations for f_α ($r+c_\alpha \Delta t, t+\Delta t$) are the same as Eq. (53) or (54), and the above-mentioned simulation procedure is also directly applicable to the D3Q19 model [76].

1.2 APPLICATIONS

1.2.1 INTRODUCTION

The term aerogel was first introduced by Kistler in 1932 to designate gels in which the liquid was replaced with a gas, without collapsing the gel solid network [50]. While wet gels were previously dried by evaporation, Kistler applied a new supercritical drying technique, according to which the liquid that impregnated the gels was evacuated after being transformed to a supercritical fluid. In practice, supercritical drying consisted in heating a gel in an autoclave, until the pressure and temperature exceeded the critical temperature T_c and pressure P_c of the liquid entrapped in the gel pores. This procedure prevented the formation of liquid–vapor meniscuses at the exit of the gel pores, responsible for a mechanical tension in the liquid and a pressure on the pore walls, which induced gel shrinkage. Besides, a supercritical fluid can be evacuated as gas, which in the end lets the "dry solid skeleton" of the initial wet material. The dry samples that were obtained had a very open porous texture, similar to the one they had in their wet stage. Overall, aerogels designate dry gels with a very high relative or specific pore volume, although the value of these characteristics depends on the nature of the solid and no official convention really exists. Typically, the relative pore volume is of the order of 90% in the most frequently studied silica aerogels [12].

1.2.1.1   SILICA AEROGEL AND ITS PROPERTIES

The first aerogels were synthesized from silica gels by replacing the liquid component with a gas. Silica aerogel, a highly porous material [50], is currently being produced using sol–gel processes. One such process is supercritical drying of tetramethoxysilane (TMOS), which is hydrolyzed to form silica and methanol [12, 34]. Several desirable properties of silica aerogels include very low densities (as low as 0.003 g cm^{-3} (, surface area as large as 1000 m^2/g, a refraction index only 5% greater than that of air, the lowest thermal conductivity among all solid materials, a speed of sound approximately three times smaller than that of air, and a dielectric constant only 10% greater than that of vacuum. [34]. These properties make silica aerogels very suitable for applications such as thermal and acoustic insulation in buildings and appliances, passive solar energy collection devices and dielectrics for integrated circuits [16], but probably one of the most notable uses was in Cherenkov radiators [4] as Cherenkov counters. Aerogels were also used to thermally insulate the 2003 Mars Exploration rovers, as well as to capture comet dust [30].

1.2.1.2   PREPARATION OF AEROGELS

Experimentally, aerogels and xerogels are made by sol–gel processing which is a core starting of an aerogel. An important boundary condition for the initial solution is the solubility of the reactants in the applied solvent. Therefore, it is important to study the miscibility diagram of the system under investigation (Fig. 1.10) or to perform corresponding tests to make sure that full solubility is assured [71]. Otherwise partial precipitation or inhomogeneous gels will result.

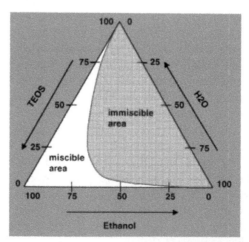

FIGURE 1.10   Miscibility diagram for the system tetraethoxysilane (TEOS), ethanol, and water [14].

The sol–gel process generally comprises two major steps on the molecular level: (1) the creation or activation of reactive sites and (2) the condensation of monomers, oligomers, and clusters via cross-linking of the reactive sites. The kinetics of each of these two reactions is controlled by various parameters, for example, the concentration of the reactants, the solvent, the catalyst used, the catalyst concentration (pH value), the temperature, and the steric effects, etc. [14].

The relative kinetics of the two steps together with reverse reactions due to solubility effects determine the morphological characteristics of the resulting gel, for example, the size of the particles forming the solid backbone, the mass distribution of the backbone elements characterized, for example, by the so-called fractal dimension of the backbone [46], and the connectivity in the solid phase, which is seen in the mechanical properties and the electrical or thermal conductivity of the backbone structure at a given porosity. Depending on the governing reaction kinetics, the gel backbone can range from a network of colloids to interconnected, highly branched polymer-like clusters (Fig. 1.11). The rate of cross-linking on a molecular scale determines the macroscopically observable gel time that is the time where the viscosity of the initial solution increases by several orders of magnitude, changing its appearance from a low viscosity liquid, to a syrup-like substance, and finally to a gel body. Almost all metal oxides can be transformed into gels and corresponding aerogels. In aqueous solution, the reactants are the corresponding salts; alkoxide precursors are processed in organic solvents. [71].

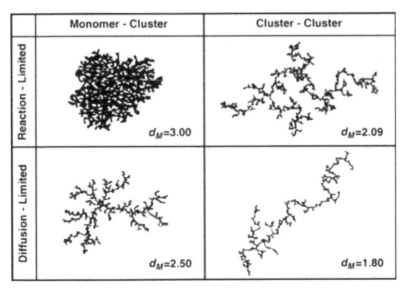

FIGURE 1.11   Morphology of the gel backbone in the case of different reaction kinetics [46]. The parameters dM are the fractal dimensions resulting from the combination of the different growth mechanisms, as indicated on the top and the left-hand side of the sketch.

The processing starts with the hydrolysis of a silica precursor, typically tetramethyl ortho silane (TMOS, Si (OCH3) 4) or tetraethyl orthosilane (TEOS, Si (xOC2H5) 4). Typically, the solvents used in these cases are ethanol or methanol and a base or acid is added to control the reaction rates (Fig. 1.12).

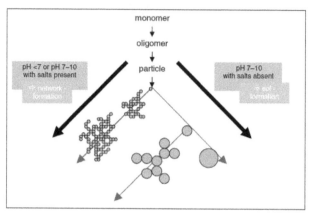

FIGURE 1.12   Impact of the pH and the presence of salts on the growth of a sol particle and the network formation upon polymerization of silica [71].

In the alcoxysilane solution, the "activation" step is the partial or complete hydrolysis of the alcoxysilane, which is favored with increasing concentration of either an acid or a base in the initial solution (reaction with TMOS, Eq. (57)). The condensation rate (Eq. 58) of the hydroxyl groups increases on both sides of the isoelectrical point (point of zero charge), but is suppressed at high Ph values [14].

$$Si\ (OCH_3)_4 + XH_2O \leftrightarrow Si\ (OCH_3)_4\text{-}X\ (OH)\ X + XCH_3OH\ \text{Hydrolysis} \qquad (57)$$

$$Si\ (OCH_3)_3\ (OH) + Si\ (OCH_3)_3\ (OH) \leftrightarrow Si_2O\ (OCH_3)_6 + H_2O\ \text{Condensation} \quad (58)$$

However other precursors can be used, too. For example, Steven Kistler, who is considered to have discovered this type of materials [8], produced the first silica aerogel using sodium silicate (Na_2SiO_3) dissolved in a solution of hydrochloridric acid (HCl) and water [19].

$$Na_2SiO_3 + 2HCl \overset{H_2O}{\leftrightarrow} [SiO_2.xH_2O] + 2NaCl \qquad (59)$$

The pH of the solution can be controlled by adding acidic or basic additives; those will react with neither the silica precursor nor the water but will affect the rate at which the hydrolysis proceeds as well as the final structure of the porous solid. In other words, the two-reaction steps hydrolysis and condensation on the pH enables a very sensitive control of the reaction, and thus the resulting gel morphology. Besides the so-called "one-step" process, in which all reactants are mixed within a short period of time, the sensitivity of the kinetics to the pH can be used to partially decouple the hydrolysis and condensation step (two-step sol–gel process) [13, 51]. For example, the reactants in a concentrated starting solution can first be prehydrolyzed under acidic conditions, subsequently diluted to match the reactant concentration for a given target solid content of the gel to be and then subject to basic conditions to accelerate the rate of cross-linking of the hydrolyzed species and to form a gel. [17, 18, 22, 71].

- Acid catalysis generally produces weakly cross-linked gels, which easily compact under drying conditions, yielding low-porosity microporous (smaller than 2 nm) xerogel structures (Fig. 1.13a).
- Conditions of neutral to basic pH result in relatively mesoporous xerogels after drying, as rigid clusters a few nanometers across pack to form mesopores. The clusters themselves may be microporous.
- Under some conditions, base-catalyzed and two-step acid-base catalyzed gels (initial polymerization under acidic conditions and further gelation under basic conditions) exhibit hierarchical structure and complex network topology (Fig. 1.13).

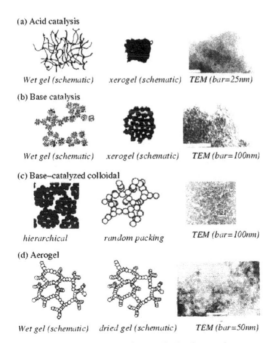

(a) Acid catalysis

Wet gel (schematic)   *xerogel (schematic)*   *TEM (bar=25nm)*

(b) Base catalysis

Wet gel (schematic)   *xerogel (schematic)*   *TEM (bar=100nm)*

(c) Base–catalyzed colloidal

hierarchical   *random packing*   *TEM (bar=100nm)*

(d) Aerogel

Wet gel (schematic)   *dried gel (schematic)*   *TEM (bar=50nm)*

FIGURE 1.13 Schematic wet and dry gel morphologies and representative transmission electron micrographs (2002 – aerogel).

After the hydrolysis, polycondensation of silica occurs and the solution becomes a gel. A large change of the viscosity indicates the gelation of the solution. At this point a nanoporous solid skeleton of silica is formed inside the solution and fills almost all its volume. If the gel is dried at room temperature and atmospheric pressure, large shrinkage of the solid structure is caused along with large reduction of the porosity. The shrinkage is mainly due to the stresses exerted on the solid branches when the liquid evaporates. Materials obtained this way are called xerogels and have porosities as high as 50%, corresponding to a density of 1.1 g/cm³. To avoid excessive shrinkage, the pressure and the temperature of the gel are increased until the critical point of the liquid phase has been exceeded. Under those conditions there is no liquid/vapor interface and no surface tension. Therefore, no large stresses are exerted on the solid skeleton and the shrinkage is highly reduced. Solids processed at supercritical conditions reach porosities up to 99.8% and are called aerogels [19]. Figure 1.14 show the process steps to synthesizing an aerogel.

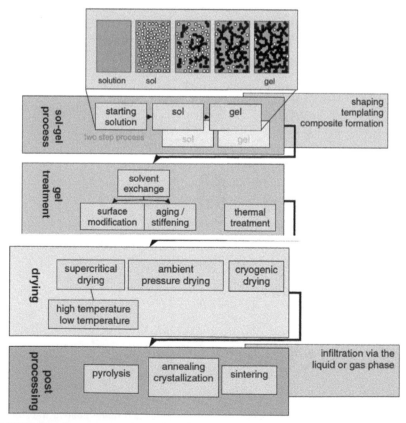

FIGURE 1.14   Process steps to synthesizing an aerogel.

1.2.1.2.1 DRYING

The removal of the liquid from the interconnected pores of a gel is the second key process step to synthesizing an aerogel. The reason why it is so tricky is the combination of small pores and thus high capillary pressures up to10 MPa (100 bar) during evaporation (Fig. 15) and a sparse, incompletely interconnected solid backbone that possesses only a low modulus of compression. These two facts typically result in a large permanent shrinkage upon drying unless specific measures are taken to prevent or at least suppress part of the effects. Possible means are:

- strengthening of the backbone to increase the stiffness of the backbone.
- reduction of the surface tension of the pore liquid by either
- replacing the initially present pore liquid with a liquid of lower surface tension and/or reducing the contact angle (toward complete wetting) via a modification of the gel surface.
- eliminating the surface tension by taking the pore liquid above the critical point of the pore liquid prior to evaporation (supercritical drying) or sublimation of the pore liquid (cryogenic drying).

Alternatively, one can allow the gel to shrink during drying; if the inner surface of the gel is chemically inert and the drying is performed slowly enough to prevent large gradients in shrinkage across the sample and thus crack formation, the gel will at least partially reexpand when the capillary forces decrease in the second stage of the drying process (Fig. 1.15) [71].

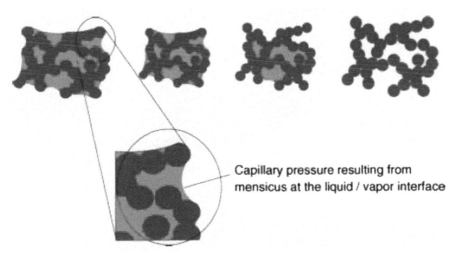

Capillary pressure resulting from mensicus at the liquid / vapor interface

FIGURE 1.15 Shrinkage of the gel body due to capillary forces upon ambient pressure drying. As soon as the pores begin to empty, the pressure is released and the gel partially re-expands.

1.2.1.2.2 SUPERCRITICAL DRYING

Supercritical drying is the extraction of a fluid in the supercritical state, thus avoiding the liquid–vapor coexistence regime and the associated capillary pressure. Figure 1.16 exemplarily shows the phase diagram of CO_2 with a possible path to be taken upon supercritical drying: In the first step, the gel containing the pore liquid is immersed in an excess of pore liquid, for example, methanol in the case of silica alcogels, to prevent partial drying; the sample in the solvent reservoir is then placed in an autoclave, pressurized and rinsed with liquid CO_2 until the initial pore liquid is completely replaced. Subsequently, the temperature is raised above the critical point of CO_2 (318°C) and the pressure is slowly released via a valve. When fully depressurized, the autoclave with the sample is allowed to cool down to room temperature. This scheme is also valid if a fluid other than CO_2 is used. Depending on the value of the critical point of the respective fluid, one distinguishes between low and high temperature supercritical drying (e.g., critical point of ethanol: 2168°C). In the case of high temperature supercritical drying, additional aging effects are observed that are mainly controlled by the solubility of the solid phase (i.e., the backbone) at the given conditions [18, 90].

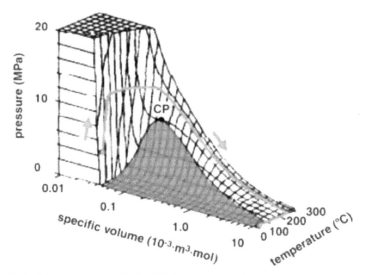

FIGURE 1.16 Phase diagram of CO_2; CP denotes the critical point. The shaded area marks the regime of coexistence of the liquid and the gas phase. The line and the arrows indicate the path taken upon supercritical drying.

In theory, it is expected that upon supercritical drying no capillary pressure and thus no shrinkage of the gel body will be observed. In practice, linear shrinkages of the order up to 10–15% percent are detected [39] (Fig. 1.17). This effect shows, for example, upon CO_2 supercritical drying when the CO_2 pressure is released and seems to result from the change in surface tension of the gel backbone upon desorption of the CO_2 surface layer. The effect has been proven to be partially reversible [38]. A rapid supercritical drying process was developed by Poco and co-workers and further characterized by Gross and co-workers [36, 37]. Here, the autoclave vessel acts as the mold for the gel and supports the backbone when tensile strains arise during depressurization. This concept enables a solution to be processed to an aerogel within 1 h. This concept was further optimized by Scherer and co-workers [71, 77].

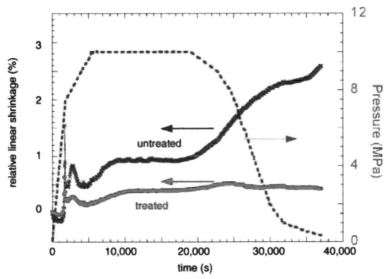

FIGURE 1.17 In situ measured relative linear shrinkage of an as prepared (untreated) and a heat-treated silica alcogel (TMOS based) upon supercritical drying in CO2.

1.2.1.2.3 AMBIENT PRESSURE DRYING

Kawaguchi and co-workers [48] and Smith and co-workers [79] were among the first to look into the drying behavior of silica gels at ambient conditions. In particular Smith and co-workers [79] studied the effects of different pore liquids and surface methylation on the drying kinetics and shrinkage of the gels in great detail. Schwertfeger and Frank applied their findings and patented a process for the drying of low-density silica aerogel granules at ambient pressure. Bhagat and co-workers [9] recently published a novel fast route for silica aerogel monoliths; hereby they use a co-precursor method for the surface modification in silica hydrogels by adding trimethylchlorosilane (TMCS, [77]) and hexamethyldisilazane (HMDS,) to the starting solution of a diluted (water glass derived) ion exchanged silicic acid. By using this approach, they have been able to prepare silica aerogels with a diameter of ~10 mm and a density as low as 0.1 g/cm^3 within ~30 h. Recently, Job and co-workers and Leonard and co-workers investigated in detail the option of applying convective drying at ambient pressure to convert monolithic organic aquagels into crack-free aerogels. The shrinkage of mesoporous aquagels upon ambient pressure drying can be significantly reduced by replacing the water in the pores by a liquid with a significantly lower surface tension, for example, acetone, isopropyl alcohol or ethanol [71].

The key to a fast crack-free drying is to avoid vapor pressure gradients across the macroscopic surface of the gel body that result in locally different rates of volume change and thus crack formation. Unexpectedly, at first sight, the fact that upon drying the first segments that is accompanied by significant shrinkage is not the most crucial phase of drying at ambient conditions. In this phase, the gel is still filled with pore liquid, and mass loss is exclusively due to liquid squeezed out of the gel by capillary forces originating from the evolution of menisci in pores at the outer surface of the sample only (Fig. 15). Unless the partial vapor pressure of the pore liquid is reduced faster than the liquid can be squeezed through the pores to the external surface of the gel, no macroscopic damage is observed. When the capillary forces are finally counterbalanced by the compressed gel body, the gel stops shrinking, the menisci invade the gel, and pores are progressively emptied. Consequently, the capillary forces decrease and the gel partially reexpands. In this phase, the drying rate is limited by transport in the gas phase and along the inner surface of the sample. The extent of reexpansion can, in theory, be 100%. In practice, however, the sample volume is only partially recovered due to the reaction of neighboring inner surfaces or creeping resulting in a plastic deformation of the gel backbone.

1.2.1.2.4. CRYOGENIC DRYING

To reduce the capillary pressure upon drying, the porefluid can also be extracted via sublimation. For this purpose, the gel has to betaken below the freezing point of the pore liquid (note that with decreasing pore size the freezing point will be shifted toward lower temperatures). Besides the temperature, the cooling rate is an important factor in this process since a low cooling rate can result in a crystal growth that can severely modify or even destroy the gel body. The second important quantity is the volume change of the pore liquid upon freezing; in particular in the case of water its density anomaly results in a 10% expansion upon freezing that is not tolerated by the gel backbone and yields a disruption of the gel network. Therefore, prior to freezing, pore water is displaced by a liquid, for example, test-butanol that preferably also possesses a high vapor pressure and thus enables faster sublimation. With respect to the mesoporosity of the resulting aerogels, freeze drying seems to be superior to ambient pressure drying at least in the case of some organic gels; however, the mesopore volume detected is still lower for freeze-dried gels compared to their super critically dried counterparts [71].

1.2.1.3   PROPERTIES OF AEROGEL

Aerogel properties are usually correlated with density, rather than related to morphological features. One reason for this is the lack of suitable representation of aerogel morphology. They are characterized by a network of three-dimensionally (3D) interconnected particles that define a system of well-accessible meso- or macropores (Fig. 1.18). Performance windows, super capacitors, heat barriers, particle traps, ultrasound probes, and ion exchange media [8, 19, 23, 24].

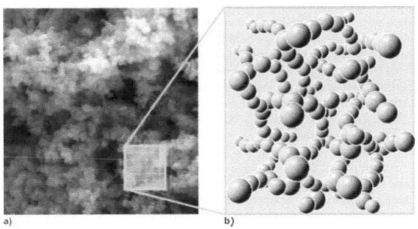

a)   b)

FIGURE 1.18   (a) The Scanning electron microscopy (SEM) image of an aerogel with a colloidal backbone morphology; (b) Corresponding sketch of the 3D interconnected skeleton [71].

Although the internal structure of aerogels and xerogels is the key factor for its excellent properties, such as large surface area and low thermal conductivity, their relatively poor mechanical response limits its final application. Hence, it is important to study and evaluate the mechanical properties of these materials.

1.2.1.3.1   MORPHOLOGICAL CHARACTERISTICS

The unique properties of aerogels are a consequence of the distribution and layout of the two phases, that is, the solid backbone and the pore network. Interestingly, the presence and properties of a void phase can significantly modify the properties of the overall material [84]. The impact of the total porosity on any macroscopic property can be predicted in a first step by using the Hashin–Shtrikman principle [84]. Here, the two extreme cases of a serial and a parallel connection of the two phases define the upper and lower bonds of the range of possible values for a physical property at a given porosity. In addition, the pore properties also play a crucial role with respect to the effective behavior of the porous medium. Aerogels per definition contain a network of pores that possess a connection to the outer surface of the gel (unless strong shrinkage takes place upon drying or sintering, resulting in the cut-off of pore connections). The main quantities characterizing the pore phase are (see Fig. 1.19): the overall porosity; the contribution of micro-, meso-, and macroporosity; the pore size distribution and the average pore size; the pore shape; the pore connectivity; and the physical and chemical properties of the solid–void interface. The highest porosities achieved for aerogels so far have even exceeded 99%.

Specific surface areas related to meso- or macropores range between a few and several hundred meters squared per gram (m^2/g). In the case of aerogels with a microporous (pores <2 nm) backbone, for example, carbon, aerogels, total specific surface areas up to ~2000 m^2/g, can be provided. The average pore sizes (not including the subnanometer micropores) are between a few nanometers and several microns; here in particular silica aerogels seem to possess a narrow pore size distribution as suggested by the comparison of quantities with a different weighting of the pore size distribution like the chord length or pore size determined from small-angle X-ray scattering and sorption isotherms, respectively, and the hydraulic radius derived from gas permeation measurements. Several orders of magnitude in mean pores sizes can easily be covered even at medium porosity with organic aerogels and their carbon counterparts [66]. In porous solids, the solid phase also has an impact on the aerogel properties through its composition, its connectivity, and the size of the backbone segments (length, diameter). The backbone characteristics are reflected in macroscopic quantities, for example, the elastomechanical behavior and the electrical or thermal transport along the solid phase (Fig. 1.19).

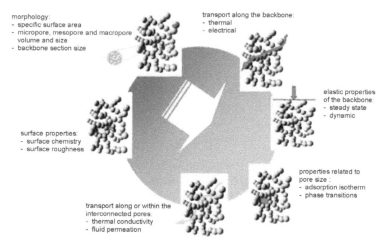

FIGURE 1.19 The morphological properties of the two phases of an aerogel, as well the characteristics of their interface (top left), determine the deduced properties, such as, for example, the transport properties in the solid and the pore phase, their mechanical behavior, and steady-state properties of the pore, and their adsorption characteristics.

Scaling Behavior: Typical for the change of the macroscopic properties with the density or porosity of an aerogel (at a given growth mechanism of the backbone) is their scaling behavior. For this reason, aerogels were popular model materials, in particular in the 1980s and 1990s, to theoretically investigate the predicted properties of percolating systems [2] and fractals. While scaling of properties with bulk density can be seen at least over a limited range for most aerogels, not all of them also exhibit fractal morphology. Porous materials with fractal morphology are characterized by a scale invariant structure, that is, they look alike under different magnifications. In practice, this behavior is only observed on a limited length scale with upper and lower bounds (Finite Size Scaling). Since a good deal of work has been performed on predicting the relationship of properties in the case of fractals, this concept will be used here to explain the intimate connection between different macroscopic properties. Similar results, however, are found for nonhierarchical, random heterogeneous media where the so-called critical exponents play a similar role as the exponents in the concept of fractal systems [84]. Important is the fact that different macroscopic properties are closely correlated for a given type of backbone and pore morphology connectivity. In some cases, the scaling concept is found to be only a coarse approximation for a small density range [31]; however, the concept is valuable for identifying trends and correlations even in these cases. Relevant quantities that describe aerogels with a fractal backbone, for example, silica aerogels, are the mass fractal dimension d_M that relates the total mass m within a sphere of radius R to the length scale R,

$$m(R) = R^{d_M} \tag{60}$$

the surface fractal dimension dS that reflects the surface roughness and thus the surface area S as a function of the scale length L

$$S(L) = L^{d_s} \tag{61}$$

and the fraction exponent or spectral dimension d0 of the material that relates the vibrational density of states $D(\omega)$ to the frequency of the vibration via

$$D(\omega) = \omega^{d'-1} \tag{62}$$

The Eqs. (60) and (61) refer to the mass distribution and roughness. These properties can directly be determined from small-angle X-ray or neutron scattering data, since the exponents in the length scale regime [see, Eqs. (60) and (61)] transfer into a scaling of the scattering intensity with the scattering vector. Typically, the mass fractal dimensions d_M of aerogels range between 1.8 and 2.5. To compare and interpret scaling exponents, one has to keep in mind that these values are based on the assumption of a fixed underlying growth mechanism (e.g., diffusion-limited aggregation, cf. Fig. 1.11) that dominates over the full range of densities considered. If the mechanism changes with density, largely deviating exponents are to be expected.

The Eq. (62) contains information about the connectivity of the gel backbone segments. From a more general point of view, the knowledge of the density of vibrational states over the full frequency range from several hertz (Hz) to terahertz (THz) provides the full spectrum of localized and nonlocalized vibrations (phonons) possible in the aerogel under investigation. All "phonon" related properties, for example, the specific heat as a function of temperature, as well as acoustical properties and the heat transfer along the backbone, can be deduced from this quantity. Typically, the density of states of an aerogel consists of four characteristic regimes (Fig. 1.20):

1. The Debye regime, where the phonon wavelength is larger than the inhomogeneities of the aerogel backbone.
2. The fracton or localized modes regime, that corresponds to vibrational motions of the chains between the knots in the disordered interconnected backbone network.
3. The particle modes regime, corresponding to the Eigen modes of the particles or struts forming the backbone.
4. The molecular regime, reflecting the motion of molecules attached to the backbone surface or in the bulk.

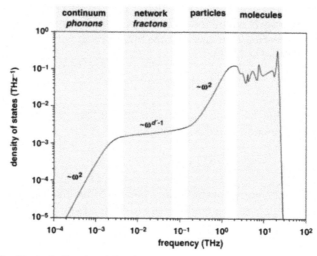

FIGURE 1.20   Typical vibrational density of states of an aerogel.

1.2.1.3.2   MACROSCOPIC PROPERTIES: MECHANICAL CHARACTERISTICS

In the case of fractals, the scaling exponents that describe the elastomechanical behavior and the transport properties along the solid backbone can be calculated from the fractal dimension d_M and the fracton scaling exponent d [87]. Scaling laws for the elastic modulus E of silica, as well as organic and carbon aerogels as a function of density ρ, were experimentally determined by Gross and co-workers:

$$E(\rho) = \rho^{\alpha} \text{ with } \alpha \text{ between } 2.0 \text{ and } 3.6 \qquad (63)$$

Theoretical assumptions yield for the scaling exponent a

$$\alpha = (5 - d_M) / (3 - d_M) \qquad (64)$$

That is, with mass fractal dimensions d_M typical for aerogels between 1.8 and 2.5, values for a between 2.7 and 5 are expected; these values roughly match the experimentally determined data. Ma and co-workers investigated the relationship between the density and the mechanical stiffness of aerogels in a very sophisticated simulation of the mechanical behavior of gel backbones; hereby the solid phase of different aerogels, generated by diffusion-limited cluster–cluster aggregation (DLCA) algorithms, mimicked the experimentally observed morphological properties [54, 55]. The goal of this study was to understand the high, experimentally derived scaling exponent for Young's modulus with density (cf. Eq. (63)). One important result was the role of the dangling bonds in the network and the decrease in their relative mass contribution with increasing bulk density; in addition, the authors were able to simulate the local strain energy distribution in the gel network as a function of gel density (Fig. 1.22). Their data revealed that at low aerogel density, only a very small percentage of the network supports an externally imposed load. Gross and co-workers showed that determining the elastic properties for low density aerogels at ambient conditions via a dynamic measurement, for example, a sound velocity measurement, results in large deviations (towards larger sound velocities) from the scaling behavior deduced from higher densities. The reason for this effect is the finite compressibility of the gas in the pores of the aerogel that dominates the properties at low backbone stiffness [37, 38]; the authors showed that the scaling behavior is fully recovered when the sample is evacuated. Typical values for the sound velocity and Young's moduli at an aerogel density of ~0.3 g/cm³ are 200–500 m/s and ~6–60 MPa, respectively (Fig. 21). The elastic properties of the backbone of aqua- and alcogels can be investigated with the beam-bending method to study, for example, the impact of additional treatments of gels prior to drying with the aim to increase the backbone stiffness [71]

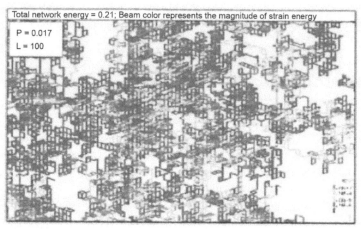

FIGURE 21 Distribution of the strain energy in a perfectly distributed, diffusion limited cluster aggregation network.

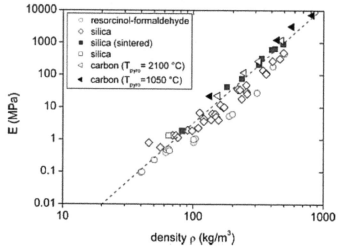

FIGURE 22 Young's modulus E versus density for different types of aerogels. The dashed line corresponds to a scaling exponent of 3.6.

The damping of sound waves in silica aerogels was studied by Zimmermann and Gross, as well as by Forest and Woignier; internal friction effects due to adsorption of molecules at the inner surface of the aerogels seem to control the damping behavior. The inverse quality factor Q^{-1} that combines the acoustic extinction coefficient a with the wavelength of the phonons under investigation l was found to be $0.3–0.5 \times 10^{-3}$ [71].

$$Q^{-1} = \alpha.\frac{\lambda}{\pi} \, with \, I(x) = I_0.\exp(-\infty.x) \tag{65}$$

Hereby I and I_0 are the acoustic intensities of the dumped and undumped wave, respectively.

1.2.1.3.3   THERMAL AND ELECTRICAL CONDUCTIVITY

Scaling theory predicts, for the conductivity of the backbone, an increase with density with a scaling exponent that is always smaller than the one for Young's modulus. For the thermal and electrical conductivity of aerogels, λ and σ, respectively, experimentally derived scaling exponents γ of ~1.5–2.2 are reported at ambient conditions. The very detailed study by Bock reveals a scaling exponent for the electrical conductivity that drops from ~3.6 at 10 K to ~1.7 at >100 K. This result is due to the temperature dependence of the charge carrier mobility. Besides porosity, the thermal, as well as the electrical backbone conductivity, are controlled by the connectivity of the solid phase and its intrinsic conductivity properties. While in the case of thermal conductivity the transport occurs via phonons and, in case of an electrically conductive material, also via electrons, the electrical conductivity takes place via charge carriers only. Because at knots the aerogel backbone is chemically connected, the conductivity of the aerogel skeleton is often higher than in a system of powder particles with the same solid content and the same chemical composition. The fact that the scaling exponent γ iswell above unity reflects the increasing amount of dangling mass with increasing porosity in aerogels [31]. The effective conductivities of aerogels can be strongly enhanced when annealing results in a phase change or molecular rearrangement of the backbone material, dangling ends are reconnected to the continuous backbone or necks at the joints of adjacent backbone particles are smeared out. Other options to affect the backbone conductivity are additives on the molecular (doping) or mesocopic (particles, fibers) scale. Pure carbon aerogels and xerogels typically show electrical conductivities at room temperatures between ~1 S/cm (for densities of ~100 kg/m³) and 50 S/cm (for densities of ~1000 kg/m³) [71].

The different methods that are typically applied to characterize the morphological and other basic properties of aerogels are summarized in Fig. 1.23. Note that some of them carry the tag "restrictions," denoting that special care has to be taken to make sure that the aerogel under investigation is neither temporarily nor permanently modified upon analysis. The restrictions for applying electron microscopy refer to the fact that the backbone units may be partially sintered due to local heating effects. Mercury porosimetry is an established technique for characterizing open porous materials with pore sizes ranging from a few nanometers to ~300 m. It makes use of the fact that mercury does not wet most substances. To force the mercury into pores, an external pressure is needed that is inversely proportional to the size of the pores to be infiltrated; to force mercury into pores of ~50 nm, a pressure of ~20 MPa (200 bar) is necessary. Especially the low-density aerogels, however, are compliant and are therefore largely compressed before the mercury actually enters the pore network.

Technique	Length scale
Electron microscopy (SEM) (restrictions!) light scattering	Millimeter
Mercury porosimetry (restrictions!)	
small angle X-ray or neutron scattering N2-Sorption (restrictions!)	
Electron microscopy (TEM) (restrictions!) X-ray diffraction	Angstrom

Further methods:	
Beam bending:	Elastic properties, fluid permeation of the aqua oralcogel
Sound velocity:	Elastic properties
Gas permeability:	Gas permeation, average pore size
Thermal conductivity:	Connectivity of the gel backbone, average pore size
Electrical conductivity:	Connectivity of the gel backbone

FIGURE 1.23   Summary of methods typically applied to characterize the morphology and other basic properties of aerogels.

A similar problem can occur upon nitrogen sorption analysis, a technique that enables the characteristics of accessible micro and mesopores to be determined. Upon nitrogen sorption, the liquid–vapor interface of nitrogen condensed in the mesopores of the sample under investigation forms a meniscus; the related capillary forces, upon both adsorption and desorption, can significantly compress the gel backbone; this results in an elastic or plastic deformation of the upon analysis and even worse a sorption isotherm that is highly affected by the compression effect rather than reflecting the pore size distribution of the uncompressed backbone as expected. One positive side effect is that, in the case of compliant aerogels, the shape of the sorption hysteresis reveals the modulus of compression and allows for a zero-order correction of the extracted pore size.

Light, X-ray, and neutron scattering are nondestructive methods that are sensitive to length scales of microns to angstroms. These methods are complementary, for example, to nitrogen sorption, since they also detect density fluctuations and closed porosity in aerogels (e.g., micropores in carbon aerogels) and even allow the morphological changes upon gelation or the morphology of a wet gel to be analyzed.

Another very valuable method to investigate the properties of the aqua- or alcogel is the beam-bending technique. Here, a rod-like shaped gel is bent in a three-point bending set-up and the force needed to keep the sample bent is measured as a function of time. Since the pores of the gel are liquid filled, the pressure gradient within the sample caused by the bending of the gel rod initiates a redistribution of the pore liquid so that the pore liquid is finally free of force. The velocity of this process is reflected in the decrease in the external force measured. Accordingly, the effective pore size can be derived from these data, if the sample geometry is well defined and the viscosity of the pore liquid and the deflection of the rod are known. The residual equilibrium force and the deflection of the gel rod finally provide the modulus of the gel backbone.

The mechanical properties of the aerogel can be quantified by measuring the time needed for an ultrasonic pulse to travel a defined path through the sample. Using the density ρ of the aerogel, Young's modulus E can be calculated from the sound velocity v via $E = v.\rho^2$. The sound velocity reflects the properties of the gel backbone and in particular its connectivity at a given porosity. It has to be noted, however, that at low sound velocities, the impact of the gas in the pores is no longer negligible and will result in a density-independent sound velocity.

The backbone connectivity can also be investigated via the thermal or the electrical conductivity of the backbone. Both sound velocity and electrical conductivity measurements (of conductive aerogels) are fast methods of extracting additional information beyond the macroscopic density. The gas pressure dependence of the thermal conductivity at pressures <0.1 MPa (1 bar) can be exploited to extract average pore sizes for pores >100 nm (162, 163) [71].

1.2.1.4 PROPERTIES OF AEROGEL

Aerogels are a promising material for a host of applications [8, 19] due to their thermal, optical and mechanical properties. [8] They are among the best thermal insulating solid materials known [6, 9, 17]. It is important to link aerogel properties to their complex internal microstructure, and to understand how such properties can be optimized for a given application [8].

After Monsanto closed down the aerogel plant in the 1970s, small companies like Airglass (Sweden) took over the production of aerogels on a small scale, mainly serving Cerenkov detector applications and research projects on window glazings. Nanopore Inc. is another company that has been continuously active in the field of aerogels over the past 20 years. Two "breakthroughs" in the 1990s triggered serious interest in aerogels from further companies: the discovery of carbon aerogels by Pekala and Kongand the patent for a new process to synthesize granular silica aerogels in 1997 by Schwertfeger and Frank. Though the potential applications of aerogels are numerous (Table 1.4) and cover almost all technical fields, it has only been in recent years that commercialization has become very active. The current worldwide promotion of nanotechnology is supporting this trend. Examples of commercially available products containing aerogel are listed in Table 1.5.

TABLE 1.4 Summery of Aerogel Application

Thermal insulations	Day lighting components:	Mechanical properties and Sensors
• Aspen Aerogels, Inc.: Technical insulation. • Corpo Nove and Aspen Aerogels, Inc.: Jacket for severe weather conditions. • Nanopore Inc.: Shipping containers. • Cabot Corporation together with Corus: HPHT pipe in pipe systems. • Toasty Feet: Shoe insoles.	• Cabot Corporation together with Wasco Products, Inc., and Kalwall • Mechanical properties: • Dunlop: Tennis racquet with aerogel composite • Sensors: • General Dynamics	• Dunlop: Tennis racquet with aerogel composite • General Dynamics
Catalyst support	**Filter**	**Electrodes**
• Aerogel Composite, LLC: Carbon aerogel-based electro catalyst for fuel cells	• Taasi aerogels: Air filters. • CDT Systems: Water purification.	• CDT Systems, Inc. • Cooper Power Systems: Supercapacitors.

TABLE 1.5   Commercially Available Products

Application of Aerogels	
Permeation or adsorption	Filter
	Gas separation
	Waste Water Treatment
	Chromatography
Mechanical or Acoustical	Shock Absorber
	Acoustic Impedance Matching
Optical	IR-reflector or –absorber
	Light Management
Electrical	Porous Electrodes
	Dielectric Layers
Carrier, Support or Matrix	For catalysts
	Drug Release
	Bio/Medical Components
	Sensors
	Explosives
Thermal	Insulation Management

Thermal insulation and shock absorber applications significantly benefited from the development of silica aerogels with respect to space applications, for example, the thermal insulation of the Pathfinder on its mission to Mars or capturing micrometeorites in the STARDUST mission (Fig. 1.24). Still of scientific and commercial interest are double-glazing systems filled with transparent silica aerogel tiles.

Current aerogel activities focus on particular catalyst supports based on carbon or other types of aerogels, molds for metal casting, preparation of aerogel particles and aerogel-comprising composites for electrodes or other functional materials. Increasingly, aerogel films and coating are synthesized and investigated to optimize them as dielectric or sensor components or to provide additional functionalities in layered systems. Recently, aerogels are also discussed in the context of hydrogen storage either as the matrix for the actual storage material or as the storage component itself.

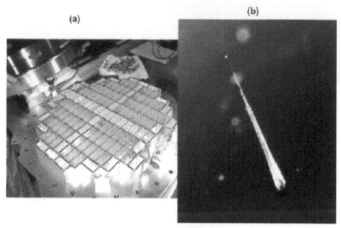

FIGURE 1.24   (a) Aerogel dust collector under construction. (b) Particle captured in aerogel.

1.2.2 CHARACTERIZATION OF AEROGELS USING MD SIMULATION

Aerogels are known to have mass and surface distributions consistent with fractal behavior over certain length scales. At sufficiently small (atomic) length scales their structure is determined by chemical considerations and is therefore not fractal, and at sufficiently large length scales they are homogeneous [31]. The evolution of this structure and the corresponding effects of relaxation, chemical kinetics, and atomic coordination have been of particular interest in modeling and simulation studies. Computer simulations and experiments are widely used for the studies of the structure and properties of aerogels and xerogels.

1.2.2.1 MODELING OF AEROGEL THROUGH DIFFERENT TECHNIQUES

In a computational model, the material structure is known exactly and completely, and so structure/property relationships can be determined and understood directly. [1]. The interatomic potential is the key component of any MD simulation. The model specifies the actual "fundamental" objects simulated, be they electrons, atoms, molecules, or sol particles, and their interactions, expressed through some kind of potential energy function [1]. It represents the most important interactions among atoms, that is, bonding interactions, interactions with nonbonding neighbors, and the extension of those interactions. If the interatomic potential does not accurately describe the interatomic interactions, the simulation results will not be representative of the actual material [73].

Having prepared a model structure, simulation studies have generally focused on structural characterization and the relationship between aerogel structure and mechanical properties. Global measures including fractal dimensions, surface areas, and pore size distributions can be directly calculated from the model structure and compared with experimental data. Microscopic measurements such as the distributions of bond lengths and bond angles and the number of bridging oxygen's bound to each silicon atom are also measured. Finally mechanical properties, including moduli, shrinkage upon drying, and the vibrational density of states can be determined and correlated with the gel structure [8]. Other topics that have been the focus of modeling studies include the prediction and analysis of gelation kinetics and aerogel mechanical properties, for which a considerable amount of experimental data is available [61–63].

The most challenging part of such a computational study is obtaining the model structure itself. Techniques applied in the case of aerogels include both "mimetic" simulations, in which the experimental preparation of an aerogel is imitated using dynamical simulations, and reconstructions, in which available experimental data is used to generate a statistically representative structure.

Simulating the formation and properties of aerogels, like all other applications of molecular simulation, involves choosing both a model and a simulation technique. The model specifies the actual "fundamental" objects simulated, be they electrons, atoms, molecules, or sol particles, and their interactions, expressed through some kind of potential energy function. Models discussed below include both atomistic descriptions, in which each atom is treated individually, and coarse-grained descriptions, which treat larger objects. At the atomic scale, there are two classes of potential used. In quantum-mechanical potentials, calculation of the energy of a configuration of atoms is accomplished by determining the electronic wave function (or density, in the case of density functional theory [24] and associated energy [25, 26]. In empirical potentials, also known as force fields, the energy is built up as a sum over different types of interactions: core-repulsions, bond stretches and bends, torsions, columbic interactions, hydrogen bonds, dispersion forces, etc., each of which is described using a relatively simple function that has been parameterized against either quantum-mechanical results or experimental data [27–33].

Once a model has been specified, different calculations may be performed. Of particular relevance here are dynamical calculations, which generate a trajectory according to specified equations of motion. In the simplest case, Newton's equations are used. The trajectory thus generated conserves total energy and, under equilibrium conditions, samples the micro canonical ensemble. The equations of motion can be modified to enforce constant-temperature and/or constant-pressure conditions. These are collectively referred to as MD simulations. Stochastic dynamics, based on the Langevin equation (or derived results), are used in order to avoid the simulation of solvent molecules [27, 33]. In these techniques, friction terms and random impulses are used to model the interaction of solute molecules with the solvent; in dilute solutions, this can reduce the number of objects simulated by several orders of magnitude. In the particular case of sol–gel processing in aqueous media, the use of such an "implicit" solvent is problematic because water itself is both a product of and catalyst for the siloxane condensation reaction. Other, "nonsimulation" operations include energy minimization and transition-state location, and are primarily used in quantum-mechanical studies of chemical reactions [1].

Generally, the model structures have been described in Fig. 1.25.

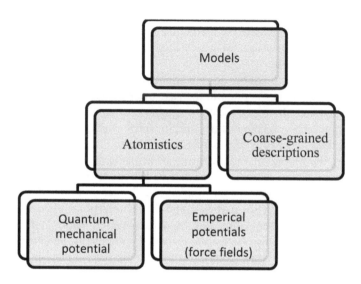

FIGURE 1.25   Modeling structure.

1. Atomistics: atom is treated individually.
    a) Quantum-mechanical potential: calculation of the energy of a configuration of atoms is accomplished by determining the electronic wave function (or density, in the case of density functional theory and associated energy.
    b) the energy is built up as a sum over different types of interactions: core-repulsions, bond stretches and bends, torsions, columbic interactions, hydrogen bonds, dispersion forces, etc., each of which is described using a relatively simple function that has been parameterized against either quantum-mechanical results or experimental data.
2. Coarse-grained descriptions: treat larger objects.

Atomistic modeling with empirical potentials is much less expensive and also scales better with increasing system size. Such calculations are therefore capable of accessing larger systems, typically as many as 105 atoms, and occasionally much larger. Likewise, they can be extended to longer times; simulations of more than a microsecond have been accomplished using commonly available computers, although most force-field-based MD simulations to date are shorter than 10 ns or so. However, in sol-gel processing due to the making and breaking of the chemical bonds biochemical simulation are unable to describe such processes. Hence, many empirical modeling required the development and validation of new potentials [1].

To access length scales larger than tens of nanometers and time scales longer than nanoseconds, simulators turn to coarse-grained models. In such simulations the "primary" particle simulated is no longer a single atom or small molecule, but a greatly simplified representation of an assembly of many atoms or molecules. These objects interact with each other through effective potentials obtained from either theoretical consideration, matching to higher-resolution atomistic simulations, or fitting against experimental data. Such models can be extended to much larger length and time scales, for several reasons. Reduction in the number of degrees of freedom in the simulation greatly reduces the computational effort required for energy and force calculations. Furthermore, because these particles are heavy (compared with atoms or small molecules) and interact through somewhat softer potentials, the equations of motion can be integrated with much larger time steps. Although many hybrids of these basic approaches are also known, the physical and chemical processes underlying aerogel preparation and properties are thought to be relatively friendly to the separation of scales and methods just described, and studies to date have exploited this. The aqueous hydrolysis/condensation chemistry underlying silicate precursor oligomerization has been studied using quantum mechanical methods. Formation of larger oligomers and complete gelation of dense aerogels has been studied with atomistic simulations of empirical force-field models. Since such simulations cannot access the time or length scales associated with the gelation of low-concentration precursors, a variety of coarse-grained models have been proposed for these processes. In this section we review these studies and discuss both the methods and models used and the conclusions reached.

An alternative approach to modeling complex porous materials is reconstruction, in which experimental data are inverted to generate a computational realization of the material structure. There is insufficient data in any experimental characterization short of high-resolution three-dimensional tomography to exactly reconstruct an amorphous material. Indeed, most experiments provide either a one-dimensional dataset, such as a structure factor or adsorption isotherm, or a two-dimensional dataset, such as an electron microscope image. Stochastic reconstruction generates representative structures that fit such limited input data but are otherwise random and isotropic. There are in general many possible model structures that will fit the input data equally well, which is known as the "problem of uniqueness." Although we do not review these techniques or their applications in detail, we do note that they have been applied to sol–gel materials in a number of recent studies. Quintanilla et al. applied the method of Gaussian random fields to generate stochastic reconstructions of TEOS-derived aerogels. Eschricht et al. used evolutionary optimization to reconstruct the commercially available "GelSil 200" material from a combination of small-angle neutron scattering data (SANS) and a pore-size distribution derived from gas adsorption data. Steriotis et al. used SANS data to guide a "process-based" atomic-scale reconstruction of a silica xerogel [1].

Finally, an entirely different approach to simulating gelation is the "Dynamic Monte Carlo" (DMC) method, in which chemical reactions are modeled by stochastic integration of phenomenological kinetic rate laws. This has been used successfully to understand the onset of gel formation, first-shell substitution effects, and the influence of cyclization in silicon alkoxide systems. However, this approach has not so far been extended to include the instantaneous positions and diffusion of each oligomer, which would be necessary in order for the calculation to generate an actual model of an aerogel that could be used in subsequent simulations [1].

These different computational and simulation techniques have their own advantages and disadvantages. For instance, although fully quantum-mechanical calculations can be very accurate, they can be applied only to a few molecules, and when used in dynamical calculations they can access only picosecond time-scales. As the system size increases the cost rises steeply. Hence, they impose high computational cost of calculating the total energy of an assembly of atoms using either wave function-based or density-functional techniques. This restricts such models only to small systems [1].

1.2.2.2   SIMULATION OF A SILICA AEROGEL THROUGH SOL-GEL PROCESSING

Mostly simulation of an aerogel follows this procedure:
1. Introduction of the MD simulation procedures used to prepare the aerogel and xerogel samples. Once a model has been specified, different calculations may be performed. Of particular relevance here are dynamical calculations, which generate a trajectory according to specified equations of motion. In the simplest case, Newton's equations are used.
2. Description of methods for obtaining structural characterization and the relationship between aerogel structure and the fractal dimension and the thermal and mechanical properties of the samples.
3. Discussion about the characteristics of the samples including: fractal dimensions, surface areas, pore size distributions, thermal properties, elastic modulus, and strength in relation to density and comparing the results with experimental data. Moreover, microscopic measurements such as the distributions of bond lengths and bond angles and the number of bridging oxygen's bound to each silicon atom are also measured. Finally mechanical properties, including moduli, shrinkage upon drying, and the vibrational density of states can be determined and correlated with the gel structure [1, 73].

Simulating the preparation of xerogels and aerogels involves separate treatment of gelation, aging, drying, and for nonporous materials, consolidation. Consolidation, or heating at high temperatures, is used to generate densified, nonporous materials for optics and other applications. The aqueous hydrolysis/condensation chemistry underlying silicate precursor oligomerization and reaction mechanisms and energetics has been studied using quantum mechanical methods. Formation of larger oligomers and complete gelation of dense aerogels has been studied with atomistic simulations of empirical force-field models. Since such simulations cannot access the time or length scales associated with the gelation of low-concentration precursors, a variety of coarse-grained models have been proposed for these processes [1].

1.2.2.2.1 GELATION

As we explained before, in the gelation step, alkoxide gel precursors in aqueous solution are hydrolyzed and polymerize through alcohol or water producing condensations:

$$\equiv Si - OR + H_2O \leftrightarrow \equiv Si - OH + ROH$$

$$\equiv Si - OR + OH - Si \equiv \leftrightarrow \equiv Si - O - Si \equiv | ROH$$

$$\equiv Si - OH + OH - Si \equiv \leftrightarrow \equiv Si - O - Si \equiv + H_2O$$

The gel morphology is influenced by temperature, the concentrations of each species (attention focuses on r, the water/alkoxide molar ratio, typically between 1 and 50), and especially acidity. An entirely different approach to simulating gelation is the "Dynamic Monte Carlo" (DMC) method, in which chemical reactions are modeled by stochastic integration of phenomenological kinetic rate laws [15].

1.2.2.2.2 AGING

Gel aging is an extension of the gelation step in which the gel network is reinforced through further polymerization, possibly at different temperature and solvent conditions. Simulating aging requires the use of an approach, which can access long time scales. The "activation-relaxation technique" (ART) is being implemented for this purpose.

1.2.2.2.3 DRYING

The gel drying process consists of removal of water from the gel system, with simultaneous collapse of the gel structure, under conditions of constant temperature, pressure, and humidity. In the coarse-grained model the equation of state is trivially calculable, and drying is easily modeled by choosing the solvent chemical potential to favor the vapor phase and allowing the particle positions and cell volume to slowly relax under the influence of solvent capillary forces. At the molecular scale, we can model this process using an extension of the "Gibbs Ensemble Monte Carlo" technique for binary mixtures, where the mixture consists of water and atmosphere.

1.2.2.2.4 CONSOLIDATION

Xerogels are higher in free energy than conventional amorphous silica (glass) and crystalline silica, as they have a substantial internal surface area and associated surface tension. Simulations of consolidation will use molecular models. Both isothermal conditions and constant heating rates can be accessed with standard MD simulations and ART as above.

1.2.2.2.4 AEROGELS

Aerogel systems do not collapse (much) under drying conditions and supercritical gel drying will be easier to simulate than the subcritical process. High-porosity aerogels are only accessible via the meso-scale model. The initial stages of gelation, when the average cluster size is very small, are best modeled with a purely atomistic approach. Hierarchically structured gels and low-density gels cannot be directly treated with molecular models; a meso-scale approach must be used in this case. Relatively dense gels can be modeled with either atomistic simulations or coarse-grained simulations. [25, 27, 91].

1.2.2.3 INTERATOMIC POTENTIAL

Atomistic modeling and simulation have been used both in studies of the hydrolysis/condensation chemistry underlying silica sol–gel processing, and in dynamical simulations of the formation of sol particles and even gels. Studies of reaction mechanisms and energetics have made extensive use of quantum mechanical methods, while the larger length scales and time scales of oligomerization and gelation have been handled using empirical potentials, as reviewed in this section.

Vashishta et al. [88, 89] developed an interatomic potential for amorphous silica. This potential involved terms representing the interaction between two atoms (two-body component); which, accounts for the potential energy due to the distance between them. The potential also includes the potential energy due to the change of orientation and bonding angle of triplets of atoms (three-body component). The two body and three-body components of the potential are given by

$$V_{ij}^{(2)} = \frac{H_{ij}}{r_{ij}^{\eta_{ij}}} + \frac{Z_i Z_j}{r_{ij}} exp\left(\frac{-r_{ij}}{r_{1s}}\right) - \frac{P_{ij}}{r_{ij}^A} exp\left(\frac{-r_{ij}}{r_{4s}}\right) \tag{66}$$

$$V_{jik}^{(3)} = B_{jik} f\left(r_{ij} r_{ik}\right)\left(\left(cos\theta_{jik} - cos\bar{\theta}_{jik}\right)\right)^2 \tag{67}$$

In the previous equations r_{ij} is the distance between the atoms i and j. The first term of Eq. (66) represents the steric repulsion due to the atomic size. The parameters H_{ij} and η_{ij} are the strength and exponent of steric repulsion. The second term corresponds to the Coulomb interactions between the atoms and accounts for the electric charge transfer, where Z_i is the effective charge of the i-th ion. The third term includes the charge–dipole interactions. It takes into account the electric polarizability of the atoms through the variable P_{ij}, which is given by

$$P_{ij} = \frac{1}{2}\left(\alpha_i Z_j^2 + \alpha_j Z_i^2\right)$$
(68)

where α_i is the electric polarizability of the i-th ion. The parameters r_{1s} and r_{4s} are cut-off values for the interactions.

In Eq. (67), B_{jik} is the strength of the three-body interaction; θ_{jik} is the angle between the vector position of the atoms, that is, r_{ij} and r_{ik}. The function f represents the effect of bond stretching and the component containing (cos θ_{jik}) takes into account the bending of the bonds, and $\tilde{\theta}_{jik}$ is a reference angle for the respective interaction. Although, there are six possible three-body interactions in the system, this potential only considers the most dominant ones, which are related to (Si–O–Si) and (O–Si–O) angles. More information about the potential, including the values of the parameters, is available in Refs. [88,89].

The first MD studies of oligomerization and the silica sol–gel process were carried out by Garofalini and Melman [20]. A solution consisting of water, silicic acid monomers, and silicic acid dimers was simulated. The potential consisted of a modified Born–Mayer–Huggins (BMH) model for the Si–Si, Si–O, and O–O interactions in nonwater molecules, and a modified Rahman–Stillinger–Lemberg (RSL2) potential [31] for O–O, O–H, and H–H interactions in water molecules.

In 1990, Feuston and Garofalini revised the model in a subsequent study [26]. The modified BMH potential was applied to interactions between all atoms, with H–X interactions supplemented by terms from the RSL2 potential [80]. A subsequent study by Garofalini and Martin extended this to larger systems (216 silicic acid monomers) at densities between 1.4 and 1.6 g/cm³ [87].

Martin and Garofalini have also further studied the details of the condensation reaction in this model [56], observing that when there are multiple bridging oxygen's in the intermediate, there is a tendency to break a silanol bond in creating a leaving group, rather than lose a nonbridging oxygen. This is consistent with the experimental finding that acid-catalyzed conditions favor the formation of linear clusters [14].

The systems described in all these studies were quite small (N ≤ 216) and of relatively high density (1.3 g/cm³ and above). The temperature and pressure conditions used are comparable with an autoclave process used to produce aerogels. These are not typical laboratory conditions for the preparation of sol–gel derived materials. [14, 56].

Rao and Gelb used the Feuston and Garofalini model in simulations closer to these conditions [28]. In this work, 729 silicic acid monomers were placed in water at liquid density with r between 0 and 26, with most work performed at r = 11. Simulation times were as long as 12 ns. High temperatures (1500–2500 K) were again used to promote reaction, and the system volume held fixed. A series of snapshots from one of these simulations is shown in Fig. 1.26. These studies confirmed the action of water as a catalyst, and further illustrated the processes involved in the conversion of monomer precursors to "sol" particles. At short times, monomers quickly dimerized, and further growth occurring via monomer addition. At longer times (after several nanoseconds), condensation between larger oligomers was also observed. The model developed by Feuston and Garofalini, while successfully encompassing much of the important chemistry involved in sol–gel processing, could nonetheless be improved in many ways.

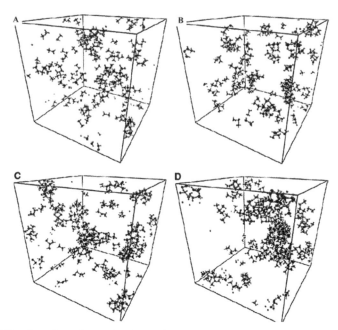

FIGURE 1.26   Snapshots from MD simulations of oligomerization in aqueous silicic acid with r = 11. Snapshots are taken from the simulation at times A. 1.0, B. 2.0, C. 3.0, and D. 4.8 ns. The evolution of small oligomers into compact sol particles is clearly visible.

Bhattacharya and Kieffer introduced a more realistic all-atom reactive "charge-transfer" potential for simulating silica sol–gel processing [10]. In this approach, a certain amount of charge is transferred between bonded atoms; this allows for more realistic electrostatic interactions between molecules, and correctly charged ions upon dissociation.

The MD simulations are used in Kieffer and Angell and Nakano et al. [49, 64, 65] to generate surrogate (computer models) of silica aerogels. Kieffer and Angell [19] propose to use MD to model silica aerogel and xerogel starting from a sample of silica glass that is gradually expanded. The expansion causes breaking of the Si–O bonds and leads to the formation of a fractal structure characterized by a fractal dimension that changes linearly with density. Silica was modeled using the Born–Mayer empirical potential, and up to 1,500 atoms were included in the simulations. The quantity of greatest interest However the linearity is not sustained for samples with densities higher than 1 g/cm³.

Very similar simulation protocols were used in subsequent work by Nakano et al. [24]. In these studies, however, simulation cells containing over 40,000 atoms were used, such that much larger length scales could be probed. An example of these large-system results is shown in Fig. 1.27.

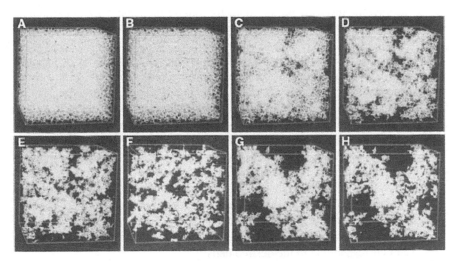

FIGURE 1.27   Snapshots of molecular dynamics of porous SiO_2 glasses A. 2.2, B. 1.6, C. 0.8, D. 0.4, E. 0.2, and F. 0.1 g/cm³ prepared at 300 K, and G. 0.2 and H. 0.1 g/cm³ prepared at 1000 K. Yellow lines represent Si–O bonds.

Comparing the results of studies, fractal dimensions found in the (low-temperature) expansion simulations of Nakano et al. [24] and Bhattacharya and Kieffer [10, 49] are not in perfect agreement, again suggesting that either or both of the potential and the simulation protocol can also affect the final network morphology.

Since the atoms are initially set at distances larger than their equilibrium distances and have high kinetic energy, they can diffuse to farther regions in the simulation volume and could potentially form any kind of structure. That is an advantage of this procedure over the gradual expansion, where the atoms can only move short distances and are bounded to remain in the same region [73].

Muralidharan et al. [61] summarizes the elastic modulus and the strength that can be obtained using four traditional potentials to model dense SiO2 systems, namely Soules potential (S-potential), Born–Meyer–Huggins potential (BMH), Feuston–Garofalini potential (FG) and van Beest–Kramer-van Santeen potential (BKS). The minimum deviation between the simulated value and the experimental one is 39% obtained using BKS-potential.

Kieffer and Angell [18] use an interaction potential of the Born–Mayer form, which accounts for two body interactions, to model a system containing between 300 and1500 particles. Nakano et al. [64] use a potential including two-body and three body interaction components to model porous silica with densities varying from 2.2 g/cm^3 to 0.2 g/cm^3 in a system composed of 41,472 particles. Similar to Kieffer and Angell [19], In studies of Nakano et al. [64] the system was gradually expanded until the desired density is reached and the dependence of the fractal dimension, internal surface area, the pore to volume ratio, pore size distribution, correlation length, and mean particle size on the density of the sample were investigated.

Campbell et al. [19], using the same potential as Nakano et al. [64], generate porous samples by placing spherical clusters of dense silica glass in a large volume and sintering the system at constant pressure and temperature. With that approach they generate samples of densities varying from 1.67 g/cm^3 to 2.2 g/cm^3, a range of density corresponding to xerogels, and study the changes on the short-range and intermediate range order of the structure with density. They also study the effect of densification on the elastic modulus of the samples, finding a power law relation between modulus and density with an exponent of 3.5±0.2.

Murillo et al. used the same potential as Nakano et al. [64] however the preparation of the samples was through different approach. The procedure to generate the porous samples analyzed in this study follows the steps listed below:

1. Placing the atoms at the crystalline sites of β-cristobalite with alattice constant corresponding to the desired density.
2. Heating up of the system to 3000 K.
3. Cooling down of the system allowing relaxation at several temperatures.

The volume of a dense crystalline sample (β-cristobalite) is expanded in one step to the desired density. Then the temperature is increased to give the atoms enough energy to diffuse through the system, and finally it is cooled down in a stepwise process. Keeping the system at high temperature, along with the cooling scheme, eliminates the effect of the initial positions of the atoms. This process resembles a diffusion-limited aggregation. The goal of this process is to create a uniform distribution of atoms across the volume. The interatomic potential was the same as the studies of Vashishta et al. This potential can accurately reproduced the structural parameters for dense silica glass. Moreover, comparing the experimental results of neutron scattering with computer simulations, this potential produces a smallest error of only 4.4%, using the factor R_x [49]. This potential also calculated the elasticity of silica glass and the results were in good agreement with experimental value reported by Muralidharan et al. [62].

Murillo et al. used an in-house Fortran program for the investigation of structural and mechanical properties of silicon aerogel and xerogel. In this code the Velocity–Verlet algorithm is used to solve the equations of motion of the particles, using a time step of 0.5 fs. The selection of a small time step ensures that the atoms do not fly off, which is important in this case because large free surfaces are generated inside the samples. Langevin dynamics is implemented and all the particles in the system are used to control its temperature accordingly [73].

1.2.2.4 SIMULATION RESULTS

The results of simulation in Murillo et al. show that increasing the temperature of the system at initial stages of the preparation process lead to an increase in the kinetic energies of the atoms. By reducing the temperature of the system, however the atoms are still able to move in the system, their numbers increases per group. At low temperature the atoms find equilibrium positions and the structure is locked. Finally the simulated samples have branched structures formed by groups of atoms interconnected by small bridges. Figure 1.28 represent the formation of clusters interconnected by small chains [73].

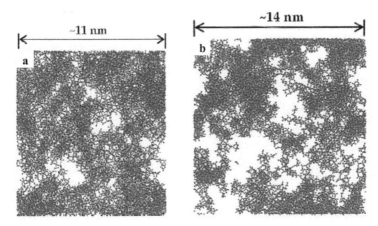

FIGURE 1.28   Simulated samples: (a) density of 0.58 g/cm³, porosity of 74%, (b) density of 0.28 g/cm³, porosity of 87%.

Yeo et al. [96] simulations are performed on the LAMMPS [67] software to attain a closer fit of the thermal conductivity. The interaction potential used is the Tersoff potential [83], re-parameterized to model interactions between silicon and oxygen [83]. Comparing the BKS potential and Tersoff potential, it is found that the Tersoff potential is more suitable for thermal conductivity studies, while, the BKS potential augmented with a "24–6" Lennard Jones potential can prevent uncontrollable dynamics at very high temperatures [89]. After producing aerogels in different density, RNEMD was used to determine the thermal conductivity at each densities [96].

Coarse-grained computer models of silica aerogels (and other gels) may be roughly divided into two categories, depending on the simulation algorithm used and type of interactions included. In hard-sphere aggregation models, which have been extensively studied, aggregates are formed out of simple particles according to one of several procedures. Nonbonded particles interact as "billiard balls," without soft attractive or repulsive forces, while bonded particles are held rigidly together at the point of contact. In flexible models, equations of motion similar to those used in atomistic simulations are applied to objects with rather more complex interactions, including soft nonbonded interactions and deformable and/or breakable bonds [1].

1.2.2.4.1 CALCULATION OF THE PAIR DISTRIBUTION FUNCTION (PDF)

After the generation of the samples using MD, the fractal dimension can be obtained based on its relation to the pair distribution function (PDF) of the samples. The PDF represents the probability of finding a pair of atoms separated by a distance r in the structure, relative to the probability expected for a randomly distributed structure having the same density. The PDF is calculated as:

$$g_{\alpha\beta}(r_1 r_2) = \frac{V^2}{N_\alpha N_\beta} \sum_{i \in \{\alpha\}}^{N_\alpha} \sum_{i \in \{\beta\}}^{N_\beta} \delta(r_1 - r_i)\delta(r_2 - r_j) \tag{69}$$

$$= \rho_\alpha^{-1} \rho_\beta^{-1} \sum_{i \in \{\alpha\}}^{N_\alpha} \sum_{i \in \{\beta\}}^{N_\beta} \delta(r_1 - r_i)\delta(r_2 - r_j)$$

where V is the volume of the system, N_α and N_β are the numbers of particles of the entities of type α and β, respectively, ρ_α and ρ_β are the corresponding densities of α and β subsystems, and the symbol $\langle\rangle$ means ensemble average. Therefore, several configurations of the system should be used to compute the PDF. The term inside the $\langle\rangle$ symbol indicates that when finding the distribution of distances between the atoms of type α and β one must locate each atom of type α and obtain the distance from it to each one of the atoms of type β. Figure 1.29 is an example of the PDF for one of the systems created. The decay of the PDF for large values of the distance between the atoms is shown in Fig. 1.29. That decay can be used to estimate the fractal dimension, d_f, of porous systems [49].

$$d_f = 3 + \frac{dLog\ (g(r))}{dLog(r)} \tag{70}$$

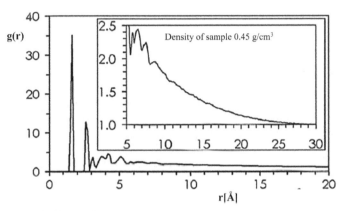

FIGURE 1.29   Pair distribution function [29].

Fractal dimension of a structure can also be obtained through a simulated scattering experiment [19, 23]. To simulate a scattering experiment one needs to calculate the scattering intensity, I, corresponding to different wavelengths of radiation shined to the sample, represented by their wave number, q. An expression to calculate I, when the positions of all the particles in the system are known, is given by Refs. [23, 26]:

$$\frac{I\ (q)}{I_0} = \sum_{ij} \frac{\sin\left(qr_{ij}\right)}{qr_{ij}} \tag{71}$$

where I is the scattering intensity, I_0 is a reference value of intensity, q is the wave number and r_{ij} is the distance between two atoms i and j. For fractal structures, the scattering intensity can be related to the wave number by a power law, where the exponent is the negative of the fractal dimension of the structure, that is, $I \propto q^{-df}$ [62, 67]. An example of the I (q) is plotted in the Fig. 1.30. For large wave numbers the scattering intensity corresponds to the individual particles and remains constant; while for small wave numbers, clusters of particles are responsible for the scattering. The limiting value of wave number in the scattering is determined by the size of the sample modeled [73].

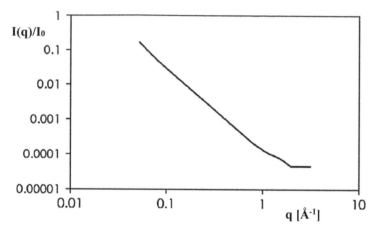

FIGURE 1.30   Log–Log plot of the scattering intensity vs. wavenumber for a sample of density of 0.45 g/cm3 [29].

Murillo et al. found that in the short-range order, features under 5 Å, the computational samples show characteristics similar to those of silica samples and the bonding distance (Si–O atoms), the nearest neighbors distances O to O atoms and Si to Si atoms are in good agreement with experimental results (Table 1.6). At higher ranges between 9 and 25 Å, approximately, the simulated samples exhibit a fractal behavior, which agrees with the fractal range found in simulations done by Nakano et al. [24], 5 to 25 Å. In aerogel samples studied by Vacher et al. the fractal range of the samples was identified extending from features as small as 4 Å to features larger than 200 Å, depending on the density of the sample. Courtens and Vacher [*] report that the fractal range of silica aerogels extends from 10 Å to 1000 Å, approximately (Table 1.7).

TABLE 1.6   The Results of Pair Distribution Function of the Samples

Bonding	Distance (Å)			
	Ref	**Simulation**	**Ref**	**Experimental's**
Si–O atoms	Murillo et al. [1]	1.63 ±0.03	[2]	1.61±0.05
O to O atoms		2.65±0.03		2.632±0.089
Si to Si atoms		3.08±0.03		2.632±0.089

TABLE 1.7   Fractal Range Found Through Simulation and Reported by Different Literatures

References	Fractal Ranges (Å)
Murillo et al. [1]	Under 5
	9 and 25
Nakano et al. [9]	5–25
Vacher et al. [5]	4–200
Courtens and Vacher [20]	10 to 1000

Murillo et al. [73] report that the fractal dimension increases with the density toward a limiting value of 3, corresponding to bulk silica. Their results are shown in Fig. 1.31. For larger densities the scattering experiments lead to fractal dimensions slightly smaller than those obtained using the decay of the PDF. For lighter samples the opposite occurs [73].

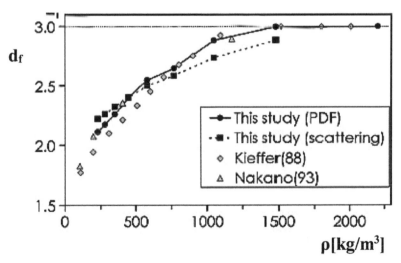

FIGURE 1.31   Fractal dimension of silica aerogels samples with different densities.

Yeo et al. [96], simulated silica aerogel on a cubic system of 52,728 atoms with densities in ranges of 0.3 to 1 g/cm³. Determination of fractal dimension was determined in accordance to Refs. [94, 95] (Fig. 1.32). In this method the total radial distributions for each density are calculated, and power-law decays are superimposed on the peak structures to determine the fractal dimensions. These results were in accordance with previous theoretical studies by Murillo et al. [73].

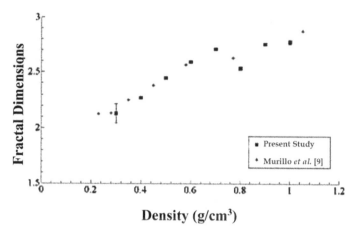

FIGURE 1.32   Decreasing fractal dimensions as density decreased.

2.2.4.2   THERMAL PROPERTIES

Experimental studies have shown that, thermal conductivity and transport mechanisms are scaled with density via a power law:

$$\lambda_S = C\rho^\alpha \qquad (72)$$

where α was approximately 1.6 for densities between 0.3 to 1.0 g/cm³ [24].

Ng et al. [42] employed negative pressure rupturing with the van Beest, Kramer and van Santen (BKS) potential [64, 65], and determined their thermal conductivities. It was found that the power-law fit of the data corresponds to experimental bulk sintered aerogel. Yeo et al. [96] studied the solid thermal conductivity of silica aerogels using reverse nonequilibrium MD (RNEMD) simulations, following the method Murillo et al. [73].

In Yeo et al. [96] studies amorphous silica samples was generated by quenching β-cristobalite, from 5000K to 300K. For validation of MD methods the thermal conductivities of increasing length of amorphous silica was compared with experimental results. They concluded that the Tersoff potential can give a much better estimation of bulk thermal properties than BKS potential. This has been shown in Fig. 1.33 that the Tersoff potential shows an almost linear dependence with increasing length scales while BKS potential overshoots the thermal conductivity of bulk amorphous silica [40]. In order to further analysis, each potential (BKS and Tersoff) were examined with vibrational density of states (vDOS) which are obtained through the discrete Fourier transform of the velocity autocorrelation function (VACF). In Tersoff potential the peaks were clearly close to those obtained from experimental and theoretical results. However, the BKS potential showed no apparent peaks. It was concluded that the reparameterized Tersoff potential is a far superior alternative in the thermal characterization of bulk amorphous [96].

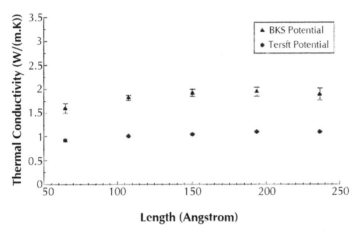

FIGURE 1.33   Amorphous silica of various lengths and their thermal conductivities.

The solid thermal conductivity of silica aerogels can be determined using reverse nonequilibrium MD (RNEMD) simulations. Yeo et al. [96] generated porous samples according to the methods Murillo et al. used [29]. Thermal conductivities were found to increase nonlinearly as the density decreased. The power-law exponent, α, of experimental bulk aerogel (1.6) in the density range of 0.3 to 1.0 g/cm³ was in accordance with α value obtained from RNEMD model (1.61). [24]. The advantage of this model is the accessibility of very low densities (lower than 0.1 g/cm³) without any adverse phenomena. However, the significant difference between thermal conductivities of simulated and experimental aerogels makes a limitation for this model. This mostly could be due to inability to attain micropores [96].

Experimentally, using small angle neutron scattering (SANS) and nuclear magnetic resonance (NMR), it has been observed that aerogels and xerogels are fractal structures characterized by a fractal dimension varying from 3 for very low porosity (almost dense glass) to 1.8 for very light specimens. The fractal dimension is slightly affected by the pH of the solution prior to gelation [23, 73, 87, 93].

Aerogel samples with densities smaller than 0.43 g/cm³ was prepared through SANS under acidic and neutral conditions have a fractal dimension around 2.4±0.03 [87]. Devreux et al. [22, 23] prepared aerogels with densities around 0.17 g/cm³ prepared under acidic conditions and a fractal dimension of 2.2. Woignier et al. [93, 94], have studied preparation of aerogel under acidic, basic and neutral pH. They found that under basic conditions aerogels have fractal dimensions close to 1.8± 0.1, while under acidic or neutral conditions the samples had a fractal dimension around 2.2±0.1 and 2.4±0.1, respectively.

2.2.4.3   MECHANICAL CHARACTERIZATION

A tension test should be simulated by the stretching the sample along one direction at strain rates larger than the one used in laboratory tests. Because at higher strain rate, the speed of the atoms, and consequently the temperature of the sample, tend to increase. Moreover, the temperature is uncontrolled when the stretching speed is near the speed of sound in the material [63]. Murillo et al. used Langevin dynamics [8] to control the speed of all the atoms in the system. Langevin dynamics is based on the following equation,

$$m_i \ddot{\vec{r}}_i = \vec{F}_i - m_i \gamma_i \dot{\vec{r}}_i + \vec{R}_i \tag{73}$$

where m_i is the mass of the atom, $\vec{r}_i, \dot{\vec{r}}_i$ and $\ddot{\vec{r}}_i$ are the position, velocity and acceleration of the particle, $\vec{F}_i$ is the force exerted on the particle by all the other particles in the system, $\vec{R}_i$ is a stochastic force applied to the particle, and γ_i is a damping coefficient. [49]

Murillo et al. [73] showed that the relation between elastic modulus and density of samples can be described by power law. The obtained exponent for this relation is 3.11 ± 0.21. Figure 1.34 shows that a power law relation between strength and density exist. The main differences in predicted strength through simulation and experimental is due to the using other potentials for silica (Table 1.4). The calculated exponent for that relation was 2.53 ± 0.15 that compares well with previously published values. Yeo et al. investigate the Young's modulus with the strain rate 0.0005 ps^{-1} for 200 ps, the tension tests are carried out on samples of different densities. The exponent of the relationship between Young's modulus and density of Yeo et al. studies was lower than the one that Murillo et al. has been mentioned. This is caused from different interactive potential [96]. The comparison of reported values is summarized in Table 1.8.

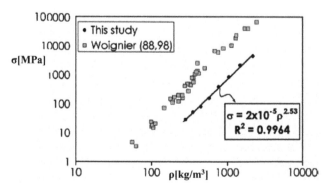

FIGURE 34   Strength vs. density for silica aerogels. Woignier [22, 24].

TABLE 1.8 Exponent for the Power Law Relation Between Elastic Modulus and Density

References	Exponent	Density [g/cm3]	Methods
Campbell et al. [20]	3.5±0.2	1.67–2.2	Computer simulation
Yeo et al. [96]	2.4313	0.3–1	Computer simulation
Murillo et al. [2]	3.11±0.21	0.23–2.2	Computer simulation
Groß et al. [10]	3.49±0.07	0.14–2.7	Exp.
Groß et al. [10]	2.97±0.05	0.08–1.2	Exp. sintered aerogels
Woignier et al. [8]	3.8±0.2	0.1–0.4 (approx. values)	Exp. pH neutral
Woignier et al. [7]	3.7±0.2	0.055–0.5 (approx. values)	Exp. pH neutral, acidic and basic
Woignier et al. [7]	3.2±0.2	0.42–2.2 (approx. values)	Exp. partially densified samples

The procedure Murillo et al. [73] used to generate the porous samples does not correspond to the events happening during the experimental gelation of a solution, nevertheless it produces samples having geometrical features that are similar to those of real aerogel and xerogel material.

TABLE 1.9 Potentials Lead to Values Strength and Elastic Modulus of Silica [61]

Potential	Strength (GPa)	Elastic Modulus (GPa)
BMH-potential	30 and 65	220
S-potential	24–35	220
BKS-potential	12 and 22	12 and 22
FG-potential	12 and 21	125

1.3   CONCLUSION

A wide variety of modeling techniques have been developed over the years, and those relevant for work at the molecular level include, in addition to MD, classical Monte Carlo, quantum based techniques and MC methods, and MD combined with discrete approaches such as lattice–Boltzmann method. MD simulation is a technique for computing the equilibrium and properties of a system and helps to better understanding the properties of assemblies of molecules in terms of their structure and the microscopic interactions between them.

Recent work at all levels of description, from quantum mechanical to empirical atomistic modeling to coarse-grained modeling, has confirmed the broad picture developed by previous studies but also illustrated the many challenges remaining in developing a complete description and understanding of silica aerogels. MD simulation has proved effective for obtaining useful structural information on sol–gel process for silica systems. The latest first-principles calculations clearly indicate that the roles of water and counter ions in the underlying chemistry are more important and more complex than previously thought. Atomistic simulations have illustrated that structure at small scales is dramatically affected by degree of hydration and that the small clusters produced by short-time oligomerization have a complex distribution of size, shape, and structure. Recent attempt at more quantitative coarse-graining indicates that quite large system sizes are required to obtain reliable results for mechanical properties of even reasonably dense aerogels.

Although atomistic empirical (or even quantum-mechanical) simulations of increasingly large scale and length would seem to be a route forward, there is no practical way that these can access the actual time- and length-scales required. Near-complete gelation was observed in many atomistic simulations, but this was only possible due to the tiny simulation cells and high densities used; with more reasonable system sizes, gelation would not have occurred on a "simulable" time-scale. In addition, important parameters, including the pH and pKas, of many of the proposed atomistic models are unknown, which makes correlation with experiment difficult. The geometrical features of the modeled samples are characterized by the fractal dimension, determined from the pair distribution function of the samples as well as from simulated scattering experiments and are found to be in good agreement with computational and experimental data. Furthermore, the mechanical properties, namely the elastic modulus and strength, of the porous samples are found to scale with density following a power law, which is expected for fractal structures.

Results showed that direct expansion of crystalline samples of β-cristobalite to reach densities between 2.2 g/cm³ and 0.23 g/cm³, along with thermal processing, leads to fractal structures which allow to investigate the properties of silica aerogels and xerogels. MD and the reparameterized Tersoff potential were used to model the porous structures of silica aerogels demonstrate that this potential is suitable for modeling thermal properties in amorphous silica. Using a quenching and expanding process, different densities of aerogel samples could be generated. Through RNEMD, their thermal conductivity is determined and the power-law fit of our data corresponds well with experimental studies.

It appears clear that further quantitative improvements in computational aerogel models are most likely to result from a multiscale approach, incorporating all three levels of descriptions studied here. Only with such combinations will modelers be able to reach the necessary system sizes and simulation times while still connected to an atomistic description of the underlying chemistry. Empirical models could be further improved by parameterization against quantum-mechanical simulation results. Large-scale atomistic simulations could be used to generate chemically realistic distributions of sol particles, even if gelation cannot be directly observed. Finally, atomistic models of all types could be used as reference systems for developing better coarse-graining techniques. Multiscale techniques are of considerable interest in the wider simulation community at this time, and developments in other areas may be expected to help further the goal of realistic simulation and modeling of aerogels.

KEYWORDS

- **Aerogels**
- **Fractals**
- **Mechanical and thermal properties**
- **Molecular dynamics**
- **Nanostructured materials**
- **Porosity**
- **Simulations**
- **Thermal insulate**

REFERENCES

1. Aegerter, M. A., & Leventis, N. (2011). Advances in Sol-gel derived Materials and Technologies, Springer.
2. Aharony, A., & Stauffer D. (2003). Introduction to Percolation Theory, Taylors & Francis.
3. Alder, B. & Wainwright T. (1957). "Phase Transitions for a Hard Sphere Systems," the Journal of Chemical Physics, *27(5)*, 1208.
4. Alemán, J., & Chadwick, A. (2007). "Definitions of Terms Relating to the Structure and Processing of Sol, Gels, Networks, and Inorganic-Organic Hybrid Materials (IUPAC Recommendations 2007)" Pure and Applied Chemistry, *79(10)*, 1801–1829.
5. Allen, M., & Tildesley, D. (1993). "*Computer Simulations in Chemical Physics*", Springer.
6. Allinger, N. L., & Chen, K. (1996). "An Improved Force Field (MM4) for Saturated Hydrocarbons" Journal of Computational Chemistry, *17(5–6)*, 642–668.
7. Allinger, N. L., & Yuh, Y. H. (1989). "Molecular Mechanics, the MM3 Force Field for Hydrocarbons 1," Journal of the American Chemical Society, *111(23)*, 8551–8566.
8. Berendsen, H. J., & Postma, J. P. M. (1984). "Molecular Dynamic with Coupling to an External Bath," The Journal of Chemical Physics, *81(8)*, 3684–3690.
9. Bhagat, S. D., & Kim, Y. H. (2007). "Rapid Synthesis of Water-Glass based Aero gels by in Situ Surface Modification of the Hydro gels," Applied Surface Science, *253(6)*, 3231–3236.
10. Bhattacharya, S., & Kieffer, J. (2005). "Fractal Dimensions of Silica Gels generated using Reactive Molecular Dynamics Simulations," The Journal of Chemical Physics, *122(9)*, 094715.
11. Bhattacharya, S., & Kieffer, J. (2008). "Molecular Dynamics Simulation Study of Growth Regimes during Poly Condensation of Silicic Acid, from Silica Nanoparticles to Porous Gels," The Journal of Physical Chemistry C, *112(6)*, 1764–1771.

12. Brinker, C., & Drotning, W. (1984). A Comparison between the Densifications Kinetics of Colloidal and Polymerics Silica Gels, MRS Proceedings, Cambridge Universal Press.
13. Brinker, C., & Keefer, K. (1982). "Sol-Gel Transition in Simple Silicates," Journal of Non-Crystalline Solids, *48(1)*, 47–64.
14. Brinker, C. J., & Scherer, G. W. (1990). "Sol-Gel Science, (1990)." New York, Academic Press.
15. Brooks, B. R., & Bruccoleri, R. E. (1983). "CHARMM, A Program for Macromolecular Energy, Minimization, and Dynamics Calculations," Journal of Computational Chemistry, *4(2)*, 187–217.
16. Buisson, P., & Pierre, A. C. (2006). "Immobilization in Quartz Fiber Felt Reinforced Silica Aero gel improves the Activity of Candida Rugosa Lipase in Organic Solvents," Journal of Molecular Catalysis B: Enzymatic, *39(1)*, 77–82.
17. Cai, W. (2005). "Handout 1, an Overview of Molecular Simulation."
18. Calas, S., & Sempere, R. (1998). "Textural Properties of Densified Aero Gels," Journal of Non-Crystalline Solids, *225*, 215–219.
19. Campbell, T., & Kalia, R. K. (1999). "Structural Correlation and Mechanical Behavior in Nano phase Silica Glasses," Physical Review Letters, *82(20)*, 4018.
20. Carofalini, S., & Melman, H. (1986). Applications of Molecular Dynamic Simulation to Sol-Gel Processing, MRS Proceeding, Cambridge Universal Press.
21. Cotter, C. J., & Reich, S. "Time Stepping Algorithms for Classical Molecular Dynamics," Computational Nanotechnology, American Scientific Publishers to appear.
22. D'Arjuzon, R. J., & Frith, W. (2003). "Brownian Dynamic Simulations of Aging colloidal Gels," Physical Review E, *67(6)*, 061404.
23. Devreux, F., Boilot, J. et al. (1990). "NMR Determinations of the Fractal Dimension in Silica Aerogels," Physical Review Letters, *65(5)*, 614.
24. Emmerling, A. & Fricke, J. (1997). "Scaling Properties and Structure of Aero Gels," Journal of Sol-Gel Science and Technology, *8(1–3)*, 781–788.
25. Feuston, B. & Garofalini, S. (1988). "Empirical Three Body Potential for Vitreous Silica," the Journal of Chemicals Physics, *89(9)*, 5818–5824.
26. Feuston, B. & Garofalini, S. (1990). "Oligomerizalion in Silica Sols," Journal of Physical Chemistry *94(13)*, 5351–5356.
27. Feuston, B., & Garofalini, S. H. (1990). "Water Induced Relaxation of the Vitreous Silica Surface," Journal of Applied Physics, *68(9)*, 4830–4836.
28. Field, M. J. (1999). A Practical Introduction to the Simulation of Molecular Systems, Cambridge University Press.
29. Frenkel, D., & Smit, B. (2001). Understanding Molecular Simulation from Algorithm to Applications, Academic Press.
30. Freundlich, H., & Hatfield, H. S. (1926). "Colloid and Capillary Chemistry."
31. Fricke, J., & Emmerling, A. (1998). "Aerogels Recent Progress in Production Techniques and Novel Applications," Journal of Sol-Gel Science and Technology, *13(1–3)*, 299–303.
32. Garcia, E., & Glaser, M. A. (1999). "HFF, Force Field for Liquid Crystal Molecules," Journal of Molecular Structure, Theoretical Chemistry, *464(1)*, 39–48.
33. Garofalini, S. H., & Martin, G. (1994). "Molecular Simulations of the Polymerization of Silica Acid Molecules and Network Formation," The Journal of Physical Chemistry, *98(4)*, 1311–1316.
34. Gash, A. E., & Tillotson, T. M. (2001). "New Sol–Gel Synthetic Route to Transition and Main-Group Metal Oxide Aerogels using Inorganic Salt Precursors," Journal of Non-Crystalline Solids, *285(1)*, 22–28.

35. Gear, C. (1966). "The Numerical Integration of Ordinary Differential Equations of various Orders (Ordinary Differential Equation Integration and Multistep Predictor-Corrector Methods for General Initial Value Problems)."

36. Gross, J., & Coronado, P. R. (1998). "Elastic Properties of Silica Aero gels from a New Rapid Supercritical Extraction Process," Journal of Non-Crystalline Solids, *225*, 282–286.

37. Gross, J., & Fricke, J. (1992). "Ultrasonic Velocity Measurements in Silica, Carbon and Organic Aero Gels," Journal of Non-Crystalline Solids, *145*, 217–222.

38. Gross, J., & Scherer, G. W. (2003). "Dynamic Pressurization Novel Method for Measuring Fluid Permeability," Journal of Non-Crystalline Solids, *325(1)*, 34–47.

39. Haereid, S., & Dahle, M. (1995). "Preparation and properties of Monolithic Silica Xerogels from TEOS-Based Alcogels Aged in Silane Solutions" Journal of Non-Crystalline Solids, *186*, 96–103.

40. Haile, J. M. (1992). Molecular Dynamics Simulation, Elementary Methods, John Wiley & Sons, Inc

41. Hairer, E., & Lubich, C. (2003). "Geometric Numerical Integration illustrated by the Störmer-Verlet Method," Acta Numerica, *12*, 399–450.

42. Himmel, B., & Bürger, H. (1995). "Structural Characterization of SiO_2 Aero gels," Journal of Non-Crystalline Solids, *185(1)*, 56–66.

43. Jabbarzadeh, A., & Tanner, R. I. (2006). "Molecular Dynamics Simulations and its Application to Nano-Rheology," Rheology Reviews 2006, 165.

44. Jones, J. E. (1924). "On the Determination of Molecular Fields, I from the Variation of the Viscosity of a Gas with Temperature," Proceedings of the Royal Society of London, Series A *106(738)*, 441–462.

45. Jorgensen, W. L., & Maxwell, D. S. (1996). "Development and Testing of the OPLS All-Atom Force Field on Conformational Energetic and Properties of Organic Liquids," Journal of the American Chemical Society, *118(45)*, 11225–11236.

46. Jullien, R., Meakin, P. et al (1992). "Three-Dimensional Models for Particle size Segregation by Shaking," Physical Review Letters, *69(4)*, 640.

47. Kallala, M., & Jullien, R. (1992). "Crossover from Gelation to Precipitation," Journal De Physique II, *2(1)*, 7–25.

48. Kawaguchi, T., & Iura, J. (1986). "Structural Changes of Monolithic Silica Gel during the Gel-to-Glass Transition," Journal of Non-Crystalline Solids, *82(1)*, 50–56.

49. Kieffer, J. & Angell, C. A. (1988). "Generation of Fractal Structures by Negative Pressure Rupturing of SiO_2 Glass," Journal of Non-Crystalline Solids, *106(1)*, 336–342.

50. Kistler, S. S. (1931). "Coherent Expanded Aero gels and Jellies," *Nature*, *127*, 741.

51. Kocon, L., & Despetis, F. (1998). "Ultralow Density Silica Aero gels by Alcohol Super Criticals Drying," Journal of Non-Crystalline Solids, *225*, 96–100.

52. Leimkuhler, B., & Reich, S. (2004). "Geometrics Numerical Methods for Hamiltonian Mechanics," Cambridge Monographs on Applied and Computational Mathematics, Cambridge University Press.

53. Lii, J. H. & Allinger, N. L. (1989). "Molecular Mechanics, The MM3 Force Field for Hydrocarbons 3, The Vander Waals' Potentials and Crystal Data for Aliphatic and Aromatic Hydrocarbons," Journal of the American Chemical Society, *111(23)*, 8576–8582.

54. Ma, H. S., & Prévost, J. H. (2001). "Computer Simulations of Mechanical Structure-Property Relationship of Aero Gels," Journal of Non-Crystalline Solids, *285(1)*, 216–221.

55. Ma, H. S., & Roberts, A. P. (2000). "Mechanical Structure-Property Relationship of Aero Gels," Journal of Non-Crystalline Solids, *277(2)*, 127–141.

56. Martin, G. E., & Garofalini, S. H. (1994). "Sol-Gel Polymerization, Analysis of Molecular Mechanisms and the Effect of Hydrogen," Journal of Non-Crystalline Solids, *171(1)*, 68–79.

57. McQuarrie, D. (1976). "Statistical Mechanics" Happers and Rows, New York.

58. Metropolis, N., & Rosenbluth, A. W. (2004). "Equation of State Calculations by Fast Computing Machines," The Journal of Chemical Physics, *21(6)*, 1087–1092.
59. Mikes, J., & Dusek, K. (1982). "Simulation of Polymer Network Formation by the Monte Carlo Method," Macromolecules, *15(1)*, 93–99.
60. Munetoh, S., & Motooka, T. (2007). "Inter Atomic Potential for Si–O Systems using Tersoff Parameterization," Computational Materials Science, *39(2)*, 334–339.
61. Muralidharan, K., & Simmons, J. (2005). "Molecular Dynamics Studies of Brittle Fracture in Vitreous Silica, Review and Recent Progress," Journal of Non-Crystalline Solids, *351(18)*, 1532–1542.
62. Murillo, J. S. R., & Barbero, E. J. (2008). Towards Toughening and Understanding of Silica Aerogels, MD Simulations, ASME 2008 International Mechanical Engineering Congress and Exposition, American Society of Mechanical Engineers.
63. Mylvaganam, K., & Zhang, L. (2004). "Important Issue in a Molecular Dynamic Simulation for Characterizing the Mechanical Properties of Carbon Nanotubes," Carbon, *42(10)*, 2025–2032.
64. Nakano, A., & Bi, L. et al. (1993). "Structural Correlations in Porous Silica, Molecular Dynamics Simulation on a Parallel Computer," Physical Review Letters, *71(1)*, 85.
65. Nakayama, T., Yakubo, K. et al. (1994). "Dynamical Properties of Fractal Networks, Scaling, Numerical Simulations, and Physical Realizations," Reviews of Modern Physics, *66(2)*, 381.
66. Petričević, R., & Reichenauer, G. et al. (1998). "Structure of Carbon Aero Gels Near the Gelation Limit of the Resorcinol–Formaldehyde Precursor," Journal of Non-Crystalline Solids, *225*, 41–45.
67. Plimpton, S. (1995). "Fast Parallel Algorithms for Short-Range Molecular Dynamics," Journal of Computational Physics, *117(1)*, 1–19.
68. Pütz, M., Kremer, K. et al. (2000). "What is the Entanglement Length in a Polymer Melt?" EPL (Europhysics Letters), *49(6)*, 735.
69. Rao, N. Z. & Gelb, L. D. (2004). "Molecular Dynamics Simulations of the Polymerization of Aqueous Silicic Acid and Analysis of the Effects of Concentration on Silica Polymorph Distributions, Growth Mechanisms, and Reaction Kinetics," The Journal of Physical Chemistry B, *108(33)*, 12418–12428.
70. Rapaport, D. C. (2004). The Art of Molecular Dynamics Simulation, Cambridge University Press.
71. Reichenauer, G. "Aerogels," Kirk-Othmer Encyclopedia of Chemical Technology.
72. Reith, D., Pütz, M. et al. (2003). "Deriving Effective Meso Scale Potentials from Atomistic Simulations," Journal of Computational Chemistry, *24(13)*, 1624–1636.
73. Rivas Murillo, J. S., Bachlechner, M. E. et al. (2010). "Structure and Mechanical Properties of Silica Aerogels and Xerogels Modeled by Molecular Dynamics Simulation," Journal of Non-Crystalline Solids, *356(25)*, 1325–1331.
74. Ryckaert, J. P. & Bellemans, A. (1975). "Molecular Dynamics of Liquid *n*-Butane near its Boiling Point," Chemical Physics Letters, *30(1)*, 123–125.
75. Saliger, R., Bock, V. et al. (1997). "Carbon Aerogels from Dilute Catalysis of Resorcinol with Formaldehyde," Journal of Non-Crystalline Solids, *221(2)*, 144–150.
76. Satoh, A. (2010). Introduction to Practice of Molecular Simulation: Molecular Dynamics, Monte Carlo, Brownian Dynamics, Lattice Boltzmann and Dissipative Particle Dynamics, Elsevier.
77. Scherer, G. W. (1998). "Characterization of Aero gels," Advances in Colloid and Interface Science, *76*, 321–339.
78. Scherer, G. W., & Gross, J. et al. (2002). "Optimization of the Rapid Supercritical Extraction Process for Aerogels" Journal of Non-Crystalline Solids, *311(3)*, 259–272.

79. Smith, D. M., & Stein, D. et al. (1995). "Preparation of Low-Density Xerogels at Ambient Pressure" Journal of Non-Crystalline Solids, *186,* 104–112.
80. Stillinger, F., & Rahman, A. (2008). "Revised Central Force Potentials for Water" Journal of Chemical Physics, *68(2),* 666–670.
81. Stone, A. (2013). Theory of Inter Molecular Forces, Oxford University Press.
82. Succi, S. (2001). The Lattice Boltzmann Equation for Fluid Dynamics and Beyond, Oxford University Press.
83. Tersoff, J. (1989). "Modeling Solid-State Chemistry, Inter Atomic Potentials for Multi Component Systems," Physical Review B, *39(8),* 5566–5568.
84. Torquato, S. (2002). Random Heterogeneous Materials, Microstructure and Macroscopic Properties, Springer.
85. Tsige, M., Curro, J. G. et al. (2003). "Molecular Dynamics Simulations and Integral Equation Theory of Alkane Chains, Comparisons of Explicit and United Atom Models," Macromolecules, *36(6),* 2158–2164.
86. Vacher, R., Courtens, E. et al. (1990). "Crossovers in the Density of States of Fractal Silica Aerogels," Physical Review Letters, *65(8),* 1008.
87. Vacher, R., Woignier, T. et al. (1988). "Structure and Self-Similarity of Silica Aero Gels," Physical Review B, *37,* 6500–6503.
88. Vashishta, P., Kalia, R. K. et al. (1996). "Molecular Dynamics Methods and Large-Scale Simulations of Amorphous Materials," Amorphous Insulators and Semiconductors, *33,* 151–213.
89. Vashishta, P., Kalia, R. K. et al. (1990). "Interactions Potential for SiO2, a Molecular Dynamics Study of Structural Correlations," Physical Review B, *41(17),* 12197.
90. Wang, P, Emmerling, A. et al. (1991). "High-temperature and Low-temperature Super Critical Drying of Acro Gels-Structural Investigations with SAXS," Journal of Applied Crystallography, *24(5),* 777–780.
91. WebbIII, E. B., & Garofalini, S. H. (1998). "Relaxation of Silica Glass Surfaces Before and After Stress Modification in a Wet and Dry Atmosphere, Molecular Dynamics Simulations," Journal of Non-Crystalline Solids, *226(1),* 47–57.
92. Weiner, S. J., Kollman, P. A. et al. (1984). "A New Force Field for Molecular Mechanical Simulation of Nucleic Acids and Proteins," Journal of the American Chemical Society *106(3),* 765–784.
93. Woignier, T., Phalippou, J. et al. (1990). "Different Kinds of Fractal Structures in Silica Aerogels," Journal of Non-Crystalline Solids, *121(1),* 198–201.
94. Woignier, T., Reynes, J. et al. (1998). "Different Kinds of Structure in Aerogels, Relationships with the Mechanical Properties," Journal of Non-Crystalline Solids, *241(1),* 45–52.
95. Wright, A. C. (1993). "The Comparison of Molecular Dynamics Simulations with Diffraction Experiments," Journal of Non-Crystalline Solids, *159(3),* 264–268.
96. Yeo, J., & Lei, J. (2013). Characterization of Mechanical and Physical Properties of Silica Aero Gels Using Molecular Dynamics Simulation, ICF13.

CHAPTER 2

BIODEGRADABLE POLYMER FILMS ON LOW DENSITY POLYETHYLENE AND CHITOSAN BASIS: A RESEARCH NOTE

M. V. BAZUNOVA and R. M. AKHMETKHANOV

Bashkir State University, 32 ZakiValidi Street, 450076 Ufa, Republic of Bashkortostan, Russia, E-mail: mbazunova@mail.ru

CONTENTS

Abstract 84
2.1   Introduction..84
2.2   Experimental Part..84
2.3   Results and Discussion 86
Keywords 88
References..88

ABSTRACT

Polymer films basing on ultra dispersed low density polyethylene powder modified by the natural polysaccharide, chitosan have been obtained under the combined effect of high pressure and shear deformation. Their water-absorbent capacity and biodegradability are estimated.

2.1 INTRODUCTION

The problem of biodegradability of well-known tonnage industrial polymers is quite urgent for modern studies. It is promising enough to use synthetic and natural polymer mixtures, which can play the roles of both filler and modifier for creating biodegradable environmentally safe polymer materials. The macromolecule fragmentation of the synthetic polymer is to be provided for due to its own biodestruction.

The synthetic polymer has been modified by the natural one under the combined effect of high-pressure and shear deformation. The usage of this method for obtaining polymer composites is sure to solve several problems at once. Firstly, the ultra dispersed powders with a high homogeneity degree of the components can be obtained under combined high pressure and shear deformation thus resulting in easing the technological process of production [1]. Secondly, the elastic deformation effects on the polymer material may lead to the chemical modification of the synthetic polymer macromolecules by the natural polymer blocks via recombination of the formed radicals. Thus, it can provide for the polymer product biodegradation. Thirdly, the choice of the best exposure conditions of high pressure and shear deformation on the polymer mixture (modification degrees, process temperature, pressure in the working zone of the dispersant, shear stress values, etc.) may lead to creating environmentally safe biodegradable polymer composite materials processed into products by conventional methods.

Therefore, the working out of the optimal method for obtaining biodegradable polymer films on the basis of ultra-dispersed powders of low-density polyethylene (LDPE) modified by the natural polymer in combined conditions of high pressure and shear deformation is quite expedient. In the paper given a polysaccharide of natural origin, chitosan (CTZ) was used as a polymer.

2.2 EXPERIMENTAL PART

LDPE 10803–020 (90,000 molecular weight, 53% crystallinity degree, and 0.917 g/sm^3 density) and CTZ samples of Bioprogress Ltd. (Russia) obtained by alkaline deacetylation of crab chitin (deacetylationdegree ~84%), and M_{sd}= 115,000 were used as components for producing biodegradable polymer films.

The initial highly dispersed powders with different mass ratio of components have been obtained by high temperature shearing under simultaneous impact of high pressure and shear deformation in an extrusion type apparatus with a screw diameter of 32 mm [2, 3]. Temperatures in kneading, compression and dispersion chambers amounted to 150°C, 150°C and 70°C, respectively.

The size of particles in powders of LDPE, CTZ and LDPE/CTZ with various mass ratio of the components were determined by "Shimadzu Salid – 7101" particle size analyzer. The film formation was carried out by roto-molding [4] at 135°C and 150°C. The film sample thickness amounted to 100 μm and 800 μm.

The absorption coefficient of the condensed vapors of volatile liquid (water, n-heptane) Kin static conditions is determined by complete saturation of the sorbent by the adsorbent vapors under standard conditions at 20°C [5] and was calculated by the formula:

$$\hat{E}' = \frac{m_{absorbed\ water}}{m_{sample}} \times 100\%,$$

where $m_{absorbed\ water}$ is weight of the saturated condensed vapors of volatile liquid, g; m_{sample} is weight of dry sample, g.

Film samples were long kept in the aqueous and enzyme media to determine the water absorption coefficient, where the absorbed water weight was calculated. The water absorption coefficient of film samples of LDPE/CTZ with different weight ratio was determined by the formula:

$$\hat{E} = \frac{m_{absorbed\ water}}{m_{sample}} \times 100\%,$$

where $m_{absorbed\ water}$ is water weight absorbed by the sample whereas m_{sample} is the sample weight. Sodium azide was added to the enzyme solution to prevent microbial contamination. Each three days both the water medium and the enzyme solution were changed. The "Liraza" agent of 1 g/dL concentration was used as an enzyme (Immunopreparat SUE, Ufa).

In experiments for determining the absorption of the condensed vapors of volatile liquid and water absorption coefficients at a confidence level of 0.95 and 5 repeated experiments, the error does not exceed 7%.

The obtained film samples were kept in soil according to the method [6] to estimate the ability to biodegradation. The soil humidity was supported on 50–60% level. The control of the soil humidity was carried out by the hygrometer ETR-310. Acidity of the soil used was close to the neutral with pH = 5.6–6.2 (Ph-meter control of 3in1 Ph). At a confidence level 0.95 and 5 repeated experiments the experiment error in determining the tensile strength and elongation does not exceed 5%.

Mechanical film properties (tensile strength (σ) and elongation (ε)) were estimated by the tensile testing machine ZWIC Z 005 at 50 mm/min tensile speed. At a confidence level 0,95 and 5 repeated experiments the experiment error in determining the tensile strength and elongation does not exceed 5%.

2.3   RESULTS AND DISCUSSION

It is well known that amorphous-crystalline polymers are subjected to high temperature grinding due to shearing impact on the polymer. For example, a good result is obtained in LDPE at high temperature shearing [2]. Despite CTZ is an infusible polymer, ultra dispersed powder with 6–60 μm particles was formed in the output of the rotary disperser after LDPE and CTZ convergence under high-pressure and shear deformation (Table 2.1). During high temperature shearing the powders of LDPE and CTZ with the latter not exceeding 60% mass were obtained. The particle distribution of LDPE/CTZ powders does not depend on the ratio of the components of the mixture and little differs from the particle distribution of the powder size of the CTZ under high temperature shearing.

The speed of the hydrolytic destruction of the polymer materials is closely connected with their ability to water absorption. Values of their absorption capacity according to water and heptane vapors were determined for a number of powder mixture samples of LDPE/CTZ (Table 2.1). It was established that the absorption coefficient of the condensed water vapors is directly proportional to the CTZ content.

TABLE 2.1   The Absorption Coefficient of the Condensed Water Vapors of Volatile Liquid (Water and n-heptane) K̈ of LDPE/CTZ Powders at 20°C.

№	LDPE/CTZ powder, mass. %	Particle size, μm	K̈ by water vapors, %	K̈ by n-heptane, %
1	0	5.5–8.0; 10.0–80.0	1.10±0.08	17±1
2	20	6.5–63.0	12.3±0.8	11.0±0.8
3	40	6.5–50.0	20±1	5.0±0.4
4	50	4.3–63.0	25±2	4.0±0.3
5	60	6.5–63.0	35±2	4.0±0.3

As the initial powders, the films with high chitosan content under rotomolding absorb water well (Table 2.2). At the same time thinner films absorb more water for a shorter period of time.

TABLE 2.2 Values of equilibrium water absorption coefficients K (%) of LDPE/CTZ films at 20°C.

№	LDPE/CTZ powder, mass %	K, %			
		Medium – water		Medium – liraza enzyme (1 g/L)	
		Film thickness 100 μm	Film thickness 800 μm	Film thickness 100 μm	Film thickness 800 μm
1	20	5.0±0.4	2.0±0.2	5.0±0.4	4.0±0.3
2	40	10.0±0.7	4.0±0.3	13.0±0.9	7.0±0.5
3	50	38±3	14±1	40±3	45±3
4	60	–	31±2	–	95.8±0.7

In case the film samples were placed into the enzyme solution, water absorption changes slightly. Firstly, the equilibrium values of the absorption coefficient of films in the enzymatic medium are higher than in water (Table 2.2). It is in the enzymatic medium usage that a longer film exposure (for more than 30–40 days) was accompanied by weight losses of the film samples. Moreover, after 40 days of testing, the film with 50%mass of chitosan and 100um thickness lost its integrity. Films of 800 μm thick and chitosan content of 50 and 60% lost their integrity after 2 months of the enzyme agent solution contact. These facts are quite logical as "Liraza" is subjected to a β-glycoside bond break in chitosan. Thus, the destruction of film integrity is caused by the biodestruction process. Higher values of the water absorption coefficient may be explained by enzyme destruction of chitosan chains as well due to some loosening in the film material structure (Table 2.2).

Tests on holding the samples in soil indicate on biodestruction of the obtained film samples either. It is found that the film weight is reduced by 7–8% during the first four months. Here the biggest weight losses are observed in samples with 50–60 mass % of chitosan.

Chitosan introduction into the polyethylene matrix is accompanied by changes in the physical and mechanical properties of the film materials (Table 2.3).

TABLE 2.3 Physical and Mechanical Properties of LDPE/CTZ Film Materials

№	Chitosan content in LDPE/CTZ, mass. %	σ, MPa		ε, %	
		Film thickness 100 μm	Film thickness 800 μm	Film thickness 100 μm	Film thickness 800 μm
1	0	13.30±0.05	40.10±0.05	460.00±0.05	125.00±0.05
2	20	5.40±0.05	22.70±0.05	24.30±0.05	13.20±0.05
3	40	7.50±0.05	25.80±0.05	12.50±0.05	7.60±0.05
4	50	11.10±0.05	29.80±0.05	6.00±0.05	6.20±0.05
5	60	11.60±0.05	30.60±0.05	5.20±0.05	4.80±0.05

As seen from Table 2.3, the polysaccharide introduction into the LDPE compounds results in slight decrease in the tensile strength of films. Wherein the number of the chitosan introduced does not affect the composition strength. However, LDPE/CTZ films obtain much less elongation values as compared with LDPE films under the same conditions. Thus, films which were obtained on the basis of ultra dispersed LDPE powders modified by chitosan possess less plasticity while retain in other satisfactory strength properties.

A method of obtaining compositions of ultra dispersed LDPE powders modified by chitosan under combined high pressure and shear deformation was worked out. The samples received obtain suitable strength properties, a good absorption ability and capability to biodegradation.

KEYWORDS

- **biodegradable polymer films**
- **chitosan**
- **low density polyethylene**

REFERENCES

1. Bazunova, M. V., Babaev, M. S., Bildanova, R. F., Protchukhan, Yu A., Kolesov, S. V., & Akhmetkhanov, R. M. (2011). Powder Polymer Technologies in Sorption-Active Composite Materials Vestn Bashkirs University, *16(3)*, 684–688.
2. Enikolopyan, N. S., Fridman, M. L., Karmilov, A. Yu., Vetsheva, A. S., & Fridman, B. M. (1987). Elastic Deformations Grindings of Thermoplastics Polymers, Reports AS USSR, *296(1)*, 134–138.
3. Akhmetkhanov, R. M., Minsker, K. S., & Zaikov, G. E. (2006). On the Mechanism of Fine dispersion of Polymer Products at Elastic Deformation Effects, Plastic Masses, *8*, 6–9.
4. Sheryshev, M. A. (1989). Formations of Polymer Sheets and Films, Braginsky, V. A. L. (Eds.) Chemistry Publishing, 120p.
5. Keltsev, N. V. (1984). Fundamental of Adsorptions Technology, Chemistry, M, 595p.
6. Ermolovitch, O. A., Makarevitch, A. V., Goncharova, E. P., & Vlasova, F. M. (2005). Estimation Methods of Biodegradation of Polymer Materials, Biotechnology, *4*, 47–54.

CHAPTER 3

A DETAILED REVIEW ON BEHAVIOR OF ETHYLENE-VINYL ACETATE COPOLYMER NANOCOMPOSITE MATERIALS

DHORALI GNANASEKARAN[1], PEDRO H. MASSINGA JR.[2], and WALTER W. FOCKE[1]

[1]Institute of Applied Materials, Department of Chemical Engineering, University of Pretoria, Pretoria 0002, South Africa.

[2]Universidade Eduardo Mondlane, Faculdade de Ciências, Campus Universitário Principal, Av. Julius Nyerere, PO Box 257, Maputo, Moçambique

CONTENTS

Abstract 90
3.1   Introduction 90
3.2   Ethylene-CO-Vinyl Acetate (EVA) 91
3.3   Classification of Nanostructured Materials (NSMS) 92
3.4   Polymer Nanocomposites 98
3.5   Ethylene-CO-Vinyl Acetate Nanocomposites 103
3.6   Effect of Equipment, Types of Clay and EVA on Nanocomposites 110
3.7   Effect of EVA Nanocomposite Structure on its Physical
      Properties and Performance 113
3.8   Conclusions 117
Acknowledgments 118
Keywords 118
References 118

ABSTRACT

The paper presents a complete conspectus of ethylene-co-vinyl acetate (EVA) nano-composites based on different types of nanostructured materials incorporated into EVA polymers. The three types of nanostructured materials, namely zero-dimensional nanostructured materials [(0DNSM) (POSS)], one-dimensional nanostructured materials [(1DNSM) (carbon nanotubes or CNTs, sepiolite)] and two-dimensional nanostructured materials [(2DNSM) (clay)] with EVA are introduced and a detailed discussion is provided on the effect of nanostructured materials on EVA polymer. Simultaneously, we discuss the recent approaches in which CNTs, sepiolite, clay and POSS nanomaterials play a vital role in EVA polymers, in addition to elucidating the influence of composite structures on the thermal, mechanical, and fire retardant properties. On the basis of this review we present the varied and versatile current research on EVA nanocomposites. The whole range of effects on polymer nanocomposite properties is covered. The great progress being made in the preparation of EVA nanocomposites offers fascinating new opportunities for materials scientists.

3.1   INTRODUCTION

There is currently immense interest in the development of nanostructured materials for a wide variety of applications and these materials offer exciting new challenges and opportunities in all the major branches of science and technology [1]. It is widely recognized that reductions in the size of components influence the interfacial interactions between them and this can, in turn, enhance the material properties to an appreciable extent [2]. Consequently, it is also possible to develop materials that are completely discontinuous, that is, that contain both organic and inorganic phases. Such materials exhibit nonlinear changes in properties with respect to composites that are made up of the same phases. This chapter focuses mainly on the classification of nanostructured materials such as zero-dimensional (polyhedral oligomeric silsesquioxane or POSS), one-dimensional (CNT, sepiolite) and two-dimensional (clay) nanostructure materials and on the recent developments in EVA nanocomposites (Fig. 3.1).

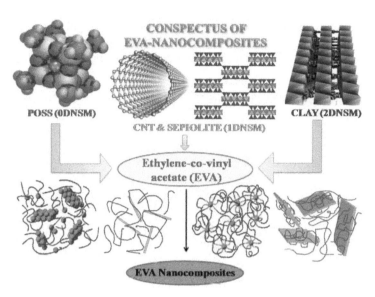

FIGURE 3.1   Overall theme of this review.

3.2   ETHYLENE-CO-VINYL ACETATE (EVA)

EVA copolymers are among the most important and widely used organic polymers, with a diversity of industrial applications such as electrical insulation [1], cable jacketing, waterproofing, corrosion protection, packaging of components [3], photovoltaic encapsulation and footwear [4]. EVA is used in paints, adhesives, coatings, textiles, wire and cable compounds, laminated safety glass, automotive plastic fuel tanks, etc. It is extremely elastomeric and accepts high filler loadings while retaining its flexible properties [5]. However, bulk EVA does not often fulfill the requirements in terms of its thermal stability and mechanical properties in some areas. In order to improve such properties, nanostructured materials are introduced as a reinforcing material. Among several polymeric materials used for polymer nanocomposites, EVA, a copolymer containing repeating units of ethylene as a nonpolar and vinyl acetate (VA) as a polar, has been newly adopted for its polymer nanostructured materials arrangement because of its potential engineering applications in the fields of packaging films and adhesives [6].

By changing the vinyl acetate content, EVA copolymers can be tailored for applications such as rubbers, thermoplastic elastomers and plastics [4]. The combinations of EVA with nanostructured materials have wide marketable applications and these applications depend on the VA contained in the main chain. As the VA content increases, the copolymer presents increasing polarity but lower crystallinity, and therefore different mechanical, thermal and morphological behaviors. The increasing polarity with increasing VA content is apparently useful in imparting a high degree of polymer nanostructured materials surface interaction, and it has been reported that there is a rise in the Young's modulus and yield strength of EVA polymeric nanocomposites [7] compared with other polymeric nanocomposites.

3.3   CLASSIFICATION OF NANOSTRUCTURED MATERIALS (NSMS)

The commencement of research into nanotechnology can be traced back over 40 years, but in the past decade hundreds of NSMs have been obtained across a variety of disciplines. NSMs are low-dimensional materials comprising building of submicron or nanoscale size in at least one direction and exhibiting size effects [8].

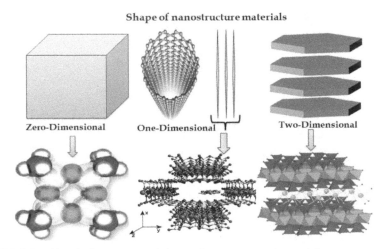

FIGURE 3.2   Chemical structure and shape of nanostructured materials.

The first part of this review can be classified into the three different types according to the dimensionality of nanomaterials (Fig 3.2), (a) 0DNSMs, such as polyhedral oligomeric silsesquioxane (POSS), which are characterized by three nanometric dimensions, (b) 1DNSMs (fibrous materials), such as carbon nanotubes and sepiolite, which are characterized by elongated structures with two nanometric dimensions, (c) 2DNSMs (layered materials), like clay (e.g., montmorillonite: MMT), which are characterized by one nanometric dimension.

3.3.1   ZERO-DIMENSIONAL NANOSTRUCTURED MATERIALS

Zero-dimensional nanostructured materials (0DNSMs) are those in which all the dimensions are measured within the nanoscale (no dimensions, or 0-D, are larger than 100 nm). The most common representation of zero-dimensional nanomaterials is nanoparticles.

Owing to their large specific surface area and other properties superior to those of their bulk counterparts arising from the quantum size effect, 0DNSMs have attracted considerable research interest and many of them have been synthesized in the past 10 years [9, 10, 11]. It is well known that the behaviors of NSMs depend strongly on their sizes, shapes, dimensionality and morphologies, which are thus the key factors giving rise to their ultimate performance and applications. It is therefore of great interest to synthesize 0DNSMs with a controlled structure and morphology. In addition, 0DNSMs are important materials due to their wide range of applications in the areas of catalysis, as magnetic materials and as electrode materials for batteries [12]. Moreover, the 0DNSMs have recently attracted intensive research interest because the nanostructures have a larger surface area and supply enough absorption sites for all involved molecules in a small space [13]. On the other hand, their having porosity in 3 dimensions could lead to better transport of molecules [14].

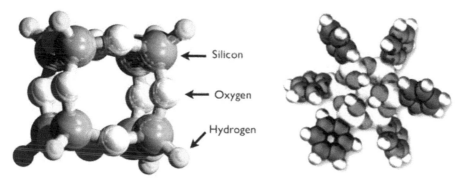

FIGURE 3.3   Chemical structure of polyhedral oligomeric silsesquioxane.

POSS is a class of organic-inorganic hybrid 0D nanostructure material constituted of an inorganic silica [15], which consists of a rigid, crystalline silica-like core, have the general formula $(RSiO_{1.5})_a (H_2O)_{0.5b}$, where R is a hydrogen atom or an organic group and a and b are integers (a = 1, 2, 3, ...; b = 1, 2, 3, ...), with $a + b = 2n$, where n is an integer (n= 1, 2, 3, ...) and $b \leq a + 2$. POSS is unique with regard to size (0.5 nm in core diameter) when compared with other nanofillers and can be functionally tailored to incorporate a wide range of reactive groups [16]. The size of a POSS molecule is nearly 1.5 nm in diameter and about 1000 D in mass; hence POSS nanostructures are approximately equivalent in size to most polymer dimensions and smaller than the polymer radii of gyration. Figure 3.3 shows the chemical skeleton of one of the POSS. POSS systems may be viewed as the smallest chemically discrete particles of silica possible, while the resins in which they are incorporated may be viewed as nanocomposites, which are intermediate between polymers and ceramics.

POSS derivatives have two unique features: (a) the chemical composition of POSS ($RSiO_{1.5}$) was found to be intermediate between that of silica (SiO_2) and si-loxane (R_2SiO), and (b) POSS compounds can be tailored to have various functional groups or solubilizing substituents that can be attached to the POSS skeleton.

One of the most popular branches of silsesquioxanes is polyhedral oligomeric silsesquioxanes, including the T_8 cage, T_{10} cage, T_{12} cage and other partial cage structures. Cubic structural compounds (completely and incompletely condensed silsesquioxanes) are commonly illustrated as T_6, T_7, T_8, T_{10} and T_{12} based on the number of silicon atoms present in the cubic structure (Fig. 3.4). The silica core of POSS is inert and rigid, whereas the surrounding organic groups provide compat-ibility with the matrix and processability. However, much more attention has been directed to the silsesquioxanes with specific cage structures designated by the ab-breviation POSS [17]. Kudo et al. [18] explored various stages of the most plausible mechanism for the synthesis of POSS, and an entire reaction scheme, including all intermediates, was considered.

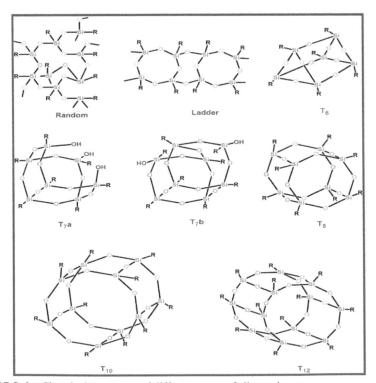

FIGURE 3.4   Chemical structures of different types of silsesquioxanes.

Conceptually, POSS may be thought of as an organic-inorganic hybrid (Fig. 3.5). Similarly, POSS is sometimes considered to be filler and sometimes a molecule. For example, POSS is rigid and inert like inorganic fillers, but unlike those conventional fillers, POSS can dissolve molecularly in a polymer.

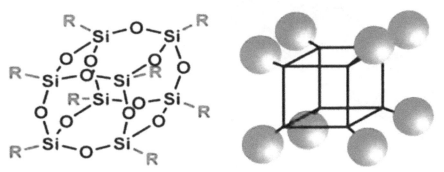

FIGURE 3.5   Silsesquioxanes Q_8 ($Q = SiO_{2/2}$); $R = H$, vinyl, epoxy, acetylene and acrylate.

3.3.2   ONE-DIMENSIONAL NANOSTRUCTURED MATERIALS

Within the various branches of nanotechnology, one-dimensional nanostructures (1DNSMs) have paved the way for numerous advances in both fundamental and applied sciences. One-dimensional nanostructured materials have one dimension that is outside the nanoscale. This leads to needle-shaped nanomaterials. One-dimensional materials, which include nanotubes, nanorods and nanowires, with at least one dimension in nanometer size fall under the category of 1DNSM. Almost all classes of materials, that is, metals, semiconductors, ceramics and organic materials, have been used to produce 1DNSMs. However, carbon nanotubes (Fig. 3.6) have occupied the most significant place and are the most widely studied 1DNSM [19].

It is generally accepted that 1DNSMs are ideal systems for exploring a large number of novel phenomena at the nanoscale and investigating the size and dimensionality dependence of functional properties. Certain fields of 1DNSMs, such as nanotubes and sepiolite, have received significant attention.

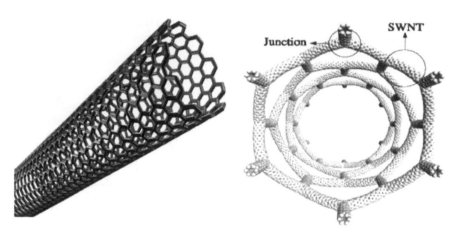

FIGURE 3.6　Schematic representation of one-dimensional nanostructured CNT.

1DNSMs have had a profound impact on nano-electronics, nanodevices and nanocomposite materials. Carbon nanotubes (CNTs) are long, slender fibers formed into tubes. The walls of the tubes are hexagonal carbon (as shown in Fig. 3.6) and the tubes are often capped at each end [20]. CNTs have been found to be effective reinforcing agents for several polymeric materials, apart from their ability to increase the electrical and thermal conductivity of these materials [21, 22]. Since Chaudhary et al. [23] first reported their existence, they have attracted increasing attention because of their high electrical and thermal conductivity, mechanical strength and chemical stability [24].

Sepiolite is a family of fibrous hydrated magnesium silicates with the theoretical half unit-cell formula $Si_{12}O_{30}Mg_8 (OH)_4 (OH_2)_4 \cdot 8H_2O$ [25]. The chemical structure of sepiolite shown in Fig. 3.7 is similar to that of the 2:1 layered structure of montmorillonite, consisting of two tetrahedral silica sheets enclosing a central sheet of octahedral magnesia, except that the layers lack continuous octahedral sheets [26]. The discontinuity of the silica sheets gives rise to the presence of silanol groups (Si–OH) at the edges of the tunnels, which are the channels opened to the external surface of the sepiolite particles [27]. The presence of silanol groups (Si–OH) can enhance interfacial interaction between sepiolite and polar polymers [28].

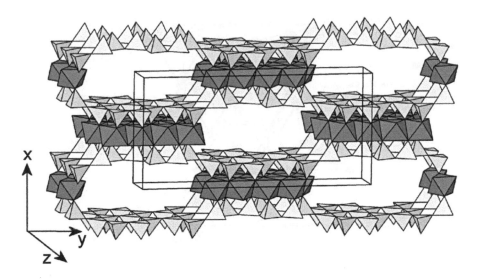

FIGURE 3.7   Schematic representation of one-dimensional nanostructured sepiolite.

3.3.3   TWO-DIMENSIONAL NANOSTRUCTURED MATERIALS WITH ETHYLENE-CO-VINYL ACETATE

Two-dimensional nanomaterials do not have two of the dimensions confined to the nanoscale, that is, 2D nanostructures (2DNSMs) have two dimensions outside of the nanometric size. 2D nanomaterials exhibit plate-like shapes and include nanolayers and nanocoatings.

In recent years, 2DNSMs have become a focus area in materials research, owing to their many low-dimensional characteristics, which differ from the bulk properties [29]. In the quest for 2DNSMs, considerable research attention has been focused over the past few years on the development of 2DNSMs. 2DNSMs with certain geometries exhibit unique shape-dependent characteristics and they are consequently utilized as building blocks and key components of nanodevices [30, 31, 32]. The 2DNSM of clay is shown in Fig. 3.8.

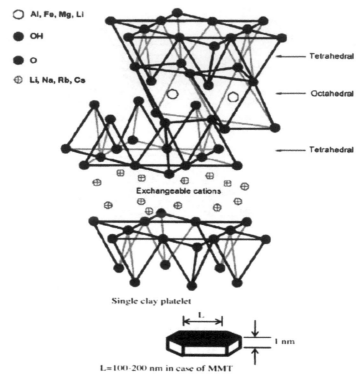

FIGURE 3.8   Graphical representation of two-dimensional nanostructured clay platelet [33].

3.4   POLYMER NANOCOMPOSITES

Nanoscale fillers, which are considered to be very important, include layered silicates (such as montmorillonite), nanotubes (mainly CNTs), fullerenes, SiO_2, metal oxides (e.g., TiO_2, Fe_2O_3, Al_2O_3), nanoparticles of metals (e.g., Au, Ag), POSS, semiconductors (e.g., PbS, CdS), carbon black, nanodiamonds, etc. Clay systems have been well developed, followed by POSS; little work has been performed using graphite-polymer and nanotube-polymer nanocomposites. In addition to clays, nanocomposites have been prepared using POSS, graphites and CNTs. The interest in such systems (organic-inorganic hybrid materials) is due to the fact that the ultrafine or nanodispersion of filler, as well as the local interactions between the matrix and the filler, lead to a higher level of properties than for equivalent micro and macrocomposites [34]. In a nanocomposite, the clay, or the nanofiller/additive, is well dispersed throughout the polymer. Polymer-clay nanocomposites are a new class of composite materials consisting of a polymer matrix with dispersed clay nanoparticles. In recent years, more attention has been given to incorporating nano-materials into polymer matrices to obtain high-performance nanocomposites. Clays typically consist of particles with a high aspect ratio, that is, their length is much longer than their width. Dispersion of the filler on a nanometer scale generally gives the polymer interesting insulation properties.

3.4.1 POSS NANOCOMPOSITES

Due to the great chemical flexibility of POSS molecules, POSS can be incorporated into polymers by copolymerization, grafting, or even blending using traditional processing methods and it can lead to a successful improvement of the flammability, thermal or polymer mechanical properties [2, 35]. Unlike traditional organic compounds, most POSS compounds release no volatile organic compounds below 300°C, are odorless, nontoxic, and reduce the toxicity of smoke upon combustion. Apart from this, the problems associated with polymer immiscibility are reduced. In the polymer-POSS nanocomposites, the POSS acts as a nanoscale building block and its interaction with the polymers on a molecular level is believed to be helpful for efficient reinforcement. Polymer-POSS nanocomposites are defined as polymers having small amounts of nanometer-size fillers (POSS), which are homogeneously dispersed by only several weight percentages (Fig. 3.9). A polymer-POSS nanocomposite with a filler having a small size leads to a dramatic increase in the interfacial area as compared with traditional composites. This interfacial area creates a significant volume fraction of interfacial polymer with properties different from those of bulk polymer even at low loadings [36, 37].

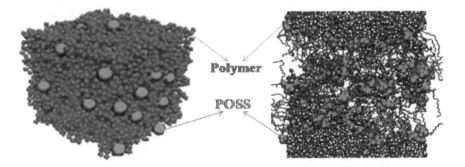

FIGURE 3.9 Systematic representation of polymer-POSS nanocomposites.

Normal fillers and especially nanofillers suffer from agglomeration. The agglomerates formed when conventional fillers are used lead to weak points in the polymer (stress concentrations) and this gives poor impact resistance and poor elongation to break. As a molecule, POSS dissolves in polymers as 1–3 nm cages and this gives performance advantages not seen with fillers. POSS increases the modulus of elastomers due to the stiffness of the cage and the high cross-link densities attainable using polyfunctional POSS cross-linkers. These include increased modulus and strength, outstanding barrier properties, improved solvent and heat resistance, and decreased flammability. This nature allows them to exhibit properties different from those of conventional microcomposites.

Organic-inorganic composite materials have been extensively studied for a long time. These may consist of two or more organic-inorganic phases in some combined form with the constraint that at least one of the phases or features must be nano-sized. Such materials combine the advantages of the organic polymer (e.g., flexible, dielectric, ductile and process able) and those of the inorganic material (e.g., rigid and thermally stable). For these reasons nanocomposites promise new applications in many fields such as for gas separation [8a, 38], in the aerospace industry [39], for electrical applications [40], as mechanically reinforced lightweight components, and in nonlinear optics, solid-state ionics, nanowires, sensors, and many others.

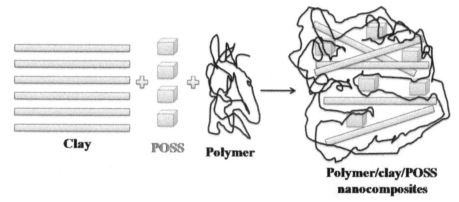

Clay   POSS   Polymer

Polymer/clay/POSS
nanocomposites

FIGURE 3.10   Formation mechanism of polymer/clay/POSS nanocomposites.

Fox et al. [41] prepared a POSS tethered imidazolium surfactant and used it to exchange montmorillonite (MMT) for the preparation of polymer nanocomposites in poly (ethylene-co-vinyl acetate) using a melt blending technique. Vaia et al. [42] have suggested that the extent of exfoliation of clay layers in a melt blended polymer composite is a result of two competing factors: enthalpic loss due to unfavorable polymer clay interactions and entropic gain due to increased polymer mobility over confinement within an intercalated clay structure. In addition, the presence of POSS interactions adds to the enthalpic barrier to exfoliation, since the POSS crystal domains must be melted. Despite these barriers to exfoliation, polymer composites exhibit small increases in exfoliation and smaller tactoids over unmodified clay. It is conceivable that the more rigid structure of POSS-imidazolium relative to other organic modifiers creates a more permanent barrier between the charged clay surface and more hydrophobic polymer chain. The model reaction of POSS/Clay composite shown in Fig. 3.10

Fox et al. [43] reported the synthesis and characterization of 1,2-dimethyl-3-(benzyl ethyl iso-butyl polyhedral oligomeric silsesquioxane) imidazolium chloride (DMIPOSS-Cl) and DMIPOSS modified montmorillonite (DMIP-MT) at several loadings of DMIPOSS-Cl. To investigated the ability of these clays to exfoliate in poly (ethylene-co-vinyl acetate) systems. In addition, they reported the effects of partial clay loading using only the POSS-modified imidazolium surfactant on the extent of exfoliation and quality of dispersion in polymer. In addition, Zhao et al. [44] have shown that although the organic groups on POSS do lower the polarity of clay surface, the surface free energy of clay exchanged with aminopropylisooctyl POSS is reduced by only half, leaving the clay surface still significantly more polar than polymer. Since clay is only partially exfoliated upon incorporation of polymers, regardless of their polarity, it is likely that the propensity for POSS to aggregate and limited space for intercalate polymer chains to move dominate the behavior in POSS exchanged clays.

3.4.2 CNT AND SEPIOLITE NANOCOMPOSITES

The one-dimensional nanostructure of CNTs, their low density, their high aspect ratio, and extraordinary properties make them particularly attractive as reinforcements in composite materials. These studies have been discussed in some excellent reviews [21, 45]. The variation of many parameters, such as CNT type, growth method, chemical pretreatment as well as polymer type and processing strategy has given some encouraging results in fabricating relatively strong CNT nanocomposites. Since the early preparation of a CNT/epoxy composite by Ref. [46], more than 30 polymer matrices have been investigated with respect to reinforcement by CNTs. The outstanding potential of CNTs as reinforcements in polymer composites is evident from the super tough composite fibers fabricated by Ref. [24]. By now, hundreds of publications have reported certain aspects of the mechanical enhancement of different polymer systems by CNTs.

CNTs have clearly demonstrated their capability as fillers in diverse multifunctional nanocomposites. The observation of an enhancement of electrical conductivity by several orders of magnitude at very low percolation thresholds (<0.1 wt.%) of CNTs in polymer matrices without compromising other performance aspects of the polymers such as their low weight, optical clarity, low melt viscosities, etc., has triggered an enormous activity world-wide in this scientific area. Nanotube-filled polymers could potentially, among the others, be used for transparent conductive coatings, electrostatic dissipation, electrostatic painting and electromagnetic interference shielding applications. A wide range of values for conductivity and percolation thresholds of CNT composites have been reported in the literature during the last decade, depending on the processing method, polymer matrix and nanotube type.

The majority of research in polymer/clay nanocomposites is focused on platelet-like clays, smectite clays such as MMT. Instead of very few works have been focused on fiber-like clays particles (sepiolite) [47]. Because of the peculiar shape, these nano-fillers are believed to be good candidates for the preparation of nanocomposites materials. In fact the dispersion of needle-like clays, compared to platelet-like clays, is favored by the relatively small contact surface area. Moreover, the reinforcement capacity of fibers in polymer nanocomposites is higher than that of platelet for uniaxial composites. These types of special reasons divert and investigate this research to sepiolite based polymer nanocomposites with different types of polymer matrices have been taken into consideration: polypropylene (PP), polyamide 6 (PA6) [48], polyurethane [49], and acrylonitrile–butadiene–styrene (ABS) [50].

3.4.3   CLAY NANOCOMPOSITES

Clays have been widely investigated and used as reinforcing agents for polymer matrices [21, 51–53]. They can dramatically enhance the mechanical performance and the barrier properties at filler loadings as low as 3–5 wt.%, without significantly changing other important characteristics such as transparency or density. In order to enhance their compatibility with polymer matrices, clays are usually modified with organic compounds (e.g., quaternary ammonium salts) but, even in this case, the properties of the polymer matrices are often not significantly improved [51]. Various researchers have reported on properties of nanoclay-EVA nanocomposites [54, 55].

IUPAC defines a composite as "a multicomponent material comprising multiple different (nongaseous) phase domains in which at least one is a continuous phase." Polymer-clay composites can be prepared in three different ways, namely by in situ polymerization [56, 57], by solution intercalation and by dispersion through melt processing. Three organoclay nanocomposite microstructures are possible, that is, phase separated, intercalated and exfoliated. The outcome is determined by which one of the interfacial interactions is favored in the system. The three main interactions are polymer-surface, polymer-surfactant and surfactant-surface [58]. The consensus is that, for thorough clay sheet dispersion, highly favorable polymer-surface interactions are essential [58]. A greater degree of exfoliation and better dispersion of layered double hydroxides (LDHs) were obtained in more polar matrices. Polyolefin nanocomposites are difficult to prepare as the low polymer polarity does not provide effective interaction with LDHs. Addition of the maleic anhydride grafted polyethylene as compatibilizer improves, to some extent, the dispersion of clays in nonpolar matrices during melt compounding [59]. The dispersed form of clay platelets is shown in Fig. 3.11.It is well established that the dispersion of particles with high aspect ratios, such as fibers and platelets, in polymeric matrices improves mechanical stiffness as well as some other properties.

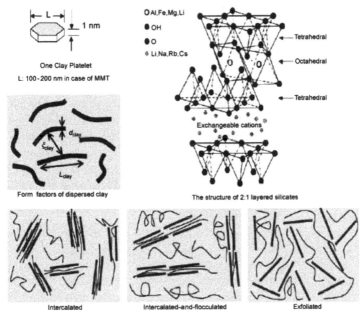

FIGURE 3.11   Schematic representation of clay with different dispersed phases [51].

However, good interfacial adhesion and a homogeneous dispersion are pre-requisites [60]. Nanostructured clays are ideal for the preparation of polymer-clay nanocomposites that possess improved gas barrier properties, better mechanical properties [61], enhanced flame retardancy [62], UV stabilization [63] or enhanced photodegradation [64], and so forth.

3.5   ETHYLENE-CO-VINYL ACETATE NANOCOMPOSITES

3.5.1   POSS-EVA NANOCOMPOSITES

EVA represents a considerable portion of the polymers currently marketed and has a wide range of applications, for example, in nonscratch films, hoses, coatings and adhesives [65]. It is commonly used in blends with polyolefins in order to improve the mechanical strength, processability, impact strength and insulation properties.

The demands for specific polymer properties for new applications have increased, including their use at higher temperatures and greater resistance to oxidation. The polymer industry has been able to keep abreast of these market demands through the use of additives, fillers and polymer blends. More recently, the diverse and entirely new chemical technology of POSS nanostructured polymers has been developed [16, 66]. This technology affords the possibility of preparing plastics that contain nanoscale reinforcements directly bound to the polymer chains.

The incorporation of POSS cages into polymer materials may result in improvements in polymer properties, including increased temperature of usage, oxidative resistance and surface hardening, resulting in better mechanical properties as well as a reduction in flammability and heat evolution [67]. In nanocomposites the addition of POSS nanoparticles can lead to the thermal stabilization of polymers during their decomposition [68]. However, in this regard the effect of the nanoparticle content is very crucial. In most cases the thermal stability enhancement takes place at a low loading (4–5 wt.%) of POSS nanoparticles, while at higher contents thermal stabilization becomes progressively lower. This is because at higher concentrations nanoparticles can form aggregates and thus the effective area of nanoparticles in contact with polymer macromolecules is lower. In this case microcomposites may be formed instead of nanocomposites and thus the protective effect of nanoparticles becomes lower.

In EVA systems the presence of 1 wt.% of POSS slightly increases the values of the stabilized and totalized torque and specific energy. This suggests that the viscosity is higher, probably due to an increase in the extent of entanglement caused by the incorporation of POSS into the polymer matrix [69]. A decrease in these values for a POSS content of 5 wt.% was also observed, suggesting that, in high concentrations, POSS forms aggregates, which are not incorporated into the polymer, leading to a decrease in viscosity. However, for EVA systems there is an increase in the stabilized and totalized torque and in the specific energy with an increase in the POSS content.

XRD analysis was carried out in order to verify the crystal structure of the POSS and of the polymer matrix. The XRD diffractograms of the POSS and the systems under study are shown in Fig. 3.1. For pure POSS, Fig. 3.1 shows peaks at $2\theta = 8°$, $8.8°$, $10.9°$, $11.7°$, $18.3°$ and $19.7°$, which are characteristic of its crystalline structure. The EVA copolymer (Fig. 3.1, 0 wt.% of POSS) shows broad peaks (2θ ranging from $21°$ to $25°$) related to the polyethylene structure of the EVA copolymer. For the other systems (Fig. 3.1) these peaks appear for POSS concentrations of 1 wt.% and above, indicating that the presence of EVA in the composite leads to the appearance of POSS aggregates at lower concentrations, probably due to the increase in the polarity of the polymer matrix.

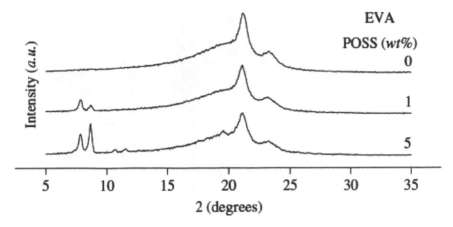

FIGURE 3.1 XRD patterns of POSS and of the composites [70].

The morphology and thermal properties of EVA composites with POSS nano-structures were analyzed and compared with those of the pure polymers. The POSS underwent aggregation at higher concentrations during composite processing, indicating a solubility limit of around 1 wt.%. The presence of EVA in the composite favors POSS aggregation due to the increase in the polarity of the polymer. These aggregates were observed in Si mapping and were characterized by the presence of a melt peak of POSS.

Further, these aggregates indicate that the polarity of the polymeric matrix plays a major role in the composite's morphology, even for immiscible systems. Figures 3.2 and 3.3 show SEM and Si mapping of 1% and 5% of POSS in EVA nanocomposites, respectively.

FIGURE 3.2 SEM (left) and Si mapping (right) of the composites with 1 wt.% of POSS [70].

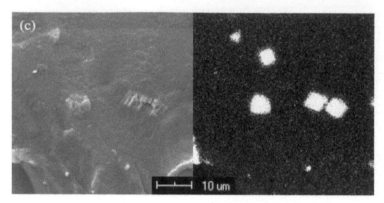

FIGURE 3.3   SEM (left) and Si mapping (right) of the composites with 5 wt.% of POSS [70].

The differential scanning calorimetry (DSC) curves of the second heating for the systems are shown in Fig. 3.4. EVA has Tm values of 84°C. The EVA systems with 5 wt.% of POSS also show a third melt peak, at approximately 56°C, which is characteristic of POSS. In the system with 1 wt.%, the POSS appears to be homogenously dispersed in the polymer matrix, minimizing the size of the domains in the samples. However, at a concentration of 5 wt.%, aggregation of the POSS takes place, which is characterized by the melt peak.

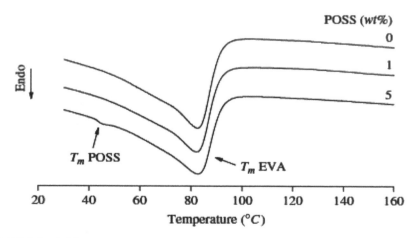

FIGURE 3.4   DSC curve of EVA–POSS nanocomposites [70].

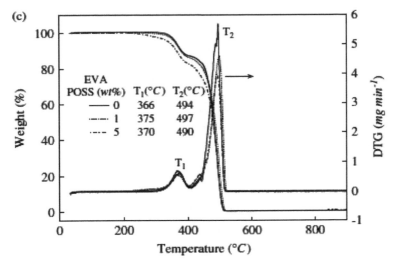

FIGURE 3.5 TGA curves of EVA-POSS nanocomposites [70].

The thermogravimetric analysis (TGA) curves of the systems are shown in Fig. 3.5. The first weight loss is observed in the temperature (T_1) range of 356 to 374°C, which is related to the elimination of the acetic acid in EVA and the degradation of POSS. In all systems, as the POSS content increases, the values of the onset temperature for degradation decrease due to the lower POSS thermal degradation temperature (285°C). Fina et al. [66] obtained similar results for the thermal degradation of polypropylene/POSS nanocomposites.

3.5.2 CNT-EVA AND SEPIOLITE-EVA NANOCOMPOSITES

The CNT is one of the stiffest materials produced commercially, having excellent mechanical, electrical and thermal properties. The reinforcement of rubbery matrices by CNTs was studied for EVA. George et al. [71] investigated the tensile strength of EVA-CNT and showed that it increased greatly (to 61%), even for very low fiber content (1.0 wt.%). The introduction of even a small number of CNTs can lead to improved performance of EVA. At 4 wt.% CNT loading, both the modulus and the tensile strength of the nanocomposite increased substantially. However, similar improvements were not observed at higher (8 wt.%) nanofiller loading due to filler agglomeration. George and Bhowmick [3] explained the effect of nanofillers and the VA content on the thermal, mechanical and conductivity properties of nanocomposites. They showed that polymers with high VA content have more affinity towards fillers due to the large free volume available, which allows easy dispersion of the nanofillers in amorphous rubbery phase, as confirmed from morphological studies. The thermal stability of nanocomposites is influenced by type of nanofiller.

Beyer [72] studied the flame retardant EVA-CNT nanocomposites and was synthesized by melt-blending. Fire property measurements by cone calorimeter revealed that the incorporation of CNT into EVA significantly reduced the peak heat release rates compared with the virgin EVA. Peak heat release rates of EVA with CNT were slightly improved compared with EVA nanocomposites based on modified layered silicates. There was also a synergistic effect by the combination of carbon nanotubes and organoclays resulting in an overall more perfect closed surface with improved heat release values. Carbon nanotubes are highly effective flame-retardants; they can also be more effective than organoclays. There was a synergistic effect between organoclays and nanotubes; the char formed during the degradation of the compound by a cone calorimeter was much less cracked compared with those from the single organoclay or carbon nanotube based compounds.

Huang et al. [56] prepared EVA-sepiolite composites and observed a remarkable change in the Young's modulus when only a small amount of sepiolite was incorporated. The fact that sepiolite has little effect on the elongation at break of the EVA/MH/SP composites, but a distinct effect on the Young's modulus, can probably be ascribed to the complex interactions between the polar EVA and the silanol groups of the high-aspect-ratio sepiolite. Hydrogen bonding may also be expected to occur between the ester groups of EVA and the characteristic silanol groups.

3.5.3 CLAY-EVA NANOCOMPOSITES

Melt intercalated nanocomposites of EVA with layered silicates have been studied over the last decade [4, 6, 8b, 23, 57, 73–81]. The method is an alternative to in situ polymerization and solution intercalation in the preparation of polymer nanocomposites. It is the most environmentally friendly method of all, versatile and compatible with current processing equipment.

Melt intercalation consists of blending molten polymer matrices with silicates. Polymer chains may diffuse from the matrix into the silicate interlayer. Conventionally, polymer chains should first intercalate into the silicate interlayers. Subsequently, the chains may push the silicate platelets further apart. Individual silicate particles may thus exfoliate into the polymer matrix. Sufficient compatibility between the surface energy of the silicate layers with that of the polymer chains is required. Silicate layers are hydrophilic, tending to be compatible only with polar polymers. Non-polar polymers appear to exhibit interfacial adhesion with organosilicates. However, compatibility between the polymer matrix and the surfactant chains of organosilicates is indispensable. It maximizes the freedom of the organosilicate chains in the polymer matrix configurationally. Hence, organosilicate chains gain entropy, which balances entropy loss owing to polymer chain confinement. It has been shown that only polar polymer chains are attached to silicate platelets, whereas nonpolar chains protrude into the bulk melt. Therefore, the interfacial adhesion with organosilicates is insufficient to yield thermodynamically stable nanocomposites.

Delaminating silicate platelets and achieving exfoliation requires strong shear forces [82]. These are attained during nanocomposite processing. Exfoliated structures are those in which the clay platelets are delaminated and individually dispersed in the polymer matrix. In exfoliated nanocomposite structures, silicate platelets are extensively interspaced. Such extensive interlayer separations disrupt coherent layer stacking and the resultant ordering of the clay platelets is not sufficient to produce a scattering peak. Hence, a featureless XRD diffraction pattern is recorded. Conversely, intercalated nanocomposites are those in which polymer chains and silicate layers alternate, in a well-ordered multilayer arrangement, with a well-defined and well-preserved interlayer distance.

Exfoliation is a key requirement for improving polymer properties. These include mechanical, thermal, barrier, flame retardant and optical properties. Exfoliation can lead to a very large surface area for the interaction of stiff silicate particles with polymer chains. EVA, $-(CH_2CH_2)_n [CH_2CH (OCOCH_3)]_m-$, is a somewhat polar copolymer due to its VA units. A higher content of VA, $-CH_2CH (OCOCH_3)-$, affords polarity for interaction with (OH groups of) pristine silicates.[77a] Good surface affinity between EVA and silicates alone is not enough to achieve exfoliation – prior break-up and expansion of intrinsic silicate stacks is required. This is accomplished through organic intercalation, using ion-exchange reactions. The stacks intrinsic to silicates are explained as follows: layered silicates comprise particles with thicknesses in the nanometer range; their aspect ratio is very large, exceeding two hundred; there is a propensity for aggregation of such particles due to the large packing area.

Generally, mechanical shear forces break up intrinsic silicate stacks. Subsequently, intercalation of large organic species expands the silicate interlayers. This expansion facilitates the diffusion of polymer chains into the layers. This, in turn, promotes the exfoliation of silicate platelets in polymer matrices. Ultimately, a large interfacial area for interaction between polymer chains and silicate particles is thus created. Intercalation may further alter the silicate surface chemistry. The changing nature of the silicate layers from hydrophilic to organophilic renders them compatible with hydrophobic polymers. To date, the preparation of fully exfoliated EVA-layered silicate nanocomposites remains a challenge. The use of different types of processing equipment and varying conditions has yielded disparities betweenn anocomposite structures and properties. This has resulted from the use of EVA with diverse features, silicates and organic modifiers of different types, modifiers with several chain lengths and different polarities, all intercalated into silicates in dissimilar amounts. It is therefore necessary to understand the influence of materials, instruments and processing conditions on the properties of created composites.

The transmission electron microscopy (TEM) shows the composites structures. Structures probed solely by X-ray diffraction analysis (XRD) are not completely certain. When no XRD peak is observed, it could be deduced that exfoliation has taken place, whereas the presence of a small number of ordered stacks may also yield a featureless XRD diffractogram. Moreover, the presence of a high number of nonuniformly dispersed clay stacks likewise yields a featureless pattern or interlayer variations. In turn, the presence of any homogeneous distribution of the clay nano-platelets could be interpreted as intercalation. Further, the presence of few silicate stacks could be deemed to indicate conventional microcomposites. Certainly, the presence of recalcitrant XRD reflections confirms the occurrence of intercalated or unmodified silicate regions. However, exfoliated regions may perhaps coexist and even predominate. TEM is critical since it gives direct evidence of composite structures. However, it may also fall short when a few samples not representative of the whole material are used. The research work reviewed below has probably been subject to this dilemma.

3.6   EFFECT OF EQUIPMENT, TYPES OF CLAY AND EVA ON NANOCOMPOSITES

3.6.1   EFFECT OF PROCESSING EQUIPMENT AND CONDITIONS ON DISPERSION OF EVA NANOCOMPOSITES

A co-rotating twin-screw extruder is considered the most effective shear device for the dispersion of silicate platelets. This is because the screws rotate in the same direction, intermesh and pass resin over and under one screw to another. The material is thus subjected to identical amounts of shear and unlikely to become stagnant. Chaudhary et al. [23] and Pistor et al. [79b] reported similar EVA nanocomposite structures. Partial intercalation and exfoliation were the predominant characteristics reported in both communications. It was further stated that all the samples contained small tactoid fractions [23, 79b]. However, the processing equipment and conditions employed by these researchers were different.

La Mantia et al. [78a] prepared equivalent samples using two different extruders. They reported similar EVA nanocomposite structures comprising intercalated silicate domains, as detected by XRD. The d_{001} was slightly increased for samples compounded in a twin-screw extruder [78a]. Similarly, Chaudhary et al. [23] and Pistor et al. [79b] found that, overall, nanocomposite structures seemed independent of processing equipment. In addition, all nanocomposites appeared to be insensitive to different processing temperatures and shear forces.

3.6.3   EFFECT OF SILICATE TYPE ON DISPERSION OF EVA NANOCOMPOSITES

The effect of the silicate type on the dispersion of EVA composites was investigated by Zanetti et al. [73b], Riva et al. [74a] and Peeterbroeck et al. [77b]. EVA matrices were melt blended with hectorite (HT), fluorohectorite (FH), magadiite [80] and montmorillonite (MMT), all organically modified. The dispersion of hectorite-type clays was found to be better than that of MMT clays. This may be attributed to the large interfacial area available for the interaction of polymer chains with hectorite-like minerals. The interaction of silicates with a high surface area with polymer chains yields excellent dispersion.[23]Hectoritehas an aspect ratio of approximately 5000, while that of MMT is <1000. Magadiite did not yield nanocomposites, only microcomposites.

3.6.4 EFFECT OF ORGANIC MODIFICATION OF SILICATES ON DISPERSION OF EVA NANOCOMPOSITES

Nanocomposite creation was found to depend on the type of silicate modification.[73] Suitable modification should render the silicate compatible with the polymer matrix. Silicates modified with ammonium cations bearing carboxylic acid moiety yielded conventional microcomposites, independent of the co-vinyl acetate content of EVA matrices. In contrast, organosilicates with nonfunctionalised alkyl ammonium tails (one and two) displayed affinity towards EVA chains. They yielded nanocomposites whatever the VA content (12, 19 and 27 wt.%) of the matrix. Exfoliated silicate sheets were observed, together with stacks of intercalated and unmodified silicates. The nanocomposite structures were considered to be intercalated/exfoliated.[73] Higher numbers of stacks were observed in EVA12 nanocomposites. These authors speculated that such a small extent of exfoliation was due to the low polarity of the EVA12 matrix.

Similarly, organosilicates with one long alkyl chain yielded nanocomposites with relatively high numbers of stacks. Riva et al. [74a] usedMMT modified with $(CH_2CH_2OH)_2N^+CH_3$ (tallow) and EVA19. The polarity of MMT was increased, further improving its affinity towards polar EVA matrices. An exfoliated structure was reported for the nanocomposite based on EVA19. In addition, stacks of unmodified silicates were observed [74a]. Zhang et al. [8b] increased the number of long alkyl chains used to modify MMT up to three. They anticipated that such an increase would decrease the organo-MMT polarity. Nevertheless, they found it necessary in order to yield wider organo-MMT basal spacing. The aim was to obtain a proper balance between the two as this would facilitate the migration and penetration of EVA chains into the silicate layers. The morphological features of the EVA-MMT nanocomposites were then examined. They were found to depend on the basal spacing of the organically modified MMT and the polarity of the EVA. Increasing both promoted better dispersions. The chains of EVA diffused more easily into the MMT layers. Two long alkyl chains were found sufficient to yield organosilicates with wider basal spacing [8b]. These authors recorded further marginal expansion of d_{001} for triple-tailed organoclay.

3.6.5 EFFECT OF VA CONTENT ON DISPERSION OF EVA NANOCOMPOSITES

The effect of changing the matrix VA content on the dispersion of the EVA composites was assessed by several authors [4, 6, 23, 73, 74b, 76, 79a]. Increasing the EVA polarity lowers the thermodynamic energy barrier for polymer interaction with silicates and therefore the polymer chains diffuse more easily into the silicate layers. Increased matrix amorphousness (with increasing VA units) further facilitates the stabilization of polymer chains within the silicate galleries [23]. These authors claimed that higher amorphous content prevents recrystallization of polymer chains during annealing. Therefore, the chains remain diffused within the silicate layers [23].

Whatever the VA content, organosilicates with OH groups along the alkyl N substituents appeared to be well dispersed. They may be due to strong intermolecular interactions between the OH groups of the organic modifier and the acetate functions of the EVA matrix [77b]. Increasing the VA content improved the degree of organosilicate dispersion, independent of the type of silicate modification. When the number of long tails was the same, organosilicates with higher chain lengths dispersed better. This may be related to the interlayer spacing of the organosilicate. Densely packing modifier into the silicate did not aid dispersion. Chaudhary et al. [23] claimed that this reduces the number of EVA chains penetrating the interlayer spaces.

Zhang and Sundararaj [6] investigated the extent of dispersion of some double-tailed organosilicates in EVA matrices with five VA contents (6, 9, 12, 18 and 28 wt.%). It was found that all EVA matrices further expanded organoclay interlayers. Increasing the VA content from 6 to 12 wt.% expanded the silicate interlayers considerably. Above such VA contents, no further interlayer expansion was recorded. An intercalation-limiting effect of the polarity after a certain critical VA content was revealed [6]. This critical VA content was found to be approximately 15 wt.%. The degree of intercalation of EVA into double-tailed organoclay increased only at VA contents up to about 15 wt.%. Thereafter, the expansion of basal spacing ceased. The interlayer expansion was attributed to increased diffusion of EVA. Polymer diffusion depends strongly on how well it flows. The latter is determined by the melt flow index (MFI).

3.6.6 EFFECT OF MFI ON DISPERSION OF EVA NANOCOMPOSITES

The propensity of EVA with a higher VA content to diffuse into the silicate interlayers has been established [8b]. Zhang and Sundararaj [6] examined the influence of MFI on the structure of nanocomposites. They used five EVA28 matrices with different MFIs (3, 6, 25, 43 and 150 g/10 min). The effect of the MFI on the intercalation-limiting effect of EVA polarity into double-tailed organoclay was investigated. Lowering the MFI from 150 to 25 did not cause any detectable change in the basal spacing. However, further decreasing the MFI to 6 did expand the silicate interlayer. Below this MFI, the silicate interlayer collapsed. It was then concluded that effective polymer diffusion requires a suitable conjugation between its mobility and its shear force [6]. The shear force should (i) create shear tensions during nanocomposite processing; (ii) aid the breaking up of organosilicate agglomerations; (iii) disperse silicate platelets or a few tactoids throughout the matrix; and keep the silicate platelets or tactoids apart. With regard to polymer mobility, it should be sufficient to promote the diffusion and penetration of polymer into the silicate layers before layer restacking. The existence of an intercalation-limiting effect of EVA into double-tailed organoclays was confirmed, although it appears to be dependent on the MFI of the matrix rather than on its VA content. Zhang and Sundararaj [6] recorded increasing interlayer distances with increasing VA content from 3 to 15 wt.%. No further expansion was recorded for organo-MMT intercalated by EVA22 with different MFIs (2 and 3 g/10 min). Zhang and Sundararaj [6] also recorded increased interlayer spacing when the MFI of EVA28 was lowered from 25 to 6. Thus it is speculated that EVA resins conjugating good mobility and sufficient shear force will have an MFI in the range of 3- 25. Marini et al. [79a] agreed that matrix viscosity is the driving force for polymer chain mobility within clay lamellae. In addition, imposed shear tension is also responsible for causing lamellae slippage and clay dispersion. Adequate affinity between polymer matrix and organosilicate was thus confirmed as indispensable.

3.7 EFFECT OF EVA NANOCOMPOSITE STRUCTURE ON ITS PHYSICAL PROPERTIES AND PERFORMANCE

3.7.1 INFLUENCE OF COMPOSITE STRUCTURE ON ITS MECHANICAL PROPERTIES

The effect of VA content on the mechanical properties of EVA/Mg layered double hydroxide nanocomposites has been studied by various groups [8b, 83]. As expected, various nanocomposites exhibit a much higher storage modulus than pure EVA grades, especially at low temperatures, given the reinforcing effect of nanofillers on the matrix. In addition, the presence of the fillers enables the matrix to sustain high modulus values at high temperatures. Also, various nanocomposites show a reduction in tan δ peak height as compared with the heights of the respective neat elastomers. This is due to the restriction in polymer chain movements imposed by the filler-polymer interactions. The enhancements in dynamic mechanical properties indicate that the more elastomeric (VA content) matrix is, the more easily the nanofillers are dispersed due to the higher free volume.

Typically, silicate particles have higher tensile moduli than polymer matrices. With increasing concentration of nanofiller, there is an increase in the Young's modulus (stiffness) of the nanocomposites. Alexandre and DuBois [84], using 5 wt.% MMT, prepared nanocomposites with double the Young's modulus of pure EVA27. It was found that EVA19 and EVA12 increased the Young's modulus by 50%. The variation in the modulus of nanocomposites was explained on the basis of their different structures. The dispersion of individual clay platelets responsible for the large increase in modulus was higher in the EVA27 nanocomposite [84]. Apart from polymer polarity, silicate modification with a surfactant having nonfunctionalised chains compatible with polymer matrices was similarly critical. The ductility of the EVA27 nanocomposite decreased only slightly compared with that of the pure polymer. This was in spite of a large increase in nanocomposite stiffness.

Zhang and Sundararaj [6] recorded ever-increasing Young's moduli of EVA nanocomposites with increasing concentration of nanofiller. In parallel, they proposed the existence of a "platelet saturation effect." Such an effect reduces the extent of platelet dispersion in the polymer matrix. The saturation effect is explained as follows: layered silicates have a large aspect ratio, exceeding 300; interaction between them is quite strong because of the large packing area; exfoliation and dispersion of silicate layers depend mainly on two factors: EVA-silicate interaction (ε_{es}) and silicate-silicate interaction (ε_{ss}); when $\varepsilon_{es} > \varepsilon_{ss}$, exfoliation of silicate layers is possible; conversely, when $\varepsilon_{es} < \varepsilon_{ss}$, exfoliation is impossible; an increase in clay content leads to a larger ε_{ss}; this is due to a shorter distance between the silicate aggregates [6]. The effect of the interplay between EVA polarity (amorphicity) and silicate concentration (wt.%) on the Young's modulus has been evaluated. It has been accepted that platelet "randomization" characterizes exfoliated nanocomposite structures. Typically, the effective dispersion of nanosilicates suppresses the ability of the matrix to absorb energy at lower VA content in the EVA matrix.[23]Nanosilicates increase spatial hindrance for polymeric chain movement. They impart rigidity to the polymer matrix, creating a "rigid" amorphous phase. Platelet-polymer and platelet-platelet interactions tend to create a flexible silicate network structure in the matrix. Owing to polymer entanglement, such a network increases the initial resistance of polymeric chains moving under stress. The initial deformation energy is then absorbed by the silicate network. Simultaneously, the flexible network increases the modulus of the nanocomposite. With increasing VA content, the flexibility of the silicate network increases. Consequently, the resistance of the polymeric chains to movement is lowered. Thus, a "mobile" amorphous phase develops, and the network's ability to absorb deformation energy decreases. This occurs in spite of platelet-polymer and platelet-platelet interactions in the flexible silicate network. Stress is then partially transferred to the polymer chains, allowing them to absorb higher deformation energy. Hence, the modulus appears to be dominated by the extent of matrix crystallinity/amorphousness rather than by the silicate network.

Rigidity may also be imparted without the formation of a silicate network structure. There will be good interaction between silicate platelets or clusters of tactoids with the matrix where they are dispersed and suitably oriented. However, tensile strength is likely to be reduced [6]. Favourable interactions at the polymer/silicate interface are critical for efficient stress transfer. Tensile strength does not increase when polymer-clay interactions are sufficiently developed. The strength of the nanocomposite reduces with increasing flexibility of the silicate network structure. Increasing matrix polarity tends to maximize extent of diffusion of EVA into silicate layers. A higher specific surface area becomes available for polymer-silicate interactions [23].

3.7.2   INFLUENCE OF COMPOSITE STRUCTURE ON ITS STEADY SHEAR RHEOLOGICAL PROPERTIES

The degree of dispersion of silicates in a polymer matrix affects the rheological behavior of nanocomposites. Measurement of complex viscosity by oscillatory testing is useful to estimate the degree of exfoliation of composites. The viscosity of highly dispersed nanocomposites, with an exfoliated structure, increases considerably when the shear rate is changed. Conversely, the viscosity of poorly dispersed nanocomposites increases only moderately with the shear rate. At a low shear rate, exfoliated nanocomposites have the propensity to display solid-like behavior. This has been attributed to the formation of a network structure by dispersed silicate layers [77a]. Polymer chains are entrapped within the network. Because there are unable to flow, the viscosity rises. High zero-shear viscosities indicate that the network of dispersed layers remains unaffected by the imposed flow. Interactions between silicate layers and polymer chains are more pronounced in exfoliated systems than in fully intercalated ones. At the same silicate concentration, the elastic modulus is higher for exfoliated structures than for intercalated ones [74a]. Hence, solid-like behavior occurs at higher silicate loading in the latter systems. This leads to a slower relaxation of polymer chains [77a].

High shear rates breakdown the silicate network and orient the platelets in the direction of flow. For this reason nanocomposites exhibit shear-thinning behavior. The slope of curves, the so-called "shear thinning exponent," is used to estimate the extent of nanocomposite exfoliation. It has been accepted that higher absolute values of the exponent indicate higher rates of exfoliation [77a, 78b]. However, Marini et al. [79a] suggested that a significant increase in viscosity in the low shear region indicates strong matrix-organosilicate interactions rather than exfoliation. Both well-dispersed intercalated and/or exfoliated silicates can lead to a huge increase in zero-shear viscosity [79a].

La Mantia and TzankovaDintcheva [78a] stated that the intensity of matrix-organosilicate interactions increases with silicate interlayer spacing. When the basal spacing increases, the surface area available for contact with the polymeric chains also increases. Moreover, due to the larger interplatelet distances, the volume concentration of the silicate increases [78a]. High interactions between the organosilicate and the polymer chains are critical for nanocomposite creation. However, they are not sufficient on their own to guarantee effective clay dispersion and exfoliation [6, 79a]. Strong matrix-organosilicate interactions are indicated by a significant increase in zero-shear viscosity, rather than simply high zero-shear viscosity. Marini et al. [79a] recordedlarge rheological differences between EVA nanocomposites, depending on the matrix viscosity. High-viscosity EVA12 (MFI = 0.3 g/10 min) and EVA19 (MFI = 2.1 g/10 min) consisted of fairly well dispersed compact tactoids and had higher zero-shear viscosity than their respective EVA matrices. However, such viscosities were of the same order of magnitude or were only one order of magnitude different. Further, pure matrices also displayed pseudoplastic behavior. Absolute values of the "shear thinning exponent" calculated for EVA12 nanocomposite and its matrix were high. Similarly, nanocomposites produced with low-viscosity EVA18 (MFI = 150 g/10 min) and EVA28 (MFI = 25 g/10 min) exhibited higher zero-shear viscosity than their respective EVA matrices. However, such viscosities were more than one order of magnitude different. Moreover, pure matrices exhibited Newtonian behavior. It was then concluded that organosilicate dispersion was dependent on EVA matrix polarity and viscosity [79a]. On its own, a high "shear thinning exponent" does not guarantee a higher rate of exfoliation. Likewise, a high zero-shear viscosity of the nanocomposites does not, on its own, guarantee strong matrix-organosilicate interactions.

3.7.3   INFLUENCE OF COMPOSITE STRUCTURE ON ITS FIRE PROPERTIES

The effect of several parameters (nature of clay and clay loading) on the fire retardancy of the nanocomposite has been investigated. It has been observed that the nature of the cations, which compensate for the negative charge of the silicate layers, affects the fire performance, even though the fire properties were improved for both the montmorillonite-type fillers investigated. The clay loading also affects the fire properties [75]. The Stanton Redcroft Cone Calorimeter was used to carry out measurements. The conventional data, namely time to ignition (TTI, s), heat release rate (HRR, kW/m^2), peak of heat release (PHRR, kW/m^2), that is, maximum of HRR, total heat release (THR, MJ/m^2) and weight loss (WL, kg) were supplied by Polymer Laboratories software.

Huang et al. [56] explained the synergistic flame retardant effects between sepiolite and magnesium hydroxide in EVA matrices. In the light of the positive results from the loss on ignition (LOI) and UL-94 tests, not only did the cone calorimeter test data indicate a reduction in the HRR and MLR, but also a prolonged TTI and a depressed smoke release (SR) were observed during combustion. Simultaneously, the tensile strength and Young's modulus of the system were also improved by the further addition of sepiolite due to the hydrogen bonds between silanol attached to sepiolite molecules and the ester groups of EVA.

Cárdenas et al. [85] studied the mechanical and fire retardant properties of EVA/ clay/sepiolite nanocomposites. Their results suggest that the synergistic effect is greater for bentonite with silica and with sepiolite than for bentonite with ATH. This is an expected effect in the case of bentonite with sepiolite taking into account that both inorganic fillers are phyllosilicates and have analogous chemical composition. However, the differences between the pHRR may also be influenced by the specific combustion mechanism of the different inorganic fillers used (silica, sepiolite). It is worth noting that EVA-sepiolite showed the lower THR among the other composites, confirming the synergistic effect between bentonite and sepiolite explained above. In the EVA-sepiolite composite, a uniform and rigid layer was formed, so the contribution of sepiolite to forming a more rigid layer of char was very clear, possibly due to the fibril structure of this type of clay. Consequently, it is possible to conclude that a nanostructure enables better fire performance to be achieved than a microstructure. In fact the presumed "diffusion effect," which leads to such improvements, occurs in a nanostructure but not in a macrostructure.

3.8   CONCLUSIONS

Detailed accounts of the different types of nanostructured materials (NSMs) to enhance the novel properties of pristine EVA have been covered in this review. The review has been systematically structured to give a clear and detailed insight into the materials. In the introduction we reviewed recent papers on the subject and classified the NSMs into three categories according to their dimensions: 0DNSM (POSS), 1DNSM (CNT, sepiolite), and 2DNSM (clay) with EVA. Next we presented a detailed discussion on the effects of POSS, various types of silicate structure and various organic modifiers on EVA nanocomposites. In the third section we discussed recent approaches to NSMs such as CNT, sepiolite, clay and POSS.As NSMs play a vital role in EVA nanocomposites, we also elucidated the influence of composite structures on their thermal, mechanical, and fire retardant properties.

With great progress being made in the preparation of EVA nanocomposites, there are fascinating new opportunities for materials scientists. While considerable attention is being paid to particular aspects of nanostructures (for instance, 0DNSM, 1DNSM and 2DNSM), future progress will hinge on a better understanding of EVA nanocomposites, their composition, size and morphology, which affect the activity of 0DNSM, 1DNSM and 2DNSM. In addition, as greater knowledge is acquired about the physical and chemical properties of 0DNSM, 1DNSM and 2DNSM, there will be more opportunities to exploit individual characteristics in thermal, electrical, mechanical and fire retardant-based applications. Moreover, the development of 0DNSM, 1DNSM and 2DNSM will help to improve our old technologies, and further research will produce more impressive results that will benefit various industries and society. Finally, it is important to note that new types of cubic silica (POSS) nanoparticles have recently been reported and their ability to form nanocomposites with enhanced properties has been proposed.

ACKNOWLEDGEMENTS

Financial support for this research from the Institutional Research Development Program (IRDP), the South Africa/Mozambique Collaborative Program of the National Research Foundation (NRF), and the Mozambican Research Foundation (FNI) is gratefully acknowledged. We are also grateful for the Vice-Chancellor's Postdoctoral fellowship, University of Pretoria, South Africa. The authors also acknowledge technical support from the Centre of Engineering Sciences at the Martin Luther University of Halle-Wittenberg.

KEYWORDS

- clay
- CNT
- EVA
- POSS
- sepiolite

REFERENCES

1. Hull, T. R., Price, D., Liu, Y., Wills, C. L., & Brady, J. (2003). An Investigation into the Decomposition and Burning Behaviour of Ethylene-Vinyl Acetate Copolymer Nanocomposite Materials, Polymer Degradation and Stability, *82(2)*, 365–371.
2. Gnanasekaran, D., Madhavan, K., Tsibouklis, J., & Reddy, B. S. R. (2011). Ring Opening Metathesis Polymerization of Polyoctahedral Oligomeric Silsesquioxanes (POSS) incorporated Oxanorbornene-5, 6-Dicarboximide, Synthesis, Characterization, and Surface Morphology of Copolymers, Australian Journal of Chemistry, *64(3)*, 309–315.
3. George, J. J., & Bhowmick, A. K. (2009). Influence of Matrix Polarity on the Properties of Ethylene Vinyl Acetate-Carbon Nanofiller Nanocomposites, Nanoscale Research Letters, *4(7)*, 655–664.
4. Cui, L., Ma, X., & Paul, D. R. (2007). Morphology and Properties of Nanocomposites Formed from Ethylene-Vinyl Acetate Copolymers and Organoclays, Polymer, *48(21)*, 6325–6339.
5. Alex, M. Henderson, (1993). Ethylene-Vinyl Acetate (EVA) Copolymers, A General Review. IEEE Electrical Insulation Magazine, *9(1)*, 30–38.
6. Zhang, F., & Sundararaj, U. (2004). Nanocomposites of Ethylene-Vinyl Acetate Copolymer (EVA) and Organoclay prepared by Twin-Screw Melt Extrusion, Polymer Composites, *25(5)*, 535–542.

7. Chaudhary, D. S., Prasad, R., Gupta, R. K., & Bhattacharya, S. N. (2005). Morphological Influence on Mechanical Characterization of Ethylene-Vinyl Acetate Copolymer-Clay Nanocomposites, Polymer Engineering & Science, 45(7), 889–897.

8. (a) Chiu, C. W., Huang, T. K., Wang, Y. C., Alamani, B. G., & Lin, J. J. (2013). Intercalation Strategies in clay/polymer hybrids, Progress in Polymer Science, (b) Zhang, C., Tjiu, W. W., Liu, T., Lui, W. Y., Phang, I. Y., & Zhang, W. D. (2011). Dramatically Enhanced Mechanical Performance of Nylon-6 Magnetic Composites with Nanostructured Hybrid One-Dimensional Carbon Nanotube-Two-Dimensional Clay Nanoplatelet Heterostructures, The Journal of Physical Chemistry B, 115(13), 3392–3399.

9. (a) Kamarudin, S. K., Achmad, F., & Daud, W. R. W. (2009). Overview on the Application of Direct Methanol Fuel Cell (DMFC) for Portable Electronic Devices, International Journal of Hydrogen Energy, 34(16), 6902–6916, (b) Chen, H., Cong, T. N., Yang, W., Tan, C., Li, Y., & Ding, Y. (2009). Progress in Electrical Energy Storage System, A critical review. Progress in Natural Science, 19(3), 291–312.

10. Ferreira-Aparicio, P., Folgado, M. A., & Daza, L. (2009). High Surface Area Graphite as Alternative Support for Proton Exchange Membrane Fuel Cell Catalysts, Journal of Power Sources, 192(1), 57–62.

11. Jin, Y. H., Lee, S. H., Shim, H. W., Ko, K. H., & Kim, D. W. (2010). Tailoring High Surface Area Nanocrystalline TiO2 Polymorphs for High-Power Li Ion Battery Electrodes, Electrochimica Acta, 55(24), 7315–7321.

12. (a) Dong, Z., Kennedy, S. J., & Wu, Y. (2011). Electrospinning Materials for Energy-Related Applications and Devices, Journal of Power Sources, 196(11), 4886–4904, (b) Armand, M., & Tarascon, J. M. (2008), Building Better Batteries, Nature, 451(7179), 652–657, (c) Simon, P., & Gogotsi, Y. (2008). Materials for Electrochemical Capacitors, Nature Materials, 7(11), 845–854, (d) Hu, C. C., Chang, K. H., Lin, M. C., & Wu, Y. T. (2006). Design and Tailoring of the Nanotubular Arrayed Architecture of Hydrous Ruo2 for Next Generation Supercapacitors, Nano Letters, 6(12), 2690–2695.

13. Shen, Q., Jiang, L., Zhang, H., Min, Q., Hou, W., & Zhu, J. J. (2008). Three-dimensional Dendritic Pt Nanostructures, Sono Electrochemical Synthesis and Electrochemical Applications, Journal of Physical Chemistry C, 112(42), 16385–16392.

14. (a) Teng, X., Liang, X., Maksimuk, S., & Yang, H. (2006). Synthesis of Porous Platinum Nanoparticles, Small, 2(2), 249–253, (b) Lee, H., Habas, S. E., Kweskin, S., Butcher, D., Somorjai, G. A., & Yang, P. (2006). Morphological Control of Catalytically Active Platinum Nanocrystals, Angewandte Chemie, 45(46), 7824–7828.

15. Brick, C. M., Ouchi, Y., Chujo, Y., & Laine, R. M. (2005). Robust Polyaromatic Octasilsesquioxanes from Polybromophenylsilsesquioxanes, Br xOPS, via Suzuki Coupling, Macromolecules, 38(11), 4661–4665.

16. Schwab, J. J., & Lichtenhan, J. D. (1998). Polyhedral Oligomeric Silsesquioxane (POSS)-Based Polymers, Applied Organometallic Chemistry, 12(10–11), 707–713.

17. Li, G., Wang, L., Ni, H., & Pittman, Jr., C. U. (2001). Polyhedral Oligomeric Silsesquioxane (POSS) Polymers and Copolymers, A Review, Journal of Inorganic and Organometallic Polymers, 11(3), 123–154.

18. Kudo, T., Machida, K., & Gordon, M. S. (2005). Exploring the Mechanism for the Synthesis of Silsesquioxanes 4, The Synthesis of T8, Journal of Physical Chemistry A, 109(24), 5424–5429.

19. Kuchibhatla, S. V. N. T., Karakoti, A. S., Bera, D., & Seal, S. (2007). One Dimensional Nanostructured Materials, Progress in Materials Science, 52(5), 699–913.

20. Li, Z. M., Li, S. N., Xu, X. B., & Lu, A. (2007). Carbon Nanotubes can Enhance Phase Dispersion in Polymer Blends, Polymer-Plastics Technology and Engineering, 46(2), 129–134.

21. Spitalsky, Z., Tasis, D., Papagelis, K., & Galiotis, C. (2010). Carbon Nanotube-Polymer Composites, Chemistry, Processing, Mechanical and Electrical Properties, Progress in Polymer Science (Oxford), *35(3)*, 357–401.

22. Gorrasi, G., Bredeau, S., Candia, C. D., Patimo, G., Pasquale, S. D., & Dubois, P. (2012). Carbon Nanotube-Filled Ethylene/Vinylacetate Copolymers, from in Situ Catalyzed Polymerization to High-Performance Electro-Conductive Nanocomposites, Polymers for Advanced Technologies, *23(11)*, 1435–1440.

23. Chaudhary, D. S., Prasad, R., Gupta, R. K., & Bhattacharya, S. N. (2005). Clay Intercalation and Influence on Crystallinity of Eva-based Clay Nanocomposites, Thermochimica Acta, *433(1–2)*, 187–195.

24. Baughman, R. H., Zakhidov, A. A., & DeHeer, W. A. (2002). Carbon Nanotubes the Route Toward Applications, Science, *297(5582)*, 787–792.

25. Zheng, Y., & Zheng, Y. (2006). Study on Sepiolite-Reinforced Polymeric Nanocomposites, Journal of Applied Polymer Science, *99(5)*, 2163–2166.

26. Chen, H., Zheng, M., Sun, H., & Jia, Q. (2007). Characterization and Properties of Sepiolite/Polyurethane Nanocomposites, Materials Science and Engineering, A, *445–446*, 725–730.

27. (a) Tartaglione, G., Tabuani, D., Camino, G., & Moisio, M. (2008). PP and PBT Composites Filled with Sepiolite, Morphology and Thermal Behaviour, Composites Science and Technology, *68(2)*, 451–460, (b) Alkan, M., & Benlikaya, R. (2009). Poly(vinyl alcohol) Nanocomposites with Sepiolite and Heat-Treated Sepiolites, Journal of Applied Polymer Science, *112(6)*, 3764–3774.

28. Huang, N. H. (2010). Synergistic Flame Retardant Effects between Sepiolite and Magnesium Hydroxide in Ethylene-Vinyl Acetate (EVA) Matrix, Express Polymer Letters, *4(4)*, 227–233.

29. Tiwari, J. N., Tiwari, R. N., & Kim, K. S. (2012). Zero-Dimensional, One-Dimensional, Two-Dimensional and Three-Dimensional Nanostructured Materials for Advanced Electrochemical Energy Devices, Progress in Materials Science, *57(4)*, 724–803.

30. Jun, Y., Seo, J., Oh, S., & Cheon, J. (2005). Recent Advances in the Shape Control of Inorganic Nano-Building Blocks, Coordination Chemistry Reviews, *249(17–18)*, 1766–1775.

31. Kim, K. S., Zhao, Y., Jang, H., Lee, S. Y., Kim, J. M., Kim, K. S., Ahn, J. H., Kim, P., Choi, J. Y., & Hong, B. H. (2009). Large-Scale Pattern Growth of Graphene Films for Stretchable Transparent Electrodes, Nature, *457(7230)*, 706–710.

32. Bae, S., Kim, H., Lee, Y., Xu, X., Park, J. S., Zheng, Y., Balakrishnan, J., Lei, T., Ri Kim, H., Song, Y. I., Kim, Y. J., Kim, K. S., Özyilmaz, B., Ahn, J. H., Hong, B. H., & Iijima, S. (2010). Roll-to-Roll Production of 30inch Graphene Films for Transparent Electrodes, Nature Nanotechnology, *5(8)*, 574–578.

33. (a) Kurecic, M., & Sfiligoj, M. (2012). Polymer Nanocomposite Hydrogels for Water Purification, (b) Chen, B., Evans, J. R. G., Greenwell, H. C., Boulet, P., Coveney, P. V., Bowden, A. A., & Whiting, A. (2008). A Critical Appraisal of Polymer-Clay Nanocomposites, Chemical Society Reviews, *37(3)*, 568–594.

34. Chrissafis, K., & Bikiaris, D. (2011). Can Nanoparticles Really Enhance Thermal Stability of Polymers? Part I, An Overview on Thermal Decomposition of Addition Polymers, Thermochimica Acta, *523(1–2)*, 1–24.

35. Gnanasekaran, D., & Reddy, B. S. R. (2010). Synthesis and Characterization of Nanocomposites Based on Copolymers of POSS-ONDI Macromonomer and TFONDI, Effect of POSS on Thermal, Microstructure and Morphological Properties, Advanced Materials Research, *123–125*, 775–778.

36. Balazs, A. C., Emrick, T., & Russell, T. P. (2006). Nanoparticle Polymer Composites, Where two Small Worlds Meet, Science, *314(5802)*, 1107–1110.

37. (a) Krishnamoorti, R., & Vaia, R. A. (2007). Polymer Nanocomposites, Journal of Polymer Science Part B, Polymer Physics, *45(24)*, 3252–3256, (b) Schaefer, D. W., & Justice, R. S. (2007). How Nano are Nanocomposites? Macromolecules, *40(24)*, 8501–8517.

38. Gnanasekaran, D., Ajit Walter, P., Asha Parveen, A., & Reddy, B. S. R. (2013). Polyhedral Oligomeric Silsesquioxane-based Fluoroimide-Containing Poly(Urethane-Imide) Hybrid Membranes, Synthesis, Characterization and Gas-Transport Properties, Separation and Purification Technology, *111*, 108–118.
39. Yang, Y., & Heeger, A. J. (1994). A New Architecture for Polymer Transistors, Nature, *372(6504)*, 344–346.
40. Gatos, K. G., Martínez Alcázar, J. G., Psarras, G. C., Thomann, R., & Karger Kocsis, J. (2007). Polyurethane Latex/Water Dispersible Boehmite Alumina Nanocomposites, Thermal, Mechanical and Dielectrical Properties, Composites Science and Technology, *67(2)*, 157–167.
41. Fox, D. M., Harris, Jr, R. H., Bellayer, S., Gilman, J. W., Gelfer, M. Y., Hsaio, B. S., Maupin, P. H., Trulove, P. C., & DeLong, H. C. (2011). The Pillaring Effect of the 1, 2-dimethyl-3(benzyl ethyl iso-butyl POSS) Imidazolium Cation in Polymer/Montmorillonite Nanocomposites, Polymer, *52(23)*, 5335–5343.
42. Vaia, R. A., Jandt, K. D., Kramer, E. J., & Giannelis, E. P. (1995). Kinetics of Polymer Melt Intercalation, Macromolecules, *28(24)*, 8080–8085.
43. Fox, D. M., Maupin, P. H., Harris, Jr, R. H., Gilman, J. W., Eldred, D. V., Katsoulis, D., Trulove, P. C., & DeLong, H. C. (2007). Use of a Polyhedral Oligomeric Silsesquioxane (POSS)-Imidazolium Cation as an Organic Modifier for Montmorillonite, Langmuir, *23(14)*, 7707–7714.
44. Zhao, F., Bao, X., McLauchlin, A. R., Gu, J., Wan, C., & Kandasubramanian, B. (2010). Effect of Poss on Morphology and Mechanical Properties of Polyamide 12/Montmorillonite Nanocomposites, Applied Clay Science, *47(3–4)*, 249–256.
45. Tjong, S. C. (2006). Structural and Mechanical Properties of Polymer Nanocomposites, Materials Science and Engineering, R, Reports, *53(3–4)*, 73–197.
46. Ajayan, P. M., Stephan, O., Colliex, C., & Trauth, D. (1994). Aligned Carbon Nanotube Arrays Formed by Cutting a Polymer Resin-Nanotube Composites, Science, *265*, 1212–1214.
47. Park, D. H., Hwang, S. J., Oh, J. M., Yang, J. H., & Choy, J. H. (2013). Polymer-Inorganic Supramolecular Nanohybrids for Red, White, Green, and Blue Applications, Progress in Polymer Science, *38(10–11)*, 1442–1486.
48. García-López, D., Fernández, J. F., Merino, J. C., & Pastor, J. M. (2013). Influence of Organic Modifier Characteristic on the Mechanical Properties of Polyamide 6/Organosepiolite Nanocomposites, Composites Part B, Engineering, *45(1)*, 459–465.
49. (a) Keledi, G., Hari, J., & Pukanszky, B. (2012). Polymer Nanocomposites, Structure, Interaction, and Functionality. Nanoscale, *4(6)*, 1919–1938, (b) Defontaine, G., Barichard, A., Letaief, S., Feng, C., Matsuura, T., & Detellier, C. (2010). Nanoporous Polymer Clay Hybrid Membranes for Gas Separation, Journal of Colloid and Interface Science, *343(2)*, 622–627.
50. Basurto, F. C., García-López, D., Villarreal-Bastardo, N., Merino, J. C., & Pastor, J. M. (2012). Nanocomposites of ABS and sepiolite, Study of different Clay Modification Processes, Composites Part B, Engineering, *43(5)*, 2222–2229.
51. Sinha Ray, S., & Okamoto, M. (2003). Polymer/Layered Silicate Nanocomposites, A Review from Preparation to Processing, Progress in Polymer Science (Oxford), *28(11)*, 1539–1641.
52. Scaffaro, R., Botta, L., Ceraulo, M., & LaMantia, F. P. (2011). Effect of Kind and Content of Organo-Modified Clay on Properties of Pet Nanocomposites, Journal of Applied Polymer Science, *122(1)*, 384–392.
53. Scaffaro, R., Maio, A., Agnello, S., & Glisenti, A. (2012). Plasma Functionalization of Multiwalled Carbon Nanotubes and their use in the Preparation of Nylon 6-Based Nanohybrids, Plasma Processes and Polymers, *9(5)*, 503–512.
54. Pramanik, M., Srivastava, S. K., Samantaray, B. K., & Bhowmick, A. K. (2002). Synthesis and Characterization of Organosoluble, Thermoplastic Elastomer/Clay Nanocomposites, Journal of Polymer Science, Part B, Polymer Physics, *40(18)*, 2065–2072.

55. Srivastava, S. K., Pramanik, M., & Acharya, H. (2006). Ethylene/Vinyl Acetate Copolymer/ Clay Nanocomposites, Journal of Polymer Science, Part B, Polymer Physics, *44(3)*, 471–480.

56. Huang, N. H., Chen, Z. J., Yi, C. H., & Wang, J. Q. (2010). Synergistic Flame Retardant Effects between Sepiolite and Magnesium Hydroxide in Ethylene-Vinyl Acetate (Eva) Matrix, Express Polymer Letters, *4(4)*, 227–233.

57. Lee, H. M., Park, B. J., Choi, H. J., Gupta, R. K., & Bhattachary, S. N. (2007). Preparation and Rheological Characteristics of Ethylene-Vinyl Acetate Copolymer/Organoclay Nanocomposites, Journal of Macromolecular Science, Part B, Physics, *46B(2)*, 261–273.

58. Giannelis, R. A. V. a. E. P. (1997). Lattice Model of Polymer Melt Intercalation in Organically-Modified Layered Silicates, Macromolecules, *30*, 7990–7999.

59. Costache, M. C., Jiang, D. D., & Wilkie, C. A. (2005). Thermal Degradation of Ethylene-Vinyl Acetate Coplymer Nanocomposites, Polymer, *46(18)*, 6947–6958.

60. Pradhan, S., Costa, F. R., Wagenknecht, U., Jehnichen, D., Bhowmick, A. K., & Heinrich, G. (2008). Elastomer/LDH Nanocomposites, Synthesis and Studies on Nanoparticle Dispersion, Mechanical Properties and Interfacial Adhesion, European Polymer Journal, *44(10)*, 3122–3132.

61. Wang, G. A., Wang, C. C., & Chen, C. Y. (2006). The Disorderly Exfoliated LDHs/PMMA Nanocomposites Synthesized by In Situ Bulk Polymerization, The Effects of LDH-U on Thermal and Mechanical Properties, Polymer Degradation and Stability, *91(10)*, 2443–2450.

62. Zubitur, M., Gómez, M. A., & Cortázar, M. (2009). Structural Characterization and Thermal Decomposition of Layered Double Hydroxide/Poly(P-Dioxanone) Nanocomposites, Polymer Degradation and Stability, *94(5)*, 804–809.

63. Bocchini, S., Morlat-Therias, S., Gardette, J. L., & Camino, G. (2008). Influence of Nano-dispersed Hydrotalcite on Polypropylene Photooxidation, European Polymer Journal, *44(11)*, 3473–3481.

64. Magagula, B., Nhlapo, N., & Focke, W. W. (2009). Mn2Al-LDH and Co2Al-LDH Stearate as Photodegradants for LDPE Film, Polymer Degradation and Stability, *94(6)*, 947–954.

65. Norman Allen, S., Miguel Rodriguez, M. E., Cristopher Liauw, M., & Fontan, E. (2000). Aspects of the Thermal Oxidation of Ethylene Vinyl Acetate Copolymer, Polymer Degradation and Stability, *68*, 363–371.

66. Fina, A., Tabuani, D., Frache, A., & Camino, G. (2005). Polypropylene-Polyhedral Oligomeric Silsesquioxanes (POSS) Nanocomposites, Polymer, *46* (19 SPEC. ISS.), 7855–7866.

67. Zheng, L., Waddon, A. J., Farris, R. J., & Coughlin, E. B. (2002). X-ray Characterizations of Polyethylene Polyhedral Oligomeric Silsesquioxane Copolymers, Macromolecules, *35(6)*, 2375–2379.

68. Tanaka, K., Adachi, S., & Chujo, Y, (2009). Structure Property Relationship of Octa Substituted POSS in Thermal and Mechanical Reinforcements of Conventional Polymers, Journal of Polymer Science, Part A, Polymer Chemistry, *47*, 5690–5697.

69. Kopesky, E. T., Haddad, T. S., Cohen, R. E., & McKinley, G. H. (2004). Thermomechanical Properties of Poly(Methyl Methacrylate)S Containing Tethered and Untethered Polyhedral Oligomeric Silsesquioxanes, Macromolecules, *37(24)*, 8992–9004.

70. Scapini, P., Figueroa, C. A., Amorim, C. L., Machado, G., Mauler, R. S., Crespo, J. S., & Oliveira, R. V. B. (2010). Thermal and Morphological Properties of High-Density Polyethylene/ Ethylene-Vinyl Acetate Copolymer Composites with Polyhedral Oligomeric Silsesquioxane Nanostructure, Polymer International, *59(2)*, 175–180.

71. George, J., & Bhowmick, A. K. (2008). Fabrication and Properties of Ethylene Vinyl Acetate-Carbon Nanofiber Nanocomposites, Nanoscale research letters, *3(12)*, 508–515.

72. Beyer, G. (2001). Flame Retardant Properties of EVA-Nanocomposites and Improvements by Combination of Nanofillers with Aluminium Trihydrate, Fire and Materials, *25(5)*, 193–197.

73. (a) Michae Alexandre, G. B., Catherine Henrist, Rudi Cloots, Andre´ Rulmont, Robert Jerome, & Philippe Dubois, (2001). "One-Pot" Preparation of Polymer/Clay Nanocomposites

Starting from Na+Montmorillonite. 1. Melt Intercalation of Ethylene-Vinyl Acetate Copolymer, Chem. Mater., *13*, 3830–3832. (b) Zanetti, M., Camino, G., Thomann, R., & Mülhaupt, R. (2001). Synthesis and Thermal Behaviour of Layered Silicate-EVA Nanocomposites, Polymer, *42(10)*, 4501–4507.

74. (a) Riva, A., Zanetti, M., Braglia, M., Camino, G., & Falqui, L. (2002). Thermal Degradation and Rheological Behaviour of EVA/Montmorillonite Nanocomposites. Polymer Degradation and Stability, *77(2)*, 299–304, (b) Cser, F., & Bhattacharya, S. N. (2003). Study of the Orientation and the Degree of Exfoliation of Nanoparticles in Poly(Ethylene-Vinyl Acetate) Nanocomposites, Journal of Applied Polymer Science, *90(11)*, 3026–3031.

75. Duquesne, S., Le Bras, C. J. M., Delobel, R., Recourt, P., & Gloaguen, J. M. (2003). Elaboration of EVA Nanoclay Systems Characterization, Thermal Behaviour and Fire Performance, Composites Science and Technology, *63*, 1141–1148.

76. Pasanovic-Zujo, V., Gupta, R. K., & Bhattacharya, S. N. (2004). Effect of Vinyl Acetate Content and Silicate Loading on Eva Nanocomposites under Shear and Extensional Flow, Rheologica Acta, *43(2)*, 99–108.

77. (a) Gupta, R. K., Pasanovic-Zujo, V., & Bhattacharya, S. N. (2005). Shear and Extensional Rheology of EVA/Layered Silicate-Nanocomposites, Journal of Non-Newtonian Fluid Mechanics, *128(2–3)*, 116–125. (b) Peeterbroeck, S., Alexandre, M., Jérôme, R., & Dubois, P. (2005). Poly(ethylene-co-vinyl acetate)/Clay Nanocomposites, Effect of Clay Nature and Organic Modifiers on Morphology, Mechanical and Thermal Properties, Polymer Degradation and Stability, *90(2)*, 288–294.

78. (a) La Mantia, F. P., & Tzankova Dintcheva, N, (2006). Eva Copolymer-Based Nanocomposites, Rheological Behavior under Shear and Isothermal and Non-Isothermal Elongational Flow, Polymer Testing, *25(5)*, 701–708. (b) Szép, A., Szabó, A., Tóth, N., Anna, P., & Marosi, G. (2006). Role of Montmorillonite in Flame Retardancy of Ethylene-Vinyl Acetate Copolymer, Polymer Degradation and Stability, *91(3)*, 593–599.

79. (a) Marini, J., Branciforti, M. C., & Lotti, C. (2009). Effect of Matrix Viscosity on the Extent of Exfoliation in Eva, Organoclay Nanocomposites, Polymers for Advanced Technologies, n/a-n/a, (b) Pistor, V., Lizot, A., Fiorio, R., & Zattera, A. J. (2010). Influence of Physical Interaction Between Organoclay and Poly(Ethylene-Co-Vinyl Acetate) Matrix and Effect of Clay Content on Rheological Melt State, Polymer, *51(22)*, 5165–5171.

80. Filippi, S., Paci, M., Polacco, G., Dintcheva, N. T., & Magagnini, P. (2011). On the Interlayer Spacing Collapse of Cloisite® 30B Organoclay, Polymer Degradation and Stability, *96(5)*, 823–832.

81. Joseph, S., & Focke, W. W. (2011). Poly(ethylene-vinyl co-vinyl acetate)/Clay Nanocomposites, Mechanical, Morphology, and Thermal Behavior, Polymer Composites, *32(2)*, 252–258.

82. Pavlidou, S., & Papaspyrides, C. D. (2008). A Review on Polymer-Layered Silicate Nanocomposites, Progress in Polymer Science (Oxford), *33(12)*, 1119–1198.

83. Soon Suh, S. H. R., Jong Hyun Bae, & Young Wook Chang, (2004). Effects of Compatibilizer on the Layered Silicate/EthyleneVinyl Acetate Nanocomposite, Journal of Applied Polymer Science, *94*, 1057–1061.

84. Alexandre, M., & Dubois, P. (2000). Polymer-Layered Silicate Nanocomposites, Preparation, Properties and uses of a New Class of Materials, Materials Science and Engineering R, Reports, *28(1)*, 1–63.

85. Cárdenas, M. Á., Basurto, F. C., García-López, D., Merino, J. C., & Pastor, J. M. (2013). Mechanical and Fire Retardant Properties of EVA/clay/ATH Nanocomposites, Effect of Functionalization of Organoclay Nanofillers, Polymer Bulletin, *70(8)*, 2169–2179.

CHAPTER 4

THE INFLUENCE OF THE ELECTRON DENSITY DISTRIBUTION IN THE MOLECULES OF (N)-AZA-TETRABENZOPORPHYRINS ON THE PHOTOCATALYTIC PROPERTIES OF THEIR FILMS

V. A. ILATOVSKY, G. V. SINKO, G. A. PTITSYN, and G. G. KOMISSAROV

N. N. Semenov Institute of Chemical Physics, Russian Academy of Sciences, 4 Kosygin str., 119991 Moscow, Russia; E-mail: gkomiss@yandex.ru; komiss@chph.ras.ru

CONTENTS

Abstract...126
Aim and Background 126
4.1   Introduction..126
4.2   Experimental Part...128
4.3   Results and Discussion 130
4.4   Conclusion 146
Keywords 147
References...147

ABSTRACT

As the nitrogen atoms are likely to be key points of the TPC macrocycle structure, there was an intention to conduct a sequential aza-substitution of all four CH groups in the TBP molecule to form phthalocyanine, with a fairly representative group of metal complexes with different degrees of the ligand bond ionicity, and explore changing of the photoelectrochemical characteristics of the films of produced pigments. Thus, the set of considered measurements on films of aza-substituted metal-derivatives of tetrabenzoporphyrin in combination with the data of quantum-chemical calculations of the electron density and energy characteristics of the pigments, leads to several conclusions: sequential substitution of carbon atoms in meso-position by nitrogen increases the electron density in the area under consideration with the corresponding increase of charges in the p-electronic bonds; the obvious consequence of the increase of electron density in the important area of structural conjugation of macrocycle is a ring currents increase and compression of the cycle, which leads to a considerable increase of the intermolecular interaction; consequences of the strengthening of the intermolecular interactions are bathochromic red shift of the absorption maxima, the increase in the extinction coefficients and the broadening of the absorption bands in the visible part of the spectrum, as well as the hypochromic shift of the Soret band; accordingly with the changes in the optical characteristics of the pigments it takes place a reduction of the band gap, lowering of the energy level of the valence band, increasing of the depth of acceptor levels formed by the adsorbed oxygen; from a practical point of view, the redistribution of electron density in the very structurally similar molecules leads to a significant increase in chemo- and light-fastness of pigments and oxidation potentials, to an 8–10 fold increase in the photocurrent, 1.6–1.8 fold increase in photopotentials, 2–5 fold in the quantum yield on a current.

AIM AND BACKGROUND

Primary goal of our study was to provide objective criteria to predict theoretically, at least qualitatively, photoelectrochemical properties of the molecules by their structure. In our opinion, the most important factor in the selection process is the electron density distribution in molecules.

4.1　INTRODUCTION

With the improvement of photovoltaic and photoelectrochemical solar energy converters interest in organic semiconductors is becoming increasingly applied nature inherent in the transition to industrial development. At this stage it is especially important to select correctly the most promising classes of organic semiconductors, and even more important to provide objective criteria to predict theoretically, at least qualitatively, photoelectrochemical properties of the molecules by their structure. In our opinion, the most important factor in the selection process is the electron density distribution in molecules, as it determines both the individual properties of the specific compound (organic semiconductor, pigment, dye), and the intermolecular interaction, that is the character of arrangement of molecules in a solid at condensation. Good examples of powerful rearrangement of electronic structure and corresponding changes of physical and chemical properties of molecules can be seen in the well-known tetrapyrrole compounds (TPC), which, of course, belong to the group of the most promising pigments. This is largely due to the enormous variability of the structure of the TPC, typified by porphyrins and their derivatives, which are cyclic aromatic polyamines, conjugated with multi-loop system containing a 16-membered macrocycle with a closed p-conjugation system including 4–8 nitrogen atoms. By replacement of hydrogen atoms in the pyrrole rings and meso-positions for a variety of donor-acceptor groups it has already been produced more than 1000 porphyrins. To this is added the variations due to the formation of different metal complexes and the introduction of extraligands. Substitution of CH-bridges by nitrogen atom (aza-substitution) in tetrabenzoporphyrin (TBP) gives a representative group of phthalocyanines (Pc) (otherwise tetrabenzo-porphyrazines / tetra- (butadiene-1,3-ylene-1,4) tetraazaporphin) and their derivatives.

To choose the most effective working pigment from that abundance is extremely difficult, especially as the theoretical assumptions for this choice are almost none. At certain stages one has to use intuitively guided screening to detect certain patterns that will eventually lead to the creation of a more or less acceptable theory. Thus, in comparison of the photocatalytic activity (the photovoltaic Becquerel effect) of thin films of TPC with different macrocycle structure it was noted that the maximum photocurrent (I_{ph}) and photopotential (U_{ph}) are given by pigments, which macrocycle's electronic structure modification is caused by exposure to a carbon atom in meso-position [1]. For example, in tetraphenyl porphyrin (TPP) substitution of <α–, β–, γ–, δ–> hydrogen to phenyl groups leads to a strong donation of electron density into the p-conjugation circuit (and hence on the pyrrole rings) and significantly increases the photoactivity of TPP. Additional polarization arises due to direct dipole interaction of atoms in the b-positions with the phenyl group. Slightly higher photocatalytic activity was shown by phthalocyanines [2], but, unlike TPP, in Pc nitrogen directly substitutes carbon atoms in meso-position. Furthermore, Pc has more substituents – benzene rings conjugated with the pyrrole ones – so, aromaticity of this compound is much higher, and it is difficult to determine which substitutions (aza- or benzo-) causes an increase in photoactivity.

As the nitrogen atoms are likely to be key points of the TPC macrocycle structure, there was an intention to conduct a sequential aza-substitution of all four CH groups in the TBP molecule to form phthalocyanine, with a fairly representative group of metal complexes with different degrees of the ligand bond ionicity, and explore changing of the photoelectrochemical characteristics of the films of produced pigments. Hoping to get a large enough material for reflection, in parallel with the experiments there were carried out quantum-chemical calculations of the changes in the distribution of electron density, bond order, the energies of the orbitals in molecules of TPC. The calculations were performed by the program GAMESS, using Rutan molecular orbitals, by restricted Hartree-Fock method (not taking into account the correlation effects) and by density functional theory method (approximately takes into account the correlation effects). Both, experimental and calculated data have shown a good correlation with phenomenological assumptions.

4.2 EXPERIMENTAL PART

In setting up experiments comparing the PEC properties of various aza-derivatives of TBP and their metal complexes we focused on the adequacy of the conditions of measurement, reproducibility and statistical significance of the results. For each modification of a pigment there were measured parameters of 30 electrodes and performed statistical analysis. The mean square variation of parameters in the cited experimental data s = 5.6%. Moreover, given the strong dependence of photocurrents I_{ph} (pH) and photopotentials U_{ph} (pH) of pigmented electrodes on the pH of the electrolyte [3,4], for all kinds of pigments there were determined extreme points of the maximum values of these parameters. Figure 4.1 shows the complete dependence of the potentials and currents of Pt-ZnPc electrode plotted from the average data for the 30 electrodes. For other pigments there were obtained similar dependences, by which it was compared their photocatalytic activities in the reaction of oxygen reduction, wherein the first single-electron step is endothermic:

$$H^+ + O_2 + e^- \rightarrow HO_2, E_o = -0.32 \ V,$$

and requires a large activation energy. Excitons, photogenerated in the bulk of the film, migrate to the pigment – electrolyte interface. High concentrations of oxygen adsorbed on the film ($\sim 10^{15}$ cm^{-3}) provides necessary conditions for separation of electron – hole pairs arising from the collapse of the excitons due to efficient trapping of electrons with formation of the charged form of adsorption O_2^-. The increase in the concentration of hydrogen ions promotes the completion of the reduction process and the transfer of the reaction products in the electrolyte. However, as can be seen from Fig. 4.1, there is an optimal pH zone for I_{ph} (pH), U_{ph} (pH), beyond which further increase the concentration of H$^+$ leads to a decrease in the photoresponse. This is due to the different photoelectrochemical stability of pigments, particularly – the possibility of protonation in the acidic environment, which is largely determined by the redox potential and ionization potential. In this regard, the comparison was performed by photoactive maxima of I_{ph} (pH), U_{ph} (pH).

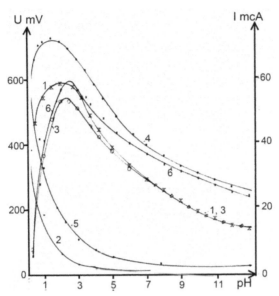

FIGURE 4.1   The dependence of currents and potentials of the electrode with Zn-Pc film on the Pt substrate on the pH of the electrolyte.

Films of tetrabenzoporphyrin and its metal derivatives with thickness of 50 nm were applied to the polished platinum substrates with a diameter of 11 mm by thermal vacuum sublimation in quasi-closed volume [5–7], using a turbo-molecular vacuum system Varian Mini-TASK (vacuum up to 10^{-9} Torr) with original design of evaporation chamber. Pigments used were previously purified twice by sublimation in vacuum 10^{-7} Torr. The amounts of impurities did not exceed 10^{-4}%. The film thickness control was carried out by frequency change of a quartz resonator located near the substrates in the evaporator system. Calibration curves were obtained by determining the thickness of pigmented films on transparent quartz substrates placed in the pigments deposition zone, with a micro-interferometer MII-11. Control of spectral parameters of the films confirmed integrity of the structure of pigment molecules. Thanks to the carefully refined method we obtained highly reproducible parameters of the films on platinum substrates. In particular, in the study of PEC parameters variations in the photocurrents did not exceed 15% and the photopotentials – 20%.

Photoactivity of the films were compared by Becquerel effect value, measuring I_{ph}, U_{ph} relatively a saturated Ag/AgCl electrode in aerated 1.0 N KCl electrolyte, changing the pH from 1 to 14 by titration with 1.0 N KOH or HCl solution. Pigmented electrodes were illuminated with an arc xenon lamp DKSSh-120 with a stabilized power supply, the light output in the plane of the electrode was 100 mW/cm². For measurement of the quantum yield on a current h pigment films with thickness d from 2 to 300 nm were applied on polished quartz substrates coated with a conductive layer of platinum. The h value was determined from a ratio,

$$h = 6.25 \cdot 10^{7} \cdot [hc\ I_{ph}/W\lambda]\ (\%),$$

where h – Planck constant, I_{ph} – photocurrent in µA, W – absorbed light power W, l – wavelength m, c – speed of light in m/s. Measurements were carried out in the region of maximum absorption of the pigments, separating from the spectrum of the lamp DKSSh-120 band width of 10 nm using interference filters at luminous flux in the plane of the electrode 10 mW/cm².

4.3   RESULTS AND DISCUSSION

The results of measurement of photoelectrochemical activity of the films of 45 metal complexes of derivatives of tetraazabenzoporphyrin with varying degrees of azasubstitution are shown in Table 4.1.

TABLE 4.1   Variations of Photopotentials, Photocurrents, Quantum Yield on a Current of Thin Film Electrodes Based on the Azasubstituted Metal Complexes of Tetrabenzo-porphyrin Depending on the Degree of Substitution and the Type of Central Atom

#	TPC	U_{ph}	pH	I_{ph}	pH	η
1	Mn-TBP	102	4	0.7	4	2.0
2	Mn-MATBP	145	3	2.8	3	4.0
3	Mn-DATBP	178	3	4.1	3	4.5
4	Mn-TATBP	198	2	5.5	2	5.0
5	Mn-Pc	230	1	6.5	1	6.0
6	Ni-TBP	53	3	0.1	3	1.1
7	Ni-MATBP	67	2	0.2	3	1.3
8	Ni-DATBP	72	1	0.2	2	1.3
9	Ni-TATB	78	1	0.3	1	1.4
10	Ni-Pc	80	0	0.3	0	1.4
11	Co-TBP	71	2	0.1	2	1.2
12	Co-MATBP	92	1	0.2	1	1.5
13	Cu-DATBP	105	1	0.3	1	1.5
14	Co-TATBP	108	1	0.4	1	1.6
15	Co-Pc	110	0	0.4	0	1.6
16	Zn-TBP	308	5	5.2	5	9.0
17	Zn-MATBP	431	4	20.0	4	13.0
18	Zn-DATBP	500	3	32.0	3	15.0

19	Zn-TATBP	540	2	46.0	2	17.0
20	Zn-Pc	570	2	54.0	2	18.0
21	Mg-TBP	160	4	4.0	3	4.0
22	Mg-MATBP	280	4	9.0	3	6.0
23	Mg-DATBP	320	3	14.0	2	7.0
24	Mg-TATBP	370	2	17.0	1.5	9.0
25	Mg-Pc	400	2	21.0	1.5	10.0
26	H_2-TBP	162	4	1.0	3	2.0
27	H_2MATBP	227	3	4.0	2	4.0
28	H_2DATBP	260	2	6.2	2	5.0
29	H_2TATBP	285	1	7.7	1	6.0
30	H_2-Pc	300	1	8.2	0	6.0
31	Fe-TBP	85	3	0.4	2	1.0
32	Fe-MATBP	140	1	1.1	1	3.0
33	Fe-DATBP	155	1	1.2	0	3.2
34	Fe-TATBP	170	0	1.3	0	3.5
35	Fe-Pc	180	0	1.4	0	4.0
36	ClFe-TBP	105	4	1.7	3	2.5
37	ClFe-MATBP	180	3	3.5	2	3.8
38	ClFe-DATBP	210	2	4.6	2	4.5
39	ClFe-TATBP	240	1	5.0	1	5.0
40	ClFe-Pc	250	1	5.4	1	5.0
41	VO-TBP	240	4	11.0	3	5.0
42	VO-MATBP	290	2	17.0	2	9.0
43	VO-DATBP	330	1	21.0	1	13.0
44	VO-TATBP	350	1	26.0	1	15.0
45	VO-Pc	350	1	28.0	0	16.0

U_{ph} – maximum values of the photopotential in mV; I_{ph} – maximum photocurrent in µA; pH – value of pH at which the maximum of U_{ph}, I_{ph}; h – quantum efficiency of a current. TBP – tetrabenzoporphyrin, MATBP – mono-tetraazabenzoporphyrin, DATBP – di-tetraazabenzoporphyrin, TATBP – tri-tetraazabenzoporphyrin, Pc – phthalocyanine (tetra-tetraazabenzoporphyrin).

For all studied compounds the same type effect of aza-substitution is observed, the most dramatic change of photocurrent (2–5 times) and photopotential (1.3–1.7 times) is the first step – the transition from the TBP to a mono-aza-substituted de-rivatives. Each subsequent step makes a smaller contribution, but the overall increase in the transition to the structure of phthalocyanine for U_{ph} reaches 185%, and for I_{ph} – 1000%. The stability of pigments changes significantly – the pH value, at which protonation of the molecules begins, decreases by 4–5 units as the proportion of nitrogen atoms in the macrocycle grows, which is associated with the increase of redox potential with aza-substitution. Changing in stability of pigments is seen in the stability of the electrodes. Photoactivity of the TBP films lost in 10–12 h of light saturating light. At the same time an output of dications of the pigments into solution was spectrally recorded. Phthalocyanine films (regardless of the nature of the central atom) did not change their characteristics after hundreds of hours of light in the same conditions. Basically, the differences in the photoactivity of pigments are caused by three factors – the efficiency of energy (charge) transfer, the band gap, the spectral range and extent of light absorption. All three are ultimately determined by the distribution of electron density in the molecules. So, for all of aza-derivatives of TBP, as a result of the transition from the structure of pure porphyrin to the phthalocyanine, there is observed an increase of photochromic properties of molecules and molecular interaction. For example, Fig. 4.2 shows the change in absorption spectrum of zinc tetrabenzoporphyrin at the sequential aza-substitution: absorption bandwidth expansion occurs simultaneously with the increase of the extinction coefficient and bathochromic shift, leading to a decrease in gap width.

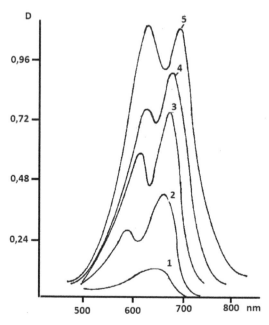

FIGURE 4.2   The absorption spectra of the films of aza-derivatives of Zn-TBP. (1 – Zn-TBP, 2 – Zn-MATBP, 3 – Zn-DATBP, 4 – Zn-TATBP, 5 – Zn-Pc).

Quantitative changes in the quantum yield on a current at sequential aza-substitution are less significant than those for the photocurrent, but nevertheless, the total increase of h reaches 300%, the greatest increase in efficiency is observed for the first step – a mono-aza-substitution. For the photocurrents that can be explained by a sharp increase in the extinction coefficient, that is, photochromic properties at the molecular level. In measurements of h this property has less influence on the final result, and increased quantum yield can be attributed to the high efficiency of energy and charge transfer between the molecules, as well as to a more advantageous arrangement of the band structure energy levels.

Lattice constants in a preferred crystal orientation plane (parallel to the substrate) are 1.98 nm for H_2-Pc and 2.19 nm for H_2-TBP, which corresponds to the axis "a" of the unit cell, connecting centers of molecules lying in one plane. Even on electron micrographs of the films pigments thickness of 5 nm (Fig. 4.3), obtained with an electron microscope JEM-100B with the ultimate resolution of 0.2 nm at the same scale is a clear difference in the interatomic distances and the changes in the lattice. This difference in interatomic distances and changes in the lattice constants are evident even in electron micrographs of pigment films with thickness of 5 nm (Fig. 4.3) received by an electron microscope JEM-100B with maximum resolution of 0.2 nm at the same scale.

 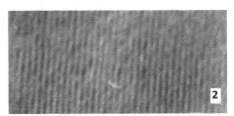

FIGURE 4.3 The change in the lattice with fourfold azazameschenii in metal-free tetrabenzporfirine (1 – tetrabenzporfirin 2 – phthalocyanine. Resolution of 0.2 nm.).

Given the almost identical molecular size (distance between the outer atoms of the benzene rings due to compression of the coordination sphere as a result of aza-substitution in Pc is of 0.13 nm smaller), the minimum distance between the nearest atoms for H_2-Pc is ~ 0.34 nm, and for H_2-TBP ~ 0.42 nm. This increases the electron affinity, decreases both ionization potential and the energy difference between the highest occupied and lowest unoccupied molecular orbitals, leading for Pc to a bathochromic shift of the first absorption band. Simultaneously, there are observed strengthening of the bond to the metal Me^{2+} at the ion radii less than 1.4 A, the appearance of dative p-bonds Me-N, change in the acid-base properties, in particular, a considerable reduction of protonation even in a strongly acidic medium. Thus, the interatomic distances are reduced by about 20%, which certainly contributes to the ring currents in the conjugation macrocycle, respectively, to strengthening of intermolecular interactions and increase of the efficiency of energy and charge transfer (tetrapyrrole compounds are characterized by "hopping" charge-transfer mechanism in which intermolecular distances are particularly important) and to an increase of the quantum yield on a current, a maximum value in the investigated group of compounds is 18%. Characteristically, the net effect depends on the nature of the central atom, whose interaction with the ligand significantly changes the electron density distribution in the macrocycle and its inner diameter.

Figure 4.4 shows the position of the energy bands for the limiting cases of substitution (no substitution – TBP, complete replacement – Pc) and the position of the acceptor level of oxygen in accordance with its redox potential.

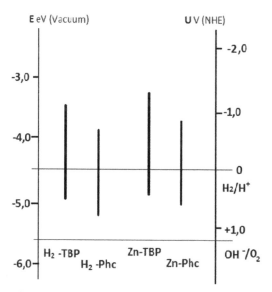

FIGURE 4.4   Energy band diagram of Me-TBP and Me-Pc.

If the surface layer is filled, that is, oxygen captures an electron to form a charge transfer complex:

$$O_2 \rightarrow O_{2\,(a)} \rightarrow [O_2^{-\delta}]_{(a)} + p^{+d},$$

then the concentration of adsorbed charged molecules is determined by conventional expression [8]:

$$\vartheta^- = \vartheta \; [1 + 1/2 \cdot exp \; (w/kT)] = \vartheta \; [1 + 1/2 \cdot exp \; ((E_a - F_s) \; /kT)],$$

where J – total concentration of adsorbed molecules, w – activation energy of the transition to the charged form, F_s – the distance from the Fermi level at the surface to E_v, E_a – the distance from the surface acceptor level to the E_v (E_v – the ceiling of the valence band). Consequently, the energy diagrams shown are initially determine increased population of acceptor levels in the dark and less photoresponse of TBP than Pc.

However, arguments based on the general concepts and physicochemical properties of the molecules do not provide sufficient certainty in assessing the impact of the electronic structure of molecules on the properties of their solid agglomerates. In connection with this there was attempted mathematical modeling of transformation of the structures by means of quantum chemical calculations of the electron density distribution and energy characteristics of the orbitals.

The calculation of the spatial and electronic structure of molecules was accomplished by the method of molecular orbitals on the program GAUSSIAN 03. In finding the molecular orbitals there were used density functional theory and the approach of Rutan in which the molecular orbitals are defined in the class of functions of the form

$$\varphi_p\left(\vec{r}\right)=\sum_{k=1}^{N_{atom}}\sum_{s=1}^{M(k)}c_{sp}^k\eta_s^{\ k}\left(\vec{r}=\vec{s}_k\right).$$

Here $h_1^k\left(\vec{r}\right),h_2^k\left(\vec{r}\right),\ldots,h_{M(k)}^k\left(\vec{r}\right)$ – a given set of functions for the k-th atom in the molecule, the position of which is determined by the vector $\vec{S}_k$. These functions, called atomic orbitals, are not assumed to be linearly independent or orthogonal, but are chosen normalized. c_{sp}^k - varying coefficients. The index s is a set of three indices (n, l, m), and the functions $\eta_s^{\ k}\left(\vec{r}\right)$ are:

$$\eta_s^{\ k}\left(\vec{r}\right)=R_{n\ell}^k\left(r\right)Y_{\ell m}\left(\Omega\right).$$

Accordingly, the sum over s is a triple sum:

$$\sum_{s=1}^{M(k)}\equiv\sum_{n=1}^{N(k)}\sum_{\ell=0}^{L(k,n)}\sum_{m=-\ell}^{\ell}$$

The set of radial parts of all atomic orbitals, used in the construction of the molecular orbitals, is the atomic basis of calculation. The basis was taken by a linear combination of Gaussian orbitals

$$R_{n\ell}\left(r\right)\approx\sum_{q=1}^{Q'}d_q^{n\ell}r^{k_q^{n\ell}}e^{-a_q^{n\ell}r^2},$$

in the form of a standard correlation consistent basis 6–31G**, which is denoted also as 6–31G (1d1pH) or cc-pVDZ. In the calculations we used two forms of the exchange-correlation functional: form PBEPBE, described in Refs. [9, 10], and the form PBE1PBE, described in Refs. [11, 12]. To determine the value of the spin in the ground state of the molecule there were carried out calculations of molecules with different spins and the total energies compared.

The results of calculation by the limited Hartree-Fock method (not taking into account correlation effects) and by the density functional theory (approximately taking into account correlation effects) are shown in Tables 4.2 and 4.3, and the corresponding numbering of atoms – in Fig. 4.5.

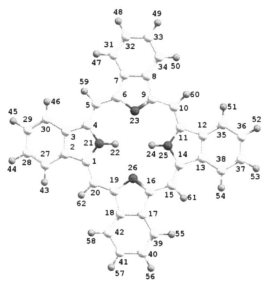

FIGURE 4.5   The numbering of the atoms in the molecules under study.

TABLE 4.2   Changes in the Distribution of Electron Density in the Molecule of Tetrabenzoporphyrin with Sequential Aza-substitution (Hartree-Fock)

Atom number	Before substitution	0	1	2-orto	2-para	3	4	After substitution
1	C	0.321051	0.437988	0.274334	0.305926	0.286533	0.584203	C
2	C	−0.08527	−0.04998	−0.03553	−0.0736	−0.06077	−0.10211	C
3	C	−0.08527	−0.12067	−0.14554	−0.11234	−0.12495	−0.1021	C
4	C	0.321052	0.750765	0.654557	0.599145	0.610537	0.584205	C
5	C	−0.16368	−0.73145	−0.68168	−0.64345	−0.65047	−0.63157	N
6	C	0.297132	0.660397	0.638233	0.592743	0.60889	0.588361	C
7	C	−0.01972	−0.12967	−0.06071	−0.05498	−0.04915	−0.04479	C
8	C	−0.10452	−0.01325	−0.10769	−0.09229	−0.11678	−0.11989	C
9	C	0.389694	0.316014	0.688339	0.411316	0.677646	0.664767	C
10	C	−0.24887	−0.17681	−0.74176	−0.24582	−0.71878	−0.70297	N

11	C	0.462781	0.336566	0.76172	0.46861	0.756837	0.746162	C
12	C	−0.07667	−0.09999	−0.13756	−0.06601	−0.10803	−0.0971	C
13	C	−0.07667	−0.07001	−0.03211	−0.10721	−0.0843	−0.0971	C
14	C	0.462781	0.298136	0.397134	0.73886	0.725921	0.746162	C
15	C	−0.24887	−0.14239	−0.18628	−0.71159	−0.68705	−0.70297	N
16	C	0.389694	0.27503	0.335136	0.643082	0.630436	0.664767	C
17	C	−0.10452	−0.01414	−0.10075	−0.13073	−0.13214	−0.11989	C
18	C	−0.01972	−0.10653	−0.0199	−0.01065	−0.00717	−0.04479	C
19	C	0.297132	0.373251	0.247478	0.291599	0.273391	0.588361	C
20	C	−0.16368	−0.2212	−0.11956	−0.15497	−0.13735	−0.63157	N
21	N	−0.80003	−0.86562	−0.78381	−0.78313	−0.78259	−0.76738	N
22	H	0.392221	0.402829	0.400495	0.397033	0.399423	0.402385	H
23	N	−0.77395	−0.77932	−0.80634	−0.78307	−0.78431	0.77959	N
24	H	0.393755	0.394374	0.409868	0.409103	0.418961	0.425386	H
25	N	−0.88939	−0.7999	−0.84798	−0.87062	−0.85203	−0.85387	N
26	N	−0.77395	−0.76184	−0.75895	−0.77188	0.76373	−0.77959	N
27	C	−0.0997	−0.12647	−0.12085	−0.10631	−0.11073	−0.09636	C
28	C	−0.16758	−0.14571	−0.14872	−0.16291	−0.15832	−0.16731	C
29	C	−0.16758	−0.16069	−0.18104	−0.1718	−0.17595	−0.16731	C
30	C	−0.0997	−0.09793	−0.08089	−0.09017	−0.08606	−0.09636	C
31	C	−0.13039	−0.10677	−0.11961	−0.11728	−0.12169	−0.1229	C
32	C	−0.14944	−0.16613	−0.14993	−0.14985	−0.14807	−0.14769	C
33	C	−0.16614	−0.1469	−0.15724	−0.15952	−0.15966	−0.16182	C
34	C	−0.11626	−0.13576	−0.11295	−0.12166	−0.11153	−0.11097	C
35	C	−0.1158	−0.09454	−0.09165	−0.12122	−0.10319	−0.10546	C
36	C	−0.1545	−0.17267	−0.16575	−0.14896	−0.15422	−0.15053	C
37	C	−0.1545	−0.16223	−0.14114	−0.15581	−0.14808	−0.15053	C
38	C	−0.1158	−0.10477	−0.12963	−0.10095	−0.10982	−0.10546	C
39	C	−0.11626	−0.13104	−0.1141	−0.10608	−0.10508	−0.11097	C
40	C	−0.16614	−0.14869	−0.17055	−0.16857	−0.17003	−0.16182	C
41	C	−0.14944	−0.16774	−0.15142	−0.14685	−0.1462	−0.14769	C
42	C	−0.13039	−0.11528	−0.12971	−0.13627	−0.1366	−0.1229	C
43	H	0.151604	0.168197	0.151732	0.149808	0.149478	0.167836	H
44	H	0.146236	0.16209	0.148571	0.144667	0.145562	0.143205	H
45	H	0.146236	0.162386	0.14674	0.14434	0.144719	0.143205	H

46	H	0.151604	0.193695	0.1753	0.16955	0.170926	0.167835	H
47	H	0.156107	0.178144	0.185657	0.185897	0.185685	0.18475	H
48	H	0.152006	0.150813	0.157821	0.158502	0.158321	0.157229	H
49	H	0.150338	0.151751	0.156449	0.155772	0.156378	0.154787	H
50	H	0.155859	0.156186	0.183034	0.160131	0.181867	0.179148	H
51	H	0.168911	0.153006	0.193815	0.170583	0.195648	0.197801	H
52	H	0.160866	0.146531	0.162843	0.164413	0.166934	0.16954	H
53	H	0.160866	0.147148	0.163625	0.165534	0.167665	0.16954	H
54	H	0.168911	0.151038	0.169237	0.196189	0.196229	0.197801	H
55	H	0.155859	0.155679	0.149019	0.174765	0.174001	0.179148	H
56	H	0.150338	0.152402	0.146604	0.148928	0.149169	0.154787	H
57	H	0.152006	0.15058	0.148584	0.150422	0.150828	0.157229	H
58	H	0.156107	0.154874	0.151201	0.154902	0.154583	0.18475	H
59	H	0.16157	0.161812					
60	H	0.150037			0.160002			
61	H	0.150037	0.165883	0.164818				
62	H	0.16157	0.158493	0.168946	0.16471	0.168241		

TABLE 4.3 Changes in the Distribution of Electron Density in the Molecule of Tetrabenzoporphyr in Sequential Aza-substitution (by Density Functional Theory on the Grid)

Atom number	Before substitution	0	1	2-orto	2-para	3	4	After substitution
1	C	0.318705	0.335082	0.324168	0.322311	0.336346	0.492694	C
2	C	0.079445	0.072616	0.082218	0.080802	0.07099	0.075752	C
3	C	0.064812	0.083751	0.072442	0.075919	0.086511	0.090177	C
4	C	0.34823	0.481363	0.490204	0.489379	0.477113	0.47302	C
5	C	−0.17246	−0.57271	−0.56948	−0.56449	−0.56676	−0.56427	N
6	C	0.297463	0.441181	0.443079	0.438523	0.443982	0.442716	C
7	C	0.083174	0.091614	0.100919	0.087589	0.098647	0.105369	C
8	C	0.100962	0.100847	0.100646	0.106798	0.099981	0.094572	C
9	C	0.286769	0.291985	0.443657	0.291054	0.442895	0.443515	C
10	C	−0.18365	−0.1751	−0.56978	−0.17955	−0.56456	−0.56052	N
11	C	0.318659	0.335855	0.489895	0.322308	0.488486	0.492695	C
12	C	0.079431	0.069906	0.072524	0.080799	0.079997	0.075719	C
13	C	0.064842	0.075938	0.081677	0.075952	0.087275	0.090224	C

14	C	0.348227	0.329671	0.324412	0.489344	0.480765	0.472982	C
15	C	−0.17247	−0.1744	−0.17218	−0.56452	−0.56436	−0.56428	N
16	C	0.297455	0.289554	0.289232	0.43853	0.439488	0.442713	C
17	C	0.08318	0.093555	0.09381	0.087571	0.092703	0.105405	C
18	C	0.100975	0.093156	0.094067	0.106829	0.100872	0.094569	C
19	C	0.286773	0.288464	0.289497	0.291053	0.293325	0.443511	C
20	C	−0.18371	−0.16973	−0.17306	−0.17962	−0.17148	−0.56055	N
21	N	−0.65905	−0.62733	−0.62116	−0.62168	−0.6195	−0.59016	N
22	H	0.294134	0.2974	0.300256	0.301075	0.30412	0.310317	H
23	N	−0.68786	−0.67021	−0.64064	−0.66216	−0.64477	−0.64205	N
24	H	0.294135	0.297583	0.300498	0.30107	0.304198	0.310326	H
25	N	−0.65905	−0.65586	−0.62216	−0.62166	−0.58796	−0.59017	N
26	N	−0.68785	−0.68467	−0.68606	−0.66224	−0.66394	−0.64213	N
27	C	−0.13023	−0.12928	−0.13376	−0.13117	−0.13039	−0.12134	C
28	C	−0.06303	−0.06185	−0.06018	−0.06119	−0.06131	−0.0614	C
29	C	−0.06419	−0.06384	−0.06413	−0.06341	−0.0622	−0.0601	C
30	C	−0.12825	0.12248	−0.11767	−0.11866	−0.12307	−0.12376	C
31	C	−0.13672	−0.13249	−0.13799	−0.13083	−0.13473	−0.13856	C
32	C	−0.0652	−0.06233	−0.06072	−0.06316	−0.06073	−0.05982	C
33	C	−0.0628	−0.06243	−0.06059	−0.06106	−0.06048	−0.06057	C
34	C	−0.13974	−0.14349	−0.13721	−0.14275	−0.13808	−0.1338	C
35	C	−0.13025	−0.12735	−0.11733	−0.13118	−0.12374	−0.12133	C
36	C	−0.06303	−0.06337	−0.06442	−0.06118	−0.0611	−0.0614	C
37	C	−0.06419	−0.06322	−0.0601	−0.06343	−0.06033	−0.0601	C
38	C	−0.12825	−0.12971	−0.13326	−0.11867	−0.12509	−0.12376	C
39	C	−0.13672	−0.13633	−0.14028	−0.13082	−0.13341	−0.13856	C
40	C	−0.0652	−0.06374	−0.06313	−0.06316	−0.06204	−0.05982	C
41	C	−0.0628	0.06362	−0.06298	−0.06105	−0.0615	−0.06058	C
42	C	−0.13974	−0.1404	−0.13993	−0.14276	−0.14191	−0.13379	C
43	H	0.065096	0.063859	0.064969	0.065592	0.066454	0.084349	H
44	H	0.064287	0.064673	0.065905	0.066126	0.067747	0.06954	H
45	H	0.064374	0.065828	0.067117	0.067146	0.06879	0.069374	H
46	H	0.065148	0.081908	0.083745	0.082705	0.084124	0.08418	H
47	H	0.056666	0.073646	0.073892	0.074406	0.07482	0.076813	H
48	H	0.058693	0.059998	0.061032	0.061571	0.062358	0.063934	H
49	H	0.05859	0.058996	0.061	0.060343	0.062486	0.063877	H
50	H	0.056809	0.056007	0.073897	0.05794	0.07516	0.076201	H
51	H	0.065097	0.066542	0.083793	0.065592	0.083283	0.084348	H

52	H	0.064288	0.066032	0.067165	0.066134	0.068048	0.06954	H
53	H	0.064374	0.066174	0.065889	0.067143	0.068044	0.069375	H
54	H	0.065148	0.066017	0.064993	0.082704	0.082491	0.084174	H
55	H	0.056666	0.056817	0.059009	0.074391	0.076035	0.076817	H
56	H	0.058693	0.059954	0.061288	0.061565	0.063045	0.063933	H
57	H	0.05859	0.060174	0.061263	0.06035	0.061879	0.063876	H
58	H	0.056808	0.058152	0.058946	0.057926	0.058215	0.076205	H
59	H	0.064099	0.067717					
60	H	0.065772			0.070909			
61	H	0.064098	0.066716	0.070527				
62	H	0.065775	0.067212	0.070551	0.070928	0.07273		

In addition, there were calculated spectrum of single-particle states and the spatial distribution of electrons in the highest occupied state (HOMO) and the lowest free state (LUMO) (Fig. 4.6). The spatial distribution of electron states in the HOMO and LUMO is illustrated in Figs. 4.7–4.8. These figures show the surface level of functions $\left| \psi(\vec{r}) \right|^2$, that is surfaces satisfying the condition $\left| \psi(\vec{r}) \right|^2 = const.$ Here $\psi(\vec{r})$ is the one-particle wave function of the electron in the states HOMO and LUMO of the corresponding molecules, constant values are the same for all molecules.

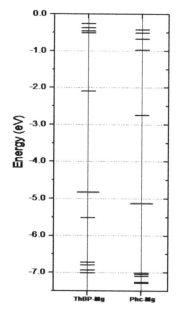

FIGURE 4.6   Spectrum of one-electron energies in the ground state of molecules Mg-tetrabenzoporphyrin and Mg-phthalocyanine (the spin of the ground state S = 0).

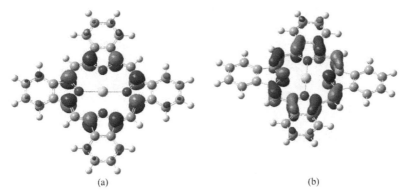

FIGURE 4.7 Spatial distribution of the electron in the one-particle states of HOMO (a) and LUMO (b) for the ground state of Mg-tetrabenzoporphyrin.

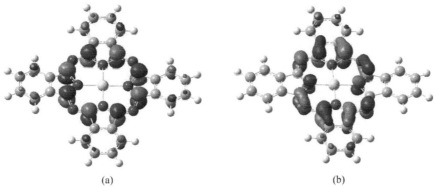

FIGURE 4.8 Spatial distribution of the electron in the one-particle states of HOMO (a) and LUMO (b) for the ground state of Mg-phthalocyanine.

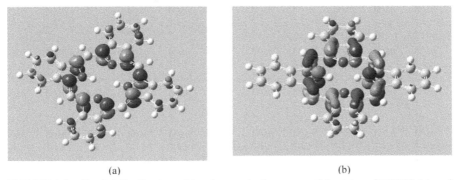

FIGURE 4.9 Spatial distribution of the electron in the one-particle states of HOMO (a) and LUMO (b) for the singlet state of the metal-free tetrabenzoporphyrin

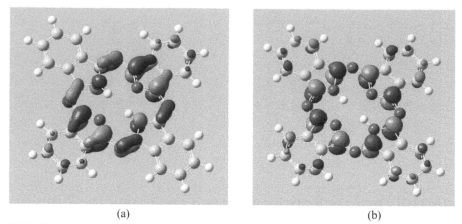

(a)   (b)

FIGURE 4.10   Spatial distribution of the electron in the one-particle states of HOMO (a) and LUMO (b) for the singlet state of the metal-free phthalocyanine

In addition, changes of the energy in sequential aza-substitution were assessed (Figs. 4.11–4.13).

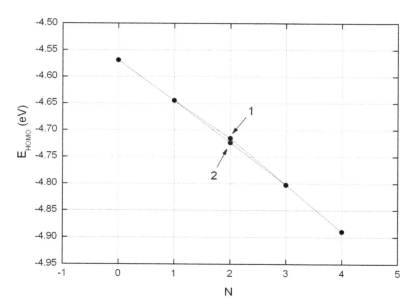

FIGURE 4.11   Dependence of the energy E_{HOMO} on the number of substitutions in the molecule of tetrabenzoporphyrin. (1 – molecule shown in Fig. 14 (c), 2 – molecule shown in Fig. 14 (d).)

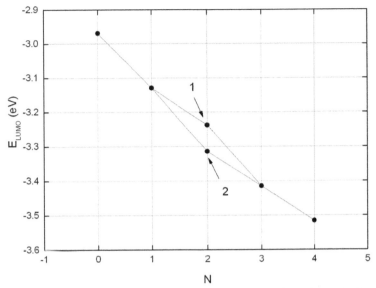

FIGURE 4.12   Dependence of the energy E_{LUMO} on the number of substitutions in the molecule of tetrabenzoporphyrin. (1 – molecule shown in Fig. 14 (c), 2 – molecule shown in Fig. 14 (d).)

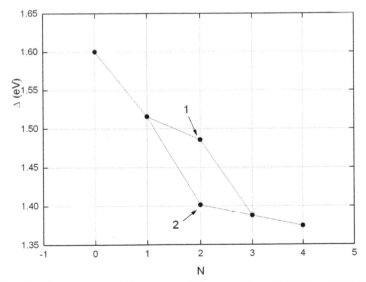

FIGURE 4.13   Dependence of the energy difference $\Delta E = ELUMO - EHOMO$ on the number of substitutions in the molecule of tetrabenzoporphyrin. (1 – molecule shown in Fig. 14 (c), 2 – molecule shown in Fig. 14 (d).)

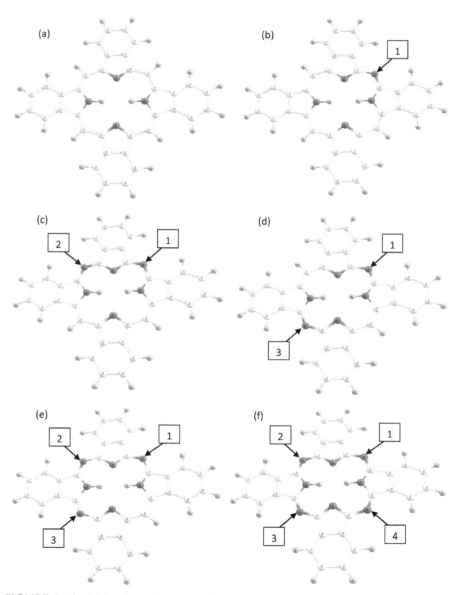

FIGURE 4.14 Molecule tetrabenzoporphyrin, arrows mark places of aza-substitution. (a) tetrabenzoporphyrin (symmetry D2h), (b) mono-substitution (symmetry C1), (c) disubstituted (symmetry D1h), (d) disubstituted (symmetry C2h), (e) triple substitution (symmetry C1), (f) phthalocyanine (symmetry D2h).

As can be seen from the above calculations, the most revealing changes occur, naturally, in the areas of substitution, with the apparent concentration of the charge in the most likely coordination of potential electron acceptor. Characteristically, the improved method of density functional on the grid clearer tracks -orto and -para substitution options. In similar calculations for other metal derivatives of tetrabenzo-porphyrin there was no significant difference, however, it should be noted a significant effect of extraligand in the case of Fe-TBP / FeCl-TBP. According to Table 4.1, the ratio of their photocurrents (0.4/1.7 = 0.24) after full aza-substitution (Fe-Pc/ClFe-Pc) is almost unchanged (1.4/5.4 = 0.26), that is a significant advantage of the pigment having extraligands on the central atom is saved. The quantum yield of the chlorinated form is 5 times higher. Therefore, looking more forward to the next subject of research, we present some of the data on these compounds (Figs. 4.15–4.18).

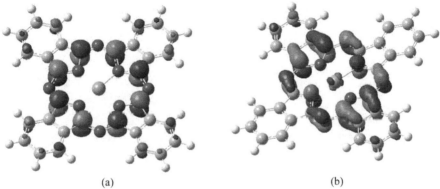

(a)   (b)

FIGURE 4.15   Spatial distribution of the electron with spin "up" in the one-particle states of HOMO (a) and LUMO (b) for the ground state of Fe-phthalocyanine. Ground-state spin S = 1.

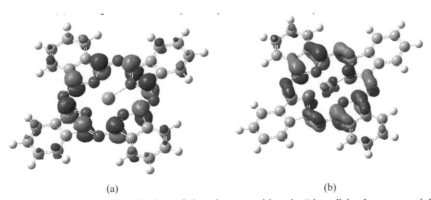

(a)   (b)

FIGURE 4.16   Spatial distribution of the electron with spin "down" in the one-particle states of HOMO (a) and LUMO (b) for the ground state of Fe-phthalocyanine. Ground-state spin S = 1.

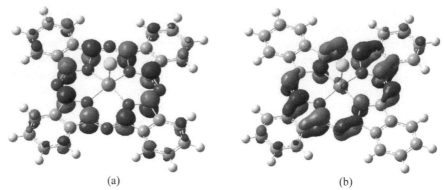

FIGURE 4.17   Spatial distribution of the electron with spin "up" in the one-particle states of HOMO (a) and LUMO (b) for the ground state of FeCl-phthalocyanine. Ground-state spin $S = 3/2$.

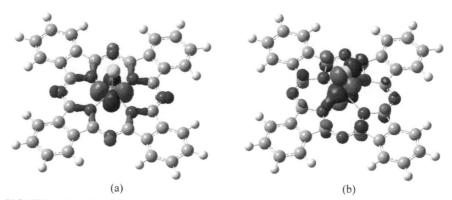

FIGURE 4.18   Spatial distribution of the electron with spin "down" in the one-particle states of HOMO (a) and LUMO (b) for the ground state of FeCl-phthalocyanine. Ground-state spin $S = 3/2$.

The picture of changes in the distribution of electron density is so obvious, that practically does not require any comment. We note only that the most impressive version of Fig. 4.18 is unlikely, since it corresponds to the band gap of about 0.8 eV, and the experimental values (about 1.5 eV) is more consistent with the spin-up states.

4.4   CONCLUSION

Thus, the set of considered measurements on films of aza-substituted metal-derivatives of tetrabenzoporphyrin in combination with the data of quantum-chemical calculations of the electron density and energy characteristics of the pigments, leads to several conclusions:

- Sequential substitution of carbon atoms in meso-position by nitrogen, that leads to a transition from the porphyrin structure to a phthalocyanine structure, increases the electron density in the area under consideration with the corresponding increase of charges in the p-electronic bonds.
- The obvious consequence of the increase of electron density in the important area of structural conjugation of macrocycle is a ring currents increase and compression of the cycle, which leads to a considerable increase of the intermolecular interaction.
- Consequences of the strengthening of the intermolecular interactions are bathochromic red shift of the absorption maxima, the increase in the extinction coefficients and the broadening of the absorption bands in the visible part of the spectrum, as well as the hypsochromic shift of the Soret band.
- Accordingly with the changes in the optical characteristics of the pigments it takes place a reduction of the band gap, lowering of the energy level of the valence band, increasing of the depth of acceptor levels formed by the adsorbed oxygen. In this case, it can be seen as a positive change, since it brings pigments to the value of the band gap of 1.5 eV, which is ideal for the converter of solar energy on the ground level.
- From a practical point of view, the redistribution of electron density in the very structurally similar molecules leads to a significant increase in chemo- and light fastness of pigments and oxidation potentials, to an 8–10 fold increase in the photocurrent, 1.6–1.8 fold increase in photopotentials, 2–5 fold in the quantum yield on a current.

KEYWORDS

- aza-substitution
- electron density distribution
- organic semiconductors
- porphyrins
- tetrapyrrole compounds

REFERENCES

1. Rudakov, V. M., Ilatovsky, V. A., & Komissarov, G. G. (1987). Photo Activity of Metal Derivatives of Tetraphenylporphyrin Khim Fizika, 6(4), 552–554 (in Russian).
2. Apresyan, E. S., Ilatovsky, V. A., & Komissarov, G. G. (1989). Photo Activity of Thins Films of Metals Derivatives of Phthalo Cyanine Zh Fiz.Himii, 63(8), 2239–2242 (in Russian).

3. Ilatovsky, V. A., Shaposhnikov, G. P., Dmitriev, I. B., Rudakov, V. M., Zhiltsov, S. L., & Komissarov, G. G. (1999). Photo Catalytic Activity of Thin Films of Aza Substituted Tetrabenzoporphyrins Zh Fiz Khimii, *73(12)*, 2240–2245 (in Russian).
4. Ilatovsky, V. A., Dmitriev, I. B., Kokorin, A. I., Ptitsyn, G. A., & Komissarov, G. G. (2009). The Influence of the Nature of the Coordinated Metal on the Photo Electrochemical Activity of Thin Films of Tetrapyrrole Compounds Khim Fiz, *28(1)*, 89–96 (in Russian).
5. Ilatovsky, V. A., Ptitsyn, G. A., & Komissarov, G. G. (2008). Influences of Molecular Structures of the Films of Tetrapyrrole Compounds on their Photo Electro chemical Characteristics at the Various Types of Sensitization Khim Fiz, *27(12)*, 66–70 (in Russian).
6. Ilatovsky, V. A., Sinko, G. V., Ptitsyn, G. A., & Komissarov, G. G. (2012). Structural Sensitizations of Pigmented Films in the Formations of Nano-Sized Mono crystal Clusters, Collections the Dynamics of Chemicals and Biological Processes, XXI century, Institutes of Chemical Physics RAS, Moscow, 173–180.
7. Ilatovsky, V. A., Apresyan, E. S., & Komissarov, G. G. (1988). Increase in Photo Activity of Phthalo Cyanines at Structural Modification of Thin Film Electrodes, Zhurn fiz khimii, *62(6)*, 1612–1617 (in Russian).
8. Wolkenstein, F. F. (1973). Physical Chemistry of Semi conductor Surface Moscow, Nauka, 398.
9. Perdew, J. P., Burke, K., & Ernzerhof, M. (1996). Physical Review Letters, *77*, 3865.
10. Perdew, J. P., Burke, K., & Ernzerhof, M. (1997). Physical Review Letters, *78*, 1396.
11. Ernzerhof, M., Perdew, J. P., & Burke, K. (1997). International Journal Quantum's Chemistry, *64*, 285.
12. Ernzerhof, M., & Scuseria, G. E. (1999). Journal Chemistry Physics, *110*, 5029.

CHAPTER 5

ON FRACTAL ANALYSIS AND POLYMERIC CLUSTER MEDIUM MODEL

G. V. KOZLOV[1], I. V. DOLBIN[1], JOZEF RICHERT[2],
O. V. STOYANOV[3], and G. E. ZAIKOV[4]

[1]Kabardino-Balkarian State University, Nal'chik – 360004, Chernyshevsky st., 173, Russian Federation; E-mail: I_dolbin@mail.ru

[2]Institut Inzynierii Materialow Polimerowych I Barwnikow, 55 M. Sklodowskiej-Curie str., 87-100 Torun, Poland; E-mail: j.richert@impib.pl

[3]Kazan National Research Technological University, Kazan, Tatarstan, Russia, E-mail: OV_Stoyanov@mail.ru

[4]N. M. Emanuel Institute of Biochemical Physics of Russian Academy of Sciences, Moscow 119334, Kosygin st., 4, Russian Federation; E-mail: Chembio@sky.chph.ras.ru

CONTENTS

Abstract 150
5.1 Introduction 150
5.2 Conclusion 158
Keywords 158
References 158

ABSTRACT

This article is review about the intercommunication of fractal analysis and polymeric cluster medium model. It is shown that close intercommunication exists between notions of local order and fractality in a glassy polymers case, having key physical grounds and expressed by the simple analytical relationship. It has also been shown, that the combined usage of these complementing one another concepts allows to broaden possibilities of polymeric mediums structure and properties analytical description. The indicated expressions will be applied repeatedly in further interpretation.

5.1　INTRODUCTION

During the last 25 years a fractal analysis methods obtained wide spreading in both theoretical physics [1] and material science [2], in particular, in physics-chemistry of polymers [3–8]. This tendency can be explained by fractal objects wide spreading in nature.

There are two main physical reasons, which define intercommunication of fractal essence and local order for solid-phase polymers: the thermodynamical nonequilibrium and dimensional periodicity of their structure. In Ref. [9] the simple relationship was obtained between thermodynamical nonequilibrium characteristic – Gibbs function change at self-assembly (cluster structure formation of polymers $\Delta \tilde{G}^{im}$ – and clusters relative fraction φ_{cl} in the form:

$$\Delta \tilde{G}^{im} \sim \phi_{cl} \tag{1}$$

This relationship graphic interpretation for amorphous glassy polymers – polycarbonate (PC) and polyarylate (PAr) – is adduced in Fig. 5.1. Since at $T = T_g (T_m)$ (where T, T_g and T_m are testing, glass transition and melting temperatures, accordingly) $\Delta \tilde{G}^{im} = 0$ [10, 11], then from the Eq. (1) it follows, that at the indicated temperatures cluster structure full decay ($\varphi_{cl} = 0$) should be occurred or transition to thermodynamically equilibrium structure.

As for the intercommunication of parameters, characterizing structure fractality and medium thermodynamical nonequilibrium, it should exist indisputably, since precisely nonequilibium processes formed fractal structures. Solid bodies fracture surfaces analysis gives evidence of such rule fulfillment – a large number of experimental papers shows their fractal structure, irrespective of the analyzed material thermodynamical state [12]. Such phenomenon is due to the fact, that the fracture process is thermodynamically nonequilibrium one [13]. Polymers structure fractality is due to the same circumstance. The experimental confirmation can be found in papers [14–16]. As for each real (physical) fractal, polymers structure fractal properties are limited by the defined linear scales. So, in papers [14, 17] these scales were determined within the range of several Ångströms (from below) up to several tens Ångströms (from above). The lower limit is connected with medium structural elements finite size and the upper one – with structure fractal dimension d_f limiting values [18]. The indicated above scale limits correspond well to cluster nanostructure specific boundary sizes: the lower – with statistical segment length l_{st}, the upper – with distance between clusters R_{cl} [19].

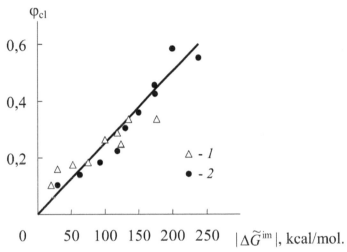

FIGURE 5.1   The dependence of clusters relative fraction ϕ_{cl} on absolute value of specific Gibbs function of nonequilibrium phase transition $|\Delta \widetilde{G}^{im}|$ for amorphous glassy polymers – polycarbonate (1) and polyarylate (2) [3].

Polymeric medium's structure fractality within the indicated above scale limits assumes the dependence of their density ρ on dimensional parameter L (see Fig. 5.2) as follows [1]:

$$\rho \sim L^{d_f - d}, \tag{2}$$

where d is dimension of Euclidean space, into which a fractal is introduced.

In Fig. 5.3, amorphous polymers nanostructure cluster model is presented. As one can see, within the limits of the indicated above dimensional periodicity scales Figs. 5.2 and 5.3 correspond each other, that is, the cluster model assumes ρ reduction as far as possible from the cluster center. Let us note that well-known Flory "felt" model [20] does not satisfy this criterion, since for it $\rho \approx$ const. Since, as it was noted above, polymeric mediums structure fractality was confirmed experimentally repeatedly [14–16], then it is obvious, that cluster model reflects real solid-phase polymers structure quite plausibly, whereas "felt" model is far from reality. It is also obvious, that opposite intercommunication is true – for density ρ finite values change of the latter within the definite limits means obligatory availability of structure periodicity.

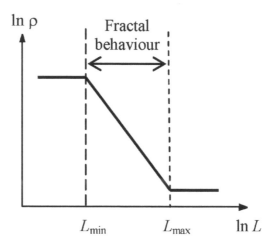

FIGURE 5.2   The schematic dependence of density ρ on structure linear scale L for solid body in logarithmic coordinates. The fractal structures are formed within the range $L_{min} - L_{max}$ [18].

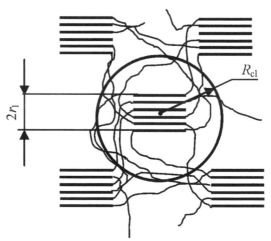

FIGURE 5.3   The model of cluster structure for amorphous polymeric media [7].

Meanwhile, we must not over look the fact that fractal analysis gives only polymers structure general mathematical description only, that is, does not identify those concrete structural units (elements), which any real polymer consists. At the same time, the physical description of polymers thermodynamically nonequilibrium structure within the frameworks of local order notions is given by the cluster model of amorphous polymers structure gives, which is capable of its elements quantitative identification [21].

The well described in literature experimental studies was served as premise for the cluster model development. As it is known (see, e.g., [22, 23]) high-molecular amorphous polymeric mediums, which are in glassy state, at their deformation above glass transition temperature can display high-elastic behavior (when high-elasticity plateau is formed), which can be described by the high-elasticity theory rules. Personally, Langievene [24, 25] and Gauss [26] equations can be used for polymers behavior description large strains. In the latter it is assumed that at deformation in plateau region polymeric chain is not stretched fully and the relationship between true stress σ^{tr} and drawing ratio λ at uniaxial tension is written as follows:

$$\sigma^{tr} = G_p (\Lambda^2 - \Lambda^{-1}), \tag{3}$$

where G_p is the so-called strain hardening modulus.

The value G_p knowledge allows formally to calculate the density v_e of macromolecular entanglements network in polymeric medium according to the well-known expressions of high-elasticity theory [23]:

$$M_e = \frac{\rho RT}{G_p}, \tag{4}$$

$$v_e = \frac{\rho N_A}{M_e}, \tag{5}$$

where ρ is polymer density, R is universal gas constant, T is testing temperature, M_e is molecular weight of chain part between entanglements, N_A is Avogadro number.

However, the attempts to estimate the value M_e (or v_e) according to the determined from Eq. (3) G_p values result to unlikely low calculated values M_e (or unreal high values v_e), contradicting to Gaussian statistics requirements, which assume availability on chain part between entanglements no less than ~ 13 monomer links [27].

As alternative the authors [28] assumed, that besides the indicated above binary hooking network in polymers glassy state another entanglements type was available, nodes of which by their structure were similar to crystallites with stretched chains (CSC). Such entanglements node possesses large enough functionality F (under node functionality emerging from it chains number is assumed [29] and it was called cluster. Cluster consists of different macromolecules segments and each such segment length is postulated equal to statistical segment length l_{st} ("stiffness part" of chain [30]). In that case the effective (real) molecular weight of chain part between clusters M_{cl}^{ef} can be calculated as follows [29]:

$$M_{cl}^{ef} = \frac{M_{cl}F}{2}, \tag{6}$$

where M_{cl} is molecular weight of the chain part between cluster, calculated according to the Eq. (4).

It is obvious, that at large enough F the reasonable values M_{cl}^{ef}, satisfied to Gaussian statistics requirements, can be obtained. Further on for parameters of entanglements cluster network and macromolecular binary hookings network distinction we will used the indices "cl" and "e," accordingly. Therefore, the offered in Ref. [28] model assumes, that amorphous polymer structure represents itself local order regions (domains), consisting of different macromolecules collinear densely packed segments (clusters), immersed in loosely packed matrix. Simultaneously clusters play the role of physical entanglements network multifunctional nodes. The value F (within the frameworks of high-elasticity theory) can be estimated as follows [31]:

$$F = \frac{2G_{\infty}}{kT\nu_{cl}} + 2, \tag{7}$$

where G_{∞} is equilibrium shear modulus, k is Boltzmann constant, ν_{cl} is cluster network density.

In Fig. 5.4, the dependences $\nu_{cl}(T)$ for polycarbonate (PC) and polyarylate (PAr) are adduced. These dependences show $_{cl}$ reduction at T growth, that assumes local order regions (clusters) thermofluctuational nature. Besides, on the indicated dependences two characteristic temperatures are found easily. The first from them, glass transition temperature T_g, defines clusters full decay (see also Fig. 5.1), the second T'_g, corresponds to the fold on curves $\nu_{cl}(T)$ and settles down on about 50 K lower T_g.

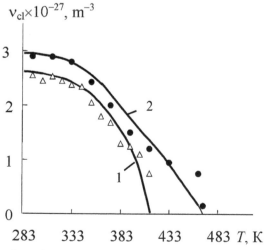

FIGURE 5.4 The dependences of macromolecular entanglements cluster network density ν_{cl} on testing temperature T for PC (1) and PAr (2) [28].

Earlier within the frameworks of local order concepts it has been shown that temperature T_g is associated with segmental mobility releasing in polymer loosely packed regions. This means, that within the frameworks of cluster model T_g can be associated with loosely packed matrix devitrification. The dependences $F(T)$ for the same polymers have a similar form (Fig. 5.5).

Two main models, describing local order structure in polymers (with folded [34] and stretched [28] chains) allows to make common conclusion: local order regions in polymer matrices play role of macromolecular entanglements physical network nodes [28, 34]. However, their reaction on mechanical deformation should be distinguished essentially, it at large strains regions with folded chains ("bundles") are capable to unfolding of folds in straightened conformations, then clusters have no such possibility and polymer matrix deformation can occur by "tie" chains (connecting only clusters) straightening, that is, by their orientation in the applied stress direction.

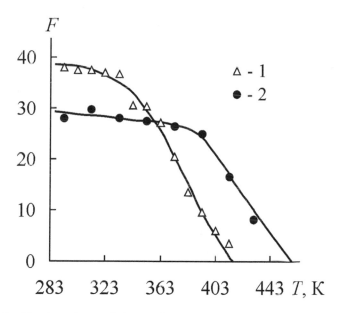

FIGURE 5.5 The dependence of clusters functionality F on testing temperature T for PC (1) and PAr (2) [28].

Turning to analogy about mediums with crystalline morphology of polymer matrices, let us note, that semicrystalline polymers (e.g., high-molecular polyethylenes) large strains, making $1000 \div 2000\%$, are realized owing to crystallites folds unfolding [35].

Proceeding from the said above and analyzing values of polymers limiting strains, one can obtain the information about local order regions type in amorphous and semicrystalline polymers. The fulfilled by the authors [36] calculations have show that the most probable type of local nanostructures in amorphous polymer matrix is an analog of crystallite with stretched chains, that is, cluster.

Let us note, completing this topic, that for the local order availability substantiation in amorphous polymer matrix (irrespective to concrete structural model of medium) strict mathematical proofs of the most common character exist. For example, according to the proved in numbers theory Ramsey's theorem any large enough quantity $N_i > R(i, j)$ of numbers, points or objects (in the considered case – statistical segments) contains without fail high-ordered system from $N_j \leq R(i, j)$ such segments. Therefore, absolute disordering of large systems (structures) is impossible [37, 48].

As it is known [39], structures, which behave themselves as fractal ones on small length scales and as homogeneous ones – on large ones, are named homogeneous fractals. Percolation clusters near percolation threshold are such fractals [1]. As it will be shown lower, cluster structure is a percolation system and in virtue of the said above – homogeneous fractal. In other words, local order availability in polymers condensed state testifies to there structure fractality [21].

The percolation system fractal dimension d_f can be expressed as follows [39]

$$d_f = d - \frac{\beta_p}{v_p}, \qquad (8)$$

where β_p and v_p are critical exponents (indices) in percolation theory.

Hence, the condition $\beta_p \neq 0$, which follows from the data of table 1.1, also testifies to also polymer matrix structure fractality. It is obvious, that polymer medium structure fractality can be determined by the dependence of φ_{cl} on temperature, that is, using analysis of local order thermofluctuational effect (see Fig. 5.4). Let us note that structure fractality and "freezing" below T_g local order (φ_{cl} = const) from the physical point of view are interexepting notions.

TABLE 5.1 A Percolation Clusters Characteristics [5]

Parameter	Experimentally determined magnitudes		Theoretical "geometrical" magnitudes
	EP-1	EP-2	
β_p	0.54	0.58	0.40
v	1.20	1.15	0.88
β_p/v_p	0.45	0.50	0.46
d_f	2.55	2.50	2.55

*EP-1 and EP-2 are epoxy polymers on the basis of resin ED-22.

The considered above principles allow to connect by analytical relationship the fractal dimension d_f and local order characteristics φ_{cl} (or v_{cl}). According to the Ref. [18] the value d_f is connected with Poisson's ratio v as follows:

$$d_f = (d-1)(1+v), \tag{9}$$

and the value v is connected, in its turn, with v_{cl} according to the relationship [21]:

$$v = 0,5 - 3,22 \times 10^{-10} \left(l_0 v_{cl} \right)^{1/2} \tag{10}$$

where the following intercommunication between φ_{cl} exists:

$$\varphi_{cl} = S l_0 C_\infty \varphi_{cl}, \tag{11}$$

where S is macromolecule cross-sectional area, l_0 is main chain skeletal length, C_∞ is characteristic ratio.

From the Eqs. (9) ÷ (11) the dependence of d_f on φ_{cl} or v_{cl} can be obtained for the most often used case $d = 3$ [21]:

$$d_f - 3 \quad 6,44 \times 10^{-10} \left(\frac{\phi_{cl}}{SC_\infty} \right)^{1/2} \tag{12}$$

or

$$d_f = 3 - 6,44 \times 10^{-10} \left(l_0 v_{cl} \right)^{1/2} \tag{13}$$

Let us note, returning to the Eq. (1), that its substation in Eq. (12) allows to obtain the following relationship:

$$\Delta \tilde{G}^{im} \sim C_\infty S \left(3 - d_f \right)^2 \tag{14}$$

The value $\Delta \tilde{G}^{im}$ (Gibbs specific function of local, for example, supra-molecular structures formation) is given for nonequilibrium phase transition "supercooled liquid → solid body" [11]. From the Eq. (14) it follows, that the condition $\Delta \tilde{G}^{im} = 0$ is achieved at $d_f = 3$, that is, at $d_f = d$ and at transition to Euclidean behavior. In other words, a fractal structures are formed only in nonequilibrium processes course, which was noted earlier [12].

Above general reasoning's about possibility of existence in polymeric media of local order regions, based on the Ramsey theorem, were presented in a simplified enough form. It can be shown similarly, that any structure, consisting of N elements, at $N > B_N$ (j) represents it self totality of finite number $k \leq j$ of put into each other self-similar structures, Hausdorff dimension of which in the general case can be different one. Therefore, any structure irrespective of its physical nature, consisting of a large enough elements number, can be represented as multifractal (in partial case as monofractal) and described by spectrum of Renyi dimensions d_q, $q = -\infty \div +$ [18].

In Ref. [40] it has been shown, that the condensed systems attainment to self-organization in scale – invariant multifractal forms is the result of key principles of open systems thermodynamics and d_q is defined by competition of short- and long-range interatomic correlations, determining volume compressibility and shear stiffness of solid bodies, accordingly.

5.2   CONCLUSION

It can be noted as a brief resume to this chapter, that close intercommunication exists between notions of local order and fractality in a glassy polymers case, having key physical grounds and expressed by the simple analytical relationship (the Eqs. (12) and (13)). It has also been shown, that the combined usage of these complementing one another concepts allows to broaden possibilities of polymeric mediums structure and properties analytical description. The indicated expressions will be applied repeatedly in further interpretation.

KEYWORDS

- fractal analysis
- intercommunication
- medium model
- molecular science
- physico-chemistry of polymers
- polymeric cluster
- theoretical physics

REFERENCES

1. Feder, E. (1989). Fractals New York, Plenum Press, 248p.
2. Ivanova, V. S., Balankin, A. S., Bunin, I. Z. H., & Oksogoev, A. A., (1994). Synergetic and Fractals in material Science, Moscow, Nauka, 383p.
3. Novikov, V. U., & Kozlov, G. V. (2000). A Macromolecules Fractal Analysis, Uspekhi Khimii, 69(4), 378–399.
4. Novikov, V. U., & Kozlov, G. V. (2000). Structure and Properties of Polymers within the Frameworks of Fractal Approach, Uspekhi Khimii, 69(6), 572–599.
5. Kozlov, G. V., & Novikov, V. U. (1998). Synergetic and Fractals Analysis of Cross-Linked Polymers, Moscow, Klassika, 112p.
6. Kozlov, G. V., Yanovskii, Yu G., & Zaikov, G. E. (2011). Synergetics and Fractal Analysis of Polymer Composites Filled with Short Fibers, New York, Nova Science Publishers, Inc, 223p.

7. Kozlov, G. V., Yanovskii, Yu G., & Zaikov, G. E. (2010). Structures and Properties of Particulate-Filled Polymers composites, the Fractal Analysis, New York, Nova Science Publishers, Inclusive, 282p.

8. Shogenov, V. N., & Kozlov, G. V. (2002). A Fractal Clusters in Physics-Chemistry of Polymers, Nal'chik, Poligrafservis I T, 268p.

9. Kozlov, G. V., & Zaikov, G. E. (2001). The Generalized Description of Local Order in Polymers, in book, Fractal and Local order in Polymeric Materials, Kozlov, G. V., Zaikov, G. E., (Ed) New York, Nova Science Publishers, Inc., 55–64.

10. Gladyshev, G. P., & Gladyshev, D. P. (1993). About Physical-Chemical Theory of Biological Evolution (Preprint), Moscow, Olimp, 24p.

11. Gladyshev, G. P., & Gladyshev, D. P. (1994). The Approximative Thermo Dynamical Equation for Nonequilibrium Phase Transitions, Zhurnal Fizicheskoi Khimii, 68(5), 790–792.

12. Hornbogen, E. (1989). Fractals in Microstructure of Metals, Interval Matter Rev, 34(6), 277–296.

13. Bessendorf, M. H. (1987). Stochastic and Fractal Analysis of Fracture Trajectories, Int J Engng Science, 25(6), 667–672.

14. Zemlyanov, M. G., Malinovskii, V. K., Novikov, V. N., Parshin, P. P., & Sokolov, A. P. (1992). The Study of Fractions in Polymers, Zhurnal Eksperiment, I Teoretich Fiziki, 101(1), 284–293.

15. Kozlov, G. V., Ozden S., Krysov, V. A., & Zaikov, G. E. (2001). The Experimental Determination of a Fractal Dimension of the Structure of Amorphous Glassy Polymers, in book, Fractals and Local Order in Polymeric Material Kozlov, G. V., Zaikov, G. E., Ed. New York, Nova Science Publishers, Inc., 83–88.

16. Kozlov, G. V., Ozden, S., & Dolbin, I. V. (2002). Small Angle X-Ray Studies of the Amorphous Polymers Fractal Structure, Russian Polymer News, 7(2), 35–38.

17. Bagryanskii, V. A., Malinovskii, V. K., Novikov, V. N., Pushchaeva, L. M., & Sokolov, A. P. (1988). In Elastic Light Diffusion on Fractal Oscillation Modes in Polymers, Fizika Tverdogo Tela, 30(8), 2360–2366.

18. Balankin, A. S. (1991). Synergetics of Deformable body, Moscow, Publishers of Defence Ministry of SSSR, 404p.

19. Kozlov, G. V., Belousov, V. N., & Mikitaev, A. K. (1998). Description of Solid Polymers as Quasitwophase Bodies, Fizika I Tekhnika Vysokikh Davlenii, 8(1), 101–107.

20. Flory, P. J. (1976). Spatial Configuration of Macromolecular Chains, Brit Polymer J, 8(1), 1–10.

21. Kozlov, G. V., Ovcharenko, E. N., & Mikitaev, A. K. (2009). Structure of the Polymers Amorphous State. Moscow, Publishers of the Mendeleev, D. I., RkhTU, 392p.

22. Haward, R. N. (1995). The Application of a Gauss-Eyring Model to Predict the behavior of Thermoplastics in Tensile Experiment, J Polymer Science, Part B, Polymer Physics, 33(8), 1481–1494.

23. Haward, R. N. (1987). The Application of a Simplified Model for the Stress–strain curves of Polymers, Polymer, 28(8), 1485–1488.

24. Boyce, M. C., Parks, D. M., & Argon, A. S. (1988). Large Inelastic Deformation in Glassy Polymers, Part I Rate Dependent Constitutive Model, Mechanical Mater, 7(1), 15–33.

25. Boyce, M. C., Parks, D. M., & Argon, A. S. (1988). Large Inelastic Deformation in Glassy Polymers, Part II Numerical Simulation of Hydrostatic Extrusion, Mechanical Mater, 7(1), 35–47.

26. Haward, R. N. (1993). Strain Hardening of Thermoplastics, Macromolecules, 26(22), 5860–5869.

27. Bartenev, G. M., Frenkel, Ya S. (1990). Physics of Polymers, leningrad, Khimiya, 432p.

28. Belousov, V. N., Kozlov, G. V., Mikitaev, A. K., & Lipatov, Yu S. (1990). Entanglements in Glassy State of Linear Amorphous Polymers, Doklady AN SSSR, *313(3)*, 630–633.
29. Flory, P. J. (1985). Molecular Theory of Rubber Elasticity, Polymer J, *17(1)*, 1–12.
30. Bernstein, V. A., & Egorov, V. M. (1990). Differential Scanning Calorimetry in physics-Chemistry of the Polymers, Leningrad, Khimiya, 256p.
31. Graessley, W. W. (1980). Linear Visco Elasticity in Gaussian Networks, Macromolecules, *13(2)*, 372–376.
32. Perepechko, I. I., & Startsev, O. V. (1973). Multiplet Temperature Transitions in Amorphous Polymers in Main Relaxation Region, Vysokomolek Soed B, *15(5)*, 321–322.
33. Belousov, V. N., Kotsev, Kh B., & Mikitaev, A. K. (1983). Two-step of Amorphous Polymers Glass Transition Doklady, AN SSSR, *270(5)*, 1145–1147.
34. Arzahakov, S. A., Bakeev, N. F., & Kabanov, V. A. (1973). Supra Molecular Structure of Amorphous Polymers, Vysokomolek Soed A, *15(5)*, 1145–1147.
35. Marisawa, Y. (1987). The Strength of Polymeric Materials, Moscow, Khimya, 400p.
36. Kozlov, G. V., Sanditov, D. S., & Serdyuk, V. D. (1993). About Supra Segmental Formations in Polymers Amorphous State, Vysokomolek, Soed, B, *35(12)*, 2067–2069.
37. Mikitaev, A. K., & Kozlov, G. V. (2008). The Fractal Mechanics of Polymeric Materials, Nal'chik, Publishers KBSU, 312p.
38. Balankin, A. S., Bugrimov, A. L., Kozlov, G. V., Mikitaev, A. K., & Sanditov, D. S. (1992). The Fractal Structure and Physical–Mechanical Properties of Amorphous Glassy Polymers, Doklady, A. N, *326(3)*, 463–466.
39. Sokolov, I. M. (1986). Dimensions and other Geometrical Critical Exponents in Percolation Theory, Uspekhi Fizicheshikh Nauk, *150(2)*, 221–256.
40. Balankin, A. S. (1991). Fractal Dynamics of Deformable Mediums, Pis'ma v ZhTF, *17(6)*, 84–89.

CHAPTER 6

POLYMERS AS NATURAL COMPOSITES: AN ENGINEERING INSIGHT

G. V. KOZLOV[1], I. V. DOLBIN[2], JOZEF RICHERT[3],
O. V. STOYANOV[4], and G. E. ZAIKOV[5]

[1]Kabardino-Balkarian State University, Nal'chik – 360004, Chernyshevsky st., 173, Russian Federation; E-mail: I_dolbin@mail.ru

[2]Institut Inzynierii Materialow Polimerowych I Barwnikow, 55 M. Sklodowskiej-Curie str., 87-100 Torun, Poland; E-mail: j.richert@impib.pl

[3]Kazan National Research Technological University, Kazan, Tatarstan, Russia; E-mail: OV_Stoyanov@mail.ru

[4]N. M. Emanuel Institute of Biochemical Physics of Russian Academy of Sciences, Moscow 119334, Kosygin st., 4, Russian Federation; E-mail: Chembio@sky.chph.ras.ru

CONTENTS

Abstract 162
6.1 Introduction 162
6.2 Natural Nanocomposites Structure 163
6.3 The Natural Nanocomposites Reinforcement 172
6.4 Intercomponent Adhesion in Natural Nanocomposites 180
6.5 The Methods of Natural Nanocomposites Nanostructure
     Regulation 195
6.6 Conclusion 203
Keywords 204
References 204

ABSTRACT

The stated in the present article results give purely practical aspect of such theoretical concepts as the cluster model of polymers amorphous state stricture and fractal analysis application for the description of structure and properties of polymers, treated as natural nanocomposites. The necessary nanostructure goal-directed making will allow to obtain polymers, not yielding (and even exceeding) by their properties to the composites, produced on their basis. Structureless (defect-free) polymers are imagined the most perspective in this respect. Such polymers can be natural replacement for a large number of elaborated at present polymer nanocomposites. The application of structureless polymers as artificial nanocomposites polymer matrix can give much larger effect. Such approach allows to obtain polymeric materials, comparable by their characteristics with metals (e.g., with aluminum).

6.1 INTRODUCTION

The idea of different classes polymers representation as composites is not new. Even 35 years ago Kardos and Raisoni [1] offered to use composite models for the description of semicrystalline polymers properties number and obtained prediction of the indicated polymers stiffness and thermal strains to a precision of ±20%. They considered semicrystalline polymer as composite, in which matrix is the amorphous and the crystallites are filler. Kardos and Raisoni [1] also supposed that other polymers, for example, hybrid polymer systems, in which two components with different mechanical properties were present, obviously can be simulated by a similar method.

In Ref. [2] it has been pointed out, that the most important consequence from works by supramolecular formation study is the conclusion that physical-mechanical properties depend in the first place on molecular structure, but are realized through supramolecular formations. At scales interval and studies methods resolving ability of polymers structure the nanoparticle size can be changed within the limits of $1 \div 100$ and more nanometers. The polymer crystallites size makes up $10 \div 20$ nm. The macromolecule can be included in several crystallites, since at molecular weight of order of 6×10^4 its length makes up more than 400 nm. These reasoning's point out, that macromolecular formations and polymer systems in virtue of their structure features are always nanostructural systems.

However, in the cited above works the amorphous glassy polymers consideration as natural composites (nanocomposites) is absent, although they are one of the most important classes of polymeric materials. This gap reason is quite enough that is, polymers amorphous state quantitative model absence. However, such model appearance lately [3–5] allows to consider the amorphous glassy polymers (both linear and cross-linked ones) as natural nanocomposites, in which local order regions (clusters) are nanofiller and surrounded them loosely packed matrix of amorphous polymers structure is matrix of nanocomposite. Proceeding from the said above, in the present chapter description of amorphous glassy polymers as natural nanocomposites, their limiting characteristics determination and practical recommendation by the indicated polymers properties improvement will be given.

6.2   NATURAL NANOCOMPOSITES STRUCTURE

The synergetics principles revealed structure adaptation mechanism to external influence and are universal ones for self-organization laws of spatial structures in dynamical systems of different nature. The structure adaptation is the reformation process of structure, which loses stability, with the new more stable structure self-organization. The fractal (multifractal) structure, which is impossible to describe within the framework of Euclidean geometry, are formed in reformation process. A wide spectrum of natural and artificial topological forms, the feature of which is self-similar hierarchically organized structure, which amorphous glassy polymers possessed [6], belongs to fractal structures.

The authors [7, 8] considered the typical amorphous glassy polymer (polycarbonate) structure change within the frameworks of solid body synergetics.

The local order region, consisting of several densely packed collinear segments of various polymer chains (for more details see previous paper) according to a signs number should be attributed to the nanoparticles (nanoclusters) [9]:

1. their size makes up $2 \div 5$ nm;
2. they are formed by self-assemble method and adapted to the external influence (e.g., temperature change results to segments number per one nanocluster change);
3. the each statistical segment represents an atoms group and boundaries between these groups are coherent owing to collinear arrangement of one segment relative to another.

The main structure parameter of cluster model-nanoclusters relative fraction φ_{cl}, which is polymers structure order parameter in strict physical sense of this tern, can be calculated according to the equation (see Chapter 5). In its turn, the polymer structure fractal dimension d_f value is determined according to the equations (see Chapter 5).

In Fig. 6.1, the dependence of φ_{cl} on testing temperature T for PC is shown, which can be approximated by the broken line, where points of folding (bifurcation points) correspond to energy dissipation mechanism change, coupling with the threshold values φ_{cl} reaching. So, in Fig. 6.1. T_1 corresponds to structure "freezing" temperature T_0 [4], T_2 to loosely packed matrix glass transition temperature T'_g [11] and T_3 to polymer glass transition temperature T_g.

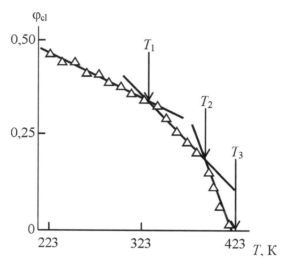

FIGURE 6.1   The dependence of nanoclusters relative fraction φ_{cl} on testing temperature T for PC. The critical temperatures of bifurcation points are indicated by arrows (explanations are given in the text) [18].

Within the frameworks of solid body synergetics it has been shown [12], that at structures self-organization the adaptation universal algorithm [12] is realized at transition from previous point of structure instability to subsequent one. The value $m = 1$ corresponds to structure minimum adaptivity and $m = m^*$ to maximum one. In Ref. [12], the table is adduced, in which values A_m, m and Δ_i are given, determined by the gold proportion rule and corresponding to spectrum of structure stability measure invariant magnitudes for the alive and lifeness nature systems. The indicated table usage facilitates determination of the interconnected by the power law stability and adaptivity of structure to external influence [12].

Using as the critical magnitudes of governing parameter the values φ_{cl} in the indicated bifurcation points T_0, T_g' and T_g (ϕ'_{cl} and T^*_{cl}, accordingly) together with the mentioned above table data [12], values A_m, Δ_i and for PC can be obtained, which are adduced in table 1. As it follows from the data of this table, systematic reduction of parameters A_m and Δ_i at the condition $m = 1 = $ const is observed. Hence, within the frameworks of solid body synergetics temperature T_g' can be characterized as bifurcation point ordering-degradation of nanostructure and T_g – as nanostructure degradation-chaos [12].

It is easy to see, that Δ_i decrease corresponds to bifurcation point critical temperature increase.

TABLE 6.1   The Critical Parameters of Nanocluster Structure State for PC [8]

The temperature range	ϕ'_{cl}	ϕ^*_{cl}	A_m	Δ_i	m	m^*
213÷333 K	0.528	0.330	0.623	0.618	1	1
333÷390 K	0.330	0.153	0.465	0.465	1	2
390÷425 K	0.153	0.049	0.324	0.324	1	8

Therefore, critical temperatures T_{cr} (T_0, T_g' and T_g) values increase should be expected at nanocluster structure stability measure Δ_i reduction. In Fig. 6.2, the dependence of T_{cr} in Δ_i reciprocal value for PC is adduced, on which corresponding values for polyarylate (PAr) are also plotted. This correlation proved to be linear one and has two characteristic points. At $\Delta_i = 1$ the linear dependence T_{cr} (Δ_i^{-1}) extrapolates to $T_{cr} = 293$K, that is, this means, that at the indicated Δ_i value glassy polymer turns into rubber-like state at the used testing temperature $T = 293$K. From the data of the determined by gold proportion law $\Delta_i = 0.213$ at $m = 1$ follows [12]. In the plot of Fig. 6.2 the greatest for polymers critical temperature $T_{cr} = T_{ll}$. (T_{ll} is the temperature of "liquid 1 to liquid 2" transition), defining the transition to "structureless liquid" [13], corresponds to this minimum Δ magnitude. For polymers this means the absence of even dynamical short-lived local order [13].

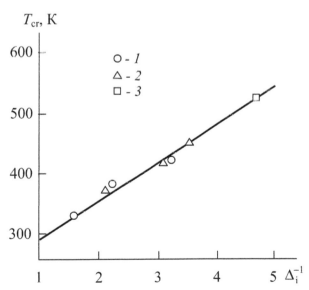

FIGURE 6.2   The dependence of critical temperatures T_{cr} on reciprocal value of nanocluster structure stability measure Δ_i for PC (1) and PAr (2), $3 - T_{ll}$ value for PC [19].

Hence, the stated above results allow giving the following interpretation of critical temperatures T_g' and T_g of amorphous glassy polymers structure within the frameworks of solid body synergetics. These temperatures correspond to governing parameter (nanocluster contents) φ_{cl} critical values, at which reaching one of the main principles of synergetics is realized-subordination principle, when a variables set is controlled by one (or several) variable, which is an order parameter. Let us also note reformations number $m = 1$ corresponds to structure formation mechanism particle-cluster [4, 5].

The authors [14, 15] considered synergetics principles application for the description of behavior of separate nanocluster structure, characterized by the integral parameter φ_{cl} nanoclusters in the system for the same amorphous glassy polymers. This aspect is very important, since, as it will be shown is subsequent sections, just separate nanoclusters characteristics define natural nanocomposites properties by critical mode. One from the criterions of nanoparticle definition has been obtained in Ref. [16]: atoms number N_{at} in it should not exceed $10^3 \div 10^4$. In Ref. [15] this criterion was applied to PC local order regions, having the greatest number of statistical segments $n_{cl} = 20$. Since nanocluster is amorphous analog of crystallite with the stretched chains and at its functionality F a number of chains emerging from it is accepted, then the value n_{cl} is determined as follows [4]:

$$n_{cl} = \frac{F}{2},$$
(1)

where the value F was calculated according to the Eq. (1.7) in previous publication.

The statistical segment volume simulated as a cylinder, is equal to $l_{st}S$ and further the volume per one atom of substance (PC) a^3 can be calculated according to the equation [17]:

$$a^3 = \frac{M}{\rho N_A p},$$
(2)

where M is repeated link molar mass, ρ is polymer density, N_A is Avogadro number, p is atoms number in a repeated link.

For PC $M = 264$ g/mol, $\rho = 1200$ kg/m^3 and $p = 37$. Then $a^3 = 9,54$ Å^3 and the value N_{at} can be estimated according to the following simple equation [17]:

$$N_{at} = \frac{l_{st} \cdot S \cdot n_{cl}}{a^3}$$
(3)

For PC $N_{at} = 193$ atoms per one nanocluster (for $n_{cl} = 20$) is obtained. It is obvious that the indicated value N_{at} corresponds well to the adduced above nanoparticle definition criterion ($N_{at} = 10^3 \div 10^4$) [9, 17].

Let us consider synergetics of nanoclusters formation in PC and PAr. Using the Eq. (3) as governing parameter critical magnitudes n_{cl} values at testing temperature

T consecutive change and the indicated above the table of the determined by gold proportion law values A_m, m and Δ_i, the dependence $\Delta(T)$ can be obtained, which is adduced in Fig. 6.3. As it follows from this figure data, the nanoclusters stability within the temperature range of $313 \div 393K$ is approximately constant and small ($\Delta_i \approx 0.232$ at minimum value $\Delta_i \approx 0.213$) and at $T > 393K$ fast growth Δ_i (nanoclusters stability enhancement) begins for both considered polymers.

This plot can be explained within the frameworks of a cluster model [3–5]. In Fig. 6.3, glass transition temperatures of loosely packed matrix T_g', which are approximately 50 K lower than polymer macroscopic glass transition temperature T_g, are indicated by vertical shaded lines. At T_g' instable nanoclusters, that is, having small n_{cl} decay occurs. At the same time stable and, hence, more steady nanoclusters remain as a structural element, that results to Δ_i growth [14].

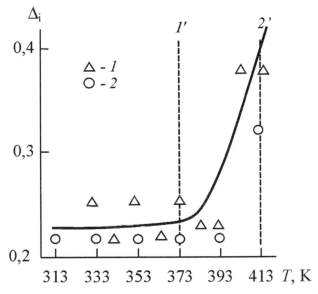

FIGURE 6.3   The dependence of nanoclusters stability measure Δ_i on testing temperature T for PC (1) and PAR (2). The vertical shaded lines indicate temperature T_g' for PC (1') and PAR (2') [14].

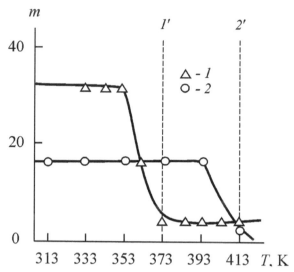

FIGURE 6.4 The dependences of reformations number m for nanoclusters on testing temperature T. The designations are the same as in Fig. 6.3 [14].

In Fig. 6.4 the dependences of reformations number m on testing temperature T for PC and PAr are adduced. At relatively low temperatures ($T < T_g'$) segments number in nanoclusters is large and segment joining (separation) to nanoclusters occurs easily enough, that explains large values m. At $T \rightarrow T_g'$ reformations number reduces sharply and at $T > T_g'$ $m \approx 4$. Since at $T > T_g'$ in the system only stable clusters remain, then it is necessary to assume, that large m at $T < T_g'$ are due to reformation of just instable nanoclusters [15].

In Fig. 6.5, the dependence of n_{cl} on m is adduced. As one can see, even small m enhancement within the range of $2 \div 16$ results to sharp increasing in segments number per one nanocluster. At $m \approx 32$ the dependence $n_{cl}(m)$ attains asymptotic branch for both studied polymers. This supposes that $n_{cl} \approx 16$ is the greatest magnitude for nanoclusters and for $m \geq 32$ this term belongs equally to both joining and separation of such segment from nanocluster.

In Fig. 6.6, the relationship of stability measure Δ_i and reformations number m for nanoclusters in PC and PAr is adduced. As it follows from the data of this figure, at $m \geq 16$ (or, according to the data of Fig. 6.5, $n_{cl} \geq 12$) Δ_i value attains its minimum asymptotic magnitude $\Delta_i = 0.213$ [12]. This means, that for the indicated n_{cl} values nanoclusters in PC and PAr structure are adopted well to the external influence change ($A_m \geq 0.91$).

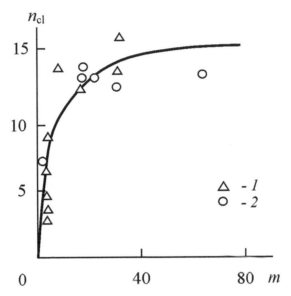

FIGURE 6.5   The dependence of segments number per one nanocluster n_{cl} on reformations number m for PC (1) and PAR (2) [14].

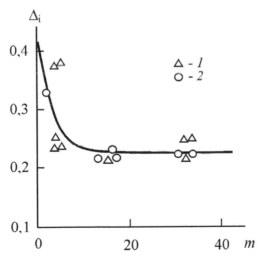

FIGURE 6.6   The dependence of stability measure Δ_i on reformation number m for PC (1) and PAR (2) [14].

Nanoclusters formation synergetics is directly connected with the studied poly-mers structure macroscopic characteristics. As it has been noted above, the fractal structure, characterized by the dimension d_p is formed as a result of nanoclusters reformations. In Fig. 6.7, the dependence $d_f (\Delta_i)$ for the considered polymers is adduced, from which d_f increase at Δ_i growth follows. This means, that the increasing of possible reformations number m, resulting to Δ_i reduction (Fig. 6.6), defines the growth of segments number in nanoclusters, the latter relative fraction φ_{cl} enhancement and, as consequence, d_f reduction [3–5].

And let us note in conclusion the following aspect, obtaining from the plot Δ_i (T) (Fig. 6.3) extrapolation to maximum magnitude $\Delta_i \approx 1.0$. The indicated Δ_i value is reached approximately at $T \approx 458$ K that corresponds to mean glass transition temperature for PC and Par. Within the frameworks of the cluster model T_g reaching means polymer nanocluster structure decay [3–5] and, in its turn, realization at T_g of the condition $\Delta_i \approx 1.0$ means, that the "degenerated" nanocluster, consisting of one statistical segment or simply statistical segment, possesses the greatest stability measure. Several such segments joining up in nanocluster mains its stability reduction (see Figs. 6.5 and 6.6), that is the cause of glassy polymers structure thermodynamical nonequilibrium [14].

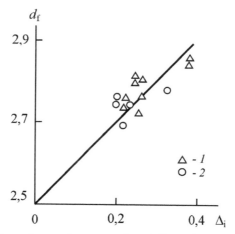

FIGURE 6.7   The dependence of structure fractal dimension d_f on stability measure of nanoclusters Δ_i for PC (1) and PAR (2) [14].

Therefore, the stated above results showed synergetics principles applicability for the description of association (dissociation) processes of polymer segments in local order domains (nanoclusters) in case of amorphous glassy polymers. Such conclusion can be a priori, since a nanoclusters are dissipative structures [6]. Testing temperature increase rises nanoclusters stability measure at the expense of possible reformations number reduction [14, 15].

As it has been shown lately, the notion "nanoparticle" (nanocluster) gets well over the limits of purely dimensional definition and means substance state specific character in sizes nanoscale. The nanoparticles, sizes of which are within the range of order of 1 ÷ 100 nm, are already not classical macroscopic objects. They represent themselves the boundary state between macro and microworld and in virtue of this they have specific features number, to which the following ones are attributed:

1. nanoparticles are self-organizing nonequilibrium structures, which submit to synergetics laws;
2. they possess very mature surface;
3. nanoparticles possess quantum (wave) properties.

For the nanoworld structures in the form of nanoparticles (nanoclusters) their size, defining the surface energy critical level, is the information parameter of feedback [19].

The first from the indicated points was considered in detail above. The authors [20, 21] showed that nanoclusters surface fractal dimension changes within the range of 2.15 ÷ 2.85 that is their well developed surface sign. And at last, let us consider quantum (wave) aspect of nanoclusters nature on the example of PC [22]. Structural levels hierarchy formation and development "scenario" in this case can be presented with the aid of iterated process [23]:

$$l_k = \langle a \rangle B_\lambda^k; \quad \lambda_k = \langle a \rangle B_\lambda^{k+1}; \quad k = 0, 1, 2,..., \qquad (4)$$

where l_k is specific spatial scale of structural changes, λ_k is length of irradiation sequence, which is due to structure reformation, k is structural hierarchy sublevel number, $B_\lambda = \lambda_b/\langle a \rangle = 2.61$ is discretely wave criterion of microfracture, λ_b is the smallest length of acoustic irradiation sequence.

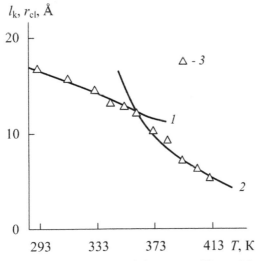

FIGURE 6.8 The dependences of structural changes specific spatial scale l_k at $B_\lambda = 1.06$ (1) and 1.19 (2) and nanoclusters radius r_{cl} (3) on testing temperature T for PC [22].

In Fig. 6.8, the dependences of l_k and nanoclusters radius r_{cl} on T are adduced, where l_k was determined according to the Eq. (4) and the value r_{cl} was calculated according to the formula (in Chapter 5). As it follows from the data of Fig. 6.8, the values l_k and r_{cl} agree within the whole studied temperatures range. Let us note, that if in Ref. [23] the value $B_\lambda = 2.61$, then for PC the indicated above agreement was obtained at $B_\lambda = 1.19$ and 1.06. This distinction confirms the thesis about distinction of synergetics laws in reference to nano-microworld objects (let us remind, that the condition $B_\lambda = 2.61$ is valid even in the case of earthquakes [14]). It is interesting to note, that B_λ change occurs at glass transition temperature of loosely packed matrix, that is, approximately at $T_g - 50$ K [11].

Hence, the stated above results demonstrated that the nanocluster possessed all nanoparticles properties, that is, they belonged to substance intermediate state-nanoworld.

And in completion of the present section let us note one more important feature of natural nanocomposites structure. In Refs. [24, 25], the interfacial regions absence in amorphous glassy polymers, treated as natural nanocomposites, was shown. This means, that such nanocomposites structure represents a nanofiller (nanoclusters), immersed in matrix (loosely packed matrix of amorphous polymer structure), that is, unlike polymer nanocomposites with inorganic nanofiller (artificial nanocomposites) they have only two structural components.

6.3　THE NATURAL NANOCOMPOSITES REINFORCEMENT

As it is well-known [26], very often a filler introduction in polymer matrix is carried out for the last stiffness enhancement. Therefore, the reinforcement degree of polymer composites, defined as a composite and matrix polymer elasticity moduli ratio, is one of their most important characteristics.

At amorphous glassy polymers as natural nanocomposites treatment the estimation of filling degree or nanoclusters relative fraction φ_{cl} has an important significance. Therefore, the authors [27] carried out the comparison of the indicated parameter estimation different methods, one of which is EPR-spectroscopy (the method of spin probes). The indicated method allows studying amorphous polymer structural heterogeneity, using radicals distribution character. As it is known [28], the method, based on the parameter d_1/d_c – the ratio of spectrum extreme components total intensity to central component intensity-measurement is the simplest and most suitable method of nitroxil radicals local concentrations determination. The value of dipole-dipole interaction ΔH_{dd} is directly proportional to spin probes concentration C_w [29]:

$$\Delta H_{dd} = A \cdot C_w, \tag{5}$$

where $A = 5 \times 10^{-20}$ Ersted·cm³ in the case of radicals chaotic distribution.

On the basis of the Eq. (5) the relationship was obtained, which allows to calculate the average distance r between two paramagnetic probes [29]:

$$r = 38(\Delta H_{dd})^{-1/3}, \text{Å} \tag{6}$$

where ΔH_{dd} is given in Ersteds.

In Fig. 6.9, the dependence of d_1/d_c on mean distance r between chaotically distributed in amorphous PC radicals-probes is adduced. For PC at $T = 77K$ the values of $d_1/d_c = 0.38 \div 0.40$ were obtained. One can make an assumption about volume fractions relation for the ordered domains (nanoclusters) and loosely packed matrix of amorphous PC. The indicated value d_1/d_c means, that in PC at probes statistical distribution 0.40 of its volume is accessible for radicals and approximately 0.60 of volume remains unoccupied by spin probes, that is, the nanoclusters relative fraction φ_{cl} according to the EPR method makes up approximately $0.60 \div 0.62$.

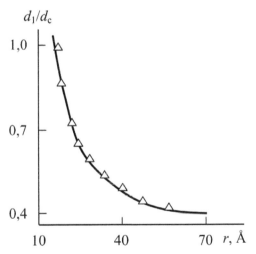

FIGURE 6.9 The dependence of parameter d_1/d_c of EPR spectrum on the value of mean distance r between radicals for PC [27].

This result corresponds well to the experimental data of Yech [30] and Perepechko [31], who obtained the values 0.60 and 0.63 for densely packed regions relative fraction in amorphous polymers.

The authors of Ref. [11] fulfilled φ_{cl} estimation with the aid of reversed gas chromatography and obtained the following magnitudes of this parameter for PC, poly (methyl methacrylate) and polysulfone: 0.70, 0.60 and 0.65, accordingly (Table 6.2).

Within the frameworks of the cluster model φ_{cl} estimation can be fulfilled by the percolation relationship (in Chapter 5) usage. Let us note, that in the given case the temperature of polymers structure quasiequilibrium state attainment, lower of which φ_{cl} value does not change, that is, T_0 [32], is accepted as testing temperature T. The calculation φ_{cl} results according to the equation (in Chapter 5) for the mentioned above polymers are adduced in Table 6.2, which correspond well to other authors estimations.

Proceeding from the circumstance, that radicals-probes are concentrated mainly in intercluster regions, the nanocluster size can be estimated, which in amorphous PC should be approximately equal to mean distance r between two paramagnetic probes, that is, ~50 Å (Fig. 6.9). This value corresponds well to the experimental data, obtained by dark-field electron microscopy method ($\approx 30 \div 100$ Å) [33].

Within the frameworks of the cluster model the distance between two neighboring nanoclusters can be estimated according to the equation (in Chapter 5) as $2R_{cl}$. The estimation $2R_{cl}$ by this mode gives the value 53.1 Å (at F = 41) that corresponds excellently to the method EPR data.

Thus, the Ref. [27] results showed, that the obtained by EPR method natural nanocomposites (amorphous glassy polymers) structure characteristics corresponded completely to both the cluster model theoretical calculations and other authors estimations. In other words, EPR data are experimental confirmation of the cluster model of polymers amorphous state structure.

The treatment of amorphous glassy polymers as natural nanocomposites allows to use for their elasticity modulus E_p (and, hence, the reinforcement degree $E_p/E_{l.m.}$, where $E_{l.m.}$ is loosely packed matrix elasticity modulus) description theories, developed for polymer composites reinforcement degree description [9, 17]. The authors [34] showed correctness of particulate-filled polymer nanocomposites reinforcement of two concepts on the example of amorphous PC. For theoretical estimation of particulate-filled polymer nanocomposites reinforcement degree E_n/E_m two equations can be used. The first from them has the look [35]:

$$\frac{E_n}{E_m} = 1 + \phi_n^{1,7} ,$$
(7)

where E_n and E_m are elasticity moduli of nanocomposites and matrix polymer, accordingly, φ_n is nanofiller volume contents.

The second equation offered by the authors of Ref. [36] is:

$$\frac{E_n}{E_m} = 1 + \frac{0,19W_n l_{st}}{D_p^{1/2}} ,$$
(8)

where W_n is nanofiller mass contents in mas.%, D_p is nanofiller particles diameter in nm.

Let us consider included in the Eqs. (7) and (8) parameters estimation methods. It is obvious, that in the case of natural nanocomposites one should accept: $E_n = E_p$, $E_m = E_{l.m.}$ and $\varphi_n = \varphi_{cl}$, the value of the latter can be estimated according to the equation (in Chapter 5).

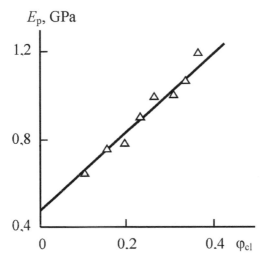

FIGURE 6.10   The dependence of elasticity modulus E_p on nanoclusters relative fraction φ_{cl} for PC [34].

The mass fraction of nanoclusters W_{cl} can be calculated as follows [37]:

$$W_{cl} = \rho \varphi_{cl}, \tag{9}$$

where ρ is nanofiller (nanoclusters) density which is equal to 1300 kg/m³ for PC.

The value $E_{l.m.}$ can be determined by the construction of E_p (φ_{cl}) plotting, which is adduced in Fig. 6.10. As one can see, this plot is approximately linear and its extrapolation to $\varphi_{cl} = 0$ gives the value $E_{l.m.}$ And at last, as it follows from the nanoclusters definition (see Chapter 1) one should accept $D_p \approx l_{st}$ for them and then the Eq. (8) accepts the following look [34]:

In Fig. 6.11, the comparison of theoretical calculation according to the Eqs. (7) and (10) with experimental values of reinforcement degree $E_p/E_{l.m.}$ for PC is adduced. As one can see, both indicated equations give a good enough correspondence with the experiment: their average discrepancy makes up 5.6% in the Eq. (7) case and 9.6% for the Eq. (10). In other words, in both cases the average discrepancy does not exceed an experimental error for mechanical tests. This means, that both considered methods can be used for PC elasticity modulus prediction. Besides, it necessary to note, that the percolation relationship (7) qualitatively describes the dependence $E_p/E_{l.m.}$ (φ_{cl}) better, than the empirical equation (10).

$$\frac{E_{n}}{E_{m}} = 1 + 0,19\rho\phi_{cl}l_{st}^{1/2} \tag{10}$$

The obtained results allowed making another important conclusion. As it is known, the percolation relationship (7) assumes, that nanofiller is percolation system (polymer composite) solid-body component and in virtue of this circumstance defines this system elasticity modulus. However, for artificial polymer particulate-filled nanocomposites, consisting of polymer matrix and inorganic nanofiller, the Eq. (7) in the cited form gives the understated values of reinforcement degree. The authors [9, 17] showed, that for such nanocomposites the sum $(\varphi_{n}+\varphi_{if})$, where φ_{if} was interfacial regions relative fraction, was a solid-body component. The correspondence of experimental data and calculation according to the Eq. (7) demonstrates, that amorphous polymer is the specific nanocomposite, in which interfacial regions are absent [24, 25]. This important circumstance is necessary to take into consideration at amorphous glassy polymers structure and properties description while simulating them as natural nanocomposites. Besides, one should note, that unlike micromechanical models the Eqs. (7) and (10) do not take into account nanofiller elasticity modulus, which is substantially differed for PC nanoclusters and inorganic nanofillers [34].

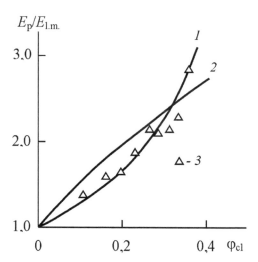

FIGURE 6.11   The dependences of reinforcement degree $E_{p}/E_{l.m}$ on nanoclusters relative fraction φ_{cl} for PC. 1 – calculation according to the equation (7); 2 – calculation according to the equation (10); 3 – the experimental data [34].

Another mode of natural nanocomposites reinforcement degree description is micromechanical models application, developed for polymer composites mechanical behavior description [1, 37–39]. So, Takayanagi and Kerner models are often used for the description of reinforcement degree on composition for the indicated materials [38, 39]. The authors [40] used the mentioned models for theoretical treatment of natural nanocomposites reinforcement degree temperature dependence on the example of PC.

Takayanagi model belongs to a micromechanical composite models group, allowing empirical description of composite response upon mechanical influence on the basis of constituent it elements properties. One of the possible expressions within the frameworks of this model has the following look [38]:

$$\frac{G_c}{G_m} = \frac{\phi_m G_m + (\alpha + \phi_f) G_f}{(1 + \alpha \phi_f) G_m + \alpha \phi_m G_f}, \tag{11}$$

where G_c, G_m and G_f are shear moduli of composite, polymer matrix and filler, accordingly, ϕ_m and ϕ_f are polymer matrix and filler relative fractions, respectively, α is a fitted parameter.

Kerner equation is identical to the Eq. (11), but for it the parameter α does not fit and has the following analytical expression [38]:

$$\alpha_m = \frac{2(4 - 5v_m)}{(7 - 5v_m)}, \tag{12}$$

where α_m and v_m are parameter α and Poisson's ratio for polymer matrix.

Let us consider determination methods of the Eqs. (11) and (12) parameters, which are necessary for the indicated equations application in the case of natural nanocomposites, Firstly, it is obvious, that in the last case one should accept: $G_c = G_p$, $G_m = G_{1.m}$, $G_f = G_{cl}$, where G_p, $G_{1.m}$ and G_{cl} are shear moduli of polymer, loosely packed matrix and nanoclusters, accordingly, and also $\phi_f = \phi_{cl}$, where ϕ_{cl} is determined according to the percolation relationship (in Chapter 5). Young's modulus for loosely packed matrix and nanoclusters can be received from the data of Fig. 6.10 by the dependence $E_p(\phi_{cl})$ extrapolation to $\phi_{cl} = 1.0$, respectively. The corresponding shear moduli were calculated according to the general equation (in Chapter 5). The value of nanoclusters fractal dimension d_f^{cl} in virtue of their dense package is accepted equal to the greatest dimension for real solids ($d_f^{cl} = 2.95$ [40]) and loosely packed matrix fractal dimension $d_f^{l.m}$ can be estimated.

However, the calculation according to the Eqs. (11) and (12) does not give a good correspondence to the experiment, especially for the temperature range of $T = 373 \div 413$ K in PC case. As it is known [38], in empirical modifications of Kerner equation it is usually supposed, that nominal concentration scale differs from mechanically effective filler fraction ϕ_f^{ef}, which can be written accounting for the designations used above for natural nanocomposites as follows [41].

$$\phi_f^{ef} = \frac{\left(G_p - G_{1.m.}\right)\left(G_{1.m.} + \alpha_{1.m.}G_{cl}\right)}{\left(G_{cl} - G_{1.m.}\right)\left(G_{1.m.} + \alpha_{1.m.}G_p\right)}, \tag{13}$$

where $\alpha_{1.m.} = \alpha_m$. The value $\alpha_{1.m.}$ can be determined according to the Eq. (12), estimating Poisson's ratio of loosely packed matrix $\nu_{1.m.}$ by the known values $d_f^{1.m}$ according to the equation (in Chapter 5).

Besides, one more empirical modification ϕ_f^{ef} exists, which can be written as follows [41]:

$$\phi_{cl_2}^{ef} = \phi_{cl} + c\left(\frac{\phi_{cl}}{2r_{cl}}\right)^{2/3}, \tag{14}$$

where c is empirical coefficient of order one, r_{cl} is nanocluster radius, determined according to the equation (in previous paper).

At the value $\phi_{cl_2}^{ef}$ calculation according to the Eq. (14) magnitude c was accepted equal to 1,0 for the temperature range of $T = 293 \div 363$ K and equal to 1.2 – for the range of $T = 373 \div 413$ K and $2r_{cl}$ is given in nm. In Fig. 12, the comparison of values ϕ_{cl}^{ef}, calculated according to the Eqs. (13) and (14) ($\phi_{cl_1}^{ef}$ and $\phi_{cl_2}^{ef}$, accordingly) is adduced. As one can see, a good enough conformity of the values ϕ_{cl}^{ef}, estimated by both methods, is obtained (the average discrepancy of $\phi_{cl_1}^{ef}$ and $\phi_{cl_2}^{ef}$ makes up slightly larger than 20%). Let us note, that the effective value ϕ_{cl} exceeds essentially the nominal one, determined according to the relationship (in Chapter 5): within the range of $T = 293 \div 363$K by about 70% and within the range of $T = 373 \div 413$K – almost in three times.

In Fig. 6.13, the comparison of experimental and calculated according to Kerner equation (the Eq. (11)) with the Eqs. (13) and (14) using values of reinforcement degree by shear modulus $G_p/G_{1.m.}$ as a function of testing temperature T for PC is adduced. As one can see, in this case at the usage of nanoclusters effective concentration scale (ϕ_{cl}^{ef} instead of ϕ_{cl}) the good conformity of theory and experiment is obtained (their average discrepancy makes up 6%).

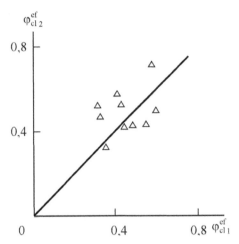

FIGURE 6.12 The comparison of nanoclusters effective concentration scale $\phi_{cl_1}^{ef}$ and $\phi_{cl_2}^{ef}$, calculated according to the Eqs. (13) and (14), respectively, for PC. A straight line shows the relation 1:1 [41].

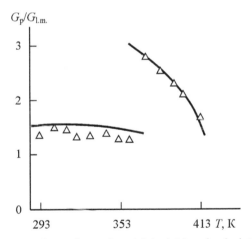

FIGURE 6.13 The comparison of experimental (points) and calculated according to the Eqs. (11), (13) and (14) (solid lines) values of reinforcement degree by shear modulus $G_p/G_{l.m.}$ as a function of testing temperature T for PC [41].

Hence, the stated above results have shown the modified Kerner equation application correctness for natural nanocomposites elastic response description. Really this fact by itself confirms the possibility of amorphous glassy polymers treatment as nanocomposites. Microcomposite models usage gives the clear notion about factors, influencing polymers stiffness.

6.4 INTERCOMPONENT ADHESION IN NATURAL NANOCOMPOSITES

Amorphous glassy polymers as natural nanocomposites puts forward to the foreground their study intercomponent interactions, that is, interactions nanoclusters – loosely packed matrix. This problem plays always one of the main roles at multiphase (multicomponent) systems consideration, since the indicated interactions or interfacial adhesion level defines to a great extent such systems properties [42]. Therefore, the authors [43] studied the physical principles of intercomponent adhesion for natural nanocomposites on the example of PC.

The authors [44] considered three main cases of the dependence of reinforcement degree E_c/E_m on φ_f. In this chapter, the authors have shown that there are the following main types of the dependences E_c/E_m (φ_f) exist:

1. the ideal adhesion between filler and polymer matrix, described by Kerner equation (perfect adhesion), which can be approximated by the following relationship:

$$\frac{E_c}{E_m} = 1 + 11,64\phi_f - 44,4\phi_f^2 + 96,3\phi_f^3 \; ; \tag{15}$$

2. zero adhesional strength at a large friction coefficient between filler and polymer matrix, which is described by the equation:

$$\frac{E_c}{E_m} = 1 + \phi_f \; ; \tag{16}$$

3. the complete absence of interaction and ideal slippage between filler and polymer matrix, when composite elasticity modulus is defined practically by polymer cross-section and connected with the filling degree by the equation:

$$\frac{E_c}{E_m} = 1 - \phi_f^{2/3} \tag{17}$$

In Fig. 6.14, the theoretical dependences $E_p/E_{1.m.}$ (φ_{cl}) plotted according to the Eqs. (15) ÷ (17), as well as experimental data (points) for PC are shown. As it follows from the adduced in Fig. 6.14 comparison at $T = 293÷363$ K the experimental data correspond well to the Eq. (16), that is, in this case zero adhesional strength at a large friction coefficient is observed. At $T = 373÷413$ K the experimental data correspond to the Eq. (15), that is, the perfect adhesion between nanoclusters and loosely packed matrix is observed. Thus, the adduced in Fig. 6.14 data demonstrated, that depending on testing temperature two types of interactions nanoclusters – loosely packed matrix are observed: either perfect adhesion or large friction between them. For quantitative estimation of these interactions it is necessary to determine their level, which can be made with the help of the parameter b_m, which is determined according to the equation [45]:

$$\sigma_f^c = \sigma_f^m K_s - b_m \phi_f, \tag{18}$$

where σ_f^c and σ_f^m are fracture stress of composite and polymer matrix, respectively, K_s is stress concentration coefficient. It is obvious, that since b_m increase results to σ_f^c reduction, then this means interfacial adhesion level decrease.

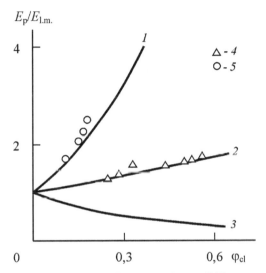

FIGURE 6.14   The dependences of reinforcement degree E_p/E_{1m} on nanoclusters relative fraction ϕ_{cl}. 1÷3 – the theoretical dependences, corresponding to the Eqs. (15) ÷ (17), accordingly; 4, 5 – the experimental data for PC within the temperature ranges: 293÷363K (4) and 373÷413K (5) [43].

The true fracture stress σ_f^{tr} for PC, taking into account sample cross-section change in a deformation process, was used as σ_f^c for natural nanocomposites, which can be determined according to the known formula:

$$\sigma_f^{tr} = \sigma_f^n \left(1 + \varepsilon_f\right), \tag{19}$$

where σ_f^n is nominal (engineering) fracture stress, ε_f is strain at fracture.

The value $\sigma_f^{1.m}$, which is accepted equal to loosely packed matrix strength $\sigma_f^{1.m}$, was determined by graphic method, namely, by the dependence $\sigma_f^{1.m}(\phi_{cl})$ plotting, which proves to be linear, and by subsequent extrapolation of it to $\phi_{cl} = 0$, that gives $\sigma_f^{1.m} = 40$ MPa [43].

And at last, the value Ks can be determined with the help of the following equation [39]:

$$\sigma_f^{tr} = \sigma_f^{l.m.} \left(1 - \phi_{cl}^{2/3}\right) K_s \qquad (20)$$

The parameter b_m calculation according to the stated above technique shows its decrease (intercomponent adhesion level enhancement) at testing temperature raising within the range of $b_m \approx 500 \div 130$.

For interactions nanoclusters – loosely packed matrix estimation within the range of $T = 293 \div 373 K$ the authors [48] used the model of Witten-Sander clusters friction, stated in Ref. [46]. This model application is due to the circumstance, that amorphous glassy polymer structure can be presented as an indicated clusters large number set [47]. According to this model, Witten-Sander clusters generalized friction coefficient t can be written as follows [46]:

$$f = \ln c + \beta \cdot \ln n_{cl}, \qquad (21)$$

where c is constant, β is coefficient, n_{cl} is statistical segments number per one nanocluster.

The coefficient β value is determined as follows [46]:

$$\beta = \left(d_f^{cl}\right)^{-1}, \qquad (22)$$

where d_f^{cl} is nanocluster structure fractal dimension, which is equal, as before, to 2.95 [40].

In Fig. 6.15, the dependence b_m (f) is adduced, which is broken down into two parts. On the first of them, corresponding to the range of $T = 293 \div 363$ K, the intercomponent interaction level is intensified at f decreasing (i.e., b_m reduction is observed and on the second one, corresponding to the range of $T = 373 \div 413$ K, b_m = const independent on value f. These results correspond completely to the data of Fig. 6.14, where in the first from the indicated temperature ranges the value $E_p/E_{l.m.}$ is defined by nanoclusters friction and in the second one by adhesion and, hence, it does not depend on friction coefficient.

As it has been shown in Ref. [48], the interfacial (or intercomponent) adhesion level depends on a number of accessible for the formation interfacial (intercomponent) bond sites (nodes) on the filler (nanocluster) particle surface N_u, which is determined as follows [49]:

$$N_u = L^{d_u}, \qquad (23)$$

where L is filler particle size, d_u is fractal dimension of accessible for contact ("non-screened") indicated particle surface.

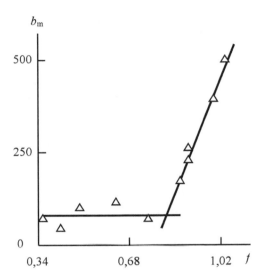

FIGURE 6.15 The dependence of parameter b_m on generalized friction coefficient f for PC [43].

One should choose the nanocluster characteristic size as L for the natural nano-composite which is equal to statistical segment l_{st}, determined according to the equation (in Chapter 5), and the dimension d_u is determined according to the following relationship [49]:

$$d_u = \left(d_{surf} - 1\right) + \left(\frac{d - d_{surf}}{d_w}\right), \tag{24}$$

where d_{surf} is nanocluster surface fractal dimension, d_w is dimension of random walk on this surface, estimated according to Aarony-Stauffer rule [49]:

$$d_w = d_{surf} + 1. \tag{25}$$

The following technique was used for the dimension d_{surf} calculation. First the nanocluster diameter $D_{cl} = 2r_{cl}$ was determined according to the equation (in Chapter 5) and then its specific surface S_u was estimated [35]:

$$S_u = \frac{6}{\rho_{cl}D_{cl}}, \tag{26}$$

where ρ_{cl} is the nanocluster density, equal to 1300 kg/m³ in the PC case.

And at last, the dimension d_{surf} was calculated with the help of the equation [20]:

$$S_u = 5,25 \times 10^3 \left(\frac{D_{cl}}{2}\right)^{d_{surf} - d} \tag{27}$$

In Fig. 6.16, the dependence b_m (N_u) for PC is adduced, which is broken down into two parts similarly to the dependence b_m (f) (Fig. 6.15). At $T = 293 \div 363$ K the value b_m is independent on N_u, since nanocluster – loosely packed matrix interactions are defined by their friction coefficient. Within the range of $T = 373 \div 413$ K intercomponent adhesion level enhancement (b_m reduction) at active sites number N_u growth is observed, as was to be expected. Thus, the data of both Figs. 6.15 and 6.16 correspond to Fig. 6.14 results.

With regard to the data of Figs. 6.15 and 6.16 two remarks should be made. Firstly, the transition from one reinforcement mechanism to another corresponds to loosely packed matrix glass transition temperature, which is approximately equal to $T_g - 50$K [11]. Secondly, the extrapolation of Fig. 6.16 plot to $b_m = 0$ gives the value $N_u \approx 71$, that corresponds approximately to polymer structure dimension $d_f = 2.86$.

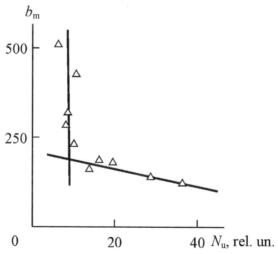

FIGURE 6.16 The dependence of parameter b_m on nanocluster surface active ("nonscreened") sites number N_u for PC [43].

In this theme completion an interesting structural aspect of intercomponent adhesion in natural nanocomposites (polymers) should be noted. Despite the considered above different mechanisms of reinforcement and nanoclusters-loosely packed matrix interaction realization the common dependence b_m (φ_{cl}) is obtained for the entire studied temperature range of 293÷413K, which is shown in Fig. 6.17. This dependence is linear, that allows to determine the limiting values $b_m \approx 970$ at $\varphi_{cl} = 1.0$ and $b_m = 0$ at $\varphi_{cl} = 0$. Besides, let us note, that the shown in Figs. 6.14÷6.16 structural transition is realized at $\varphi_{cl} \approx 0.26$ [43].

Hence, the stated above results have demonstrated, that intercomponent adhesion level in natural nanocomposites (polymers) has structural origin and is defined by nanoclusters relative fraction. In two temperature ranges two different reinforcement mechanisms are realized, which are due to large friction between nanoclusters and loosely packed matrix and also perfect (by Kerner) adhesion between them. These mechanisms can be described successfully within the frameworks of fractal analysis.

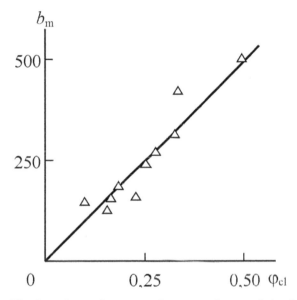

FIGURE 6.17   The dependence of parameter b_m on nanoclusters relative fraction φ_{cl} for PC [43].

The further study of intercomponent adhesion in natural nanocomposites was fulfilled in Ref. [50]. In Fig. 6.18, the dependence b_m (T) for PC is shown, from which b_m reduction or intercomponent adhesion level enhancement at testing temperature growth follows. In the same figure the maximum value b_m for nanocomposites polypropylene/Na^+-montmorillonite [9] was shown by a horizontal shaded line. As one can see, b_m values for PC within the temperature range of $T = 373 \div 413$ K by absolute value are close to the corresponding parameter for the indicated nanocomposite, that indicates high enough intercomponent adhesion level for PC within this temperature range.

Let us note an important structural aspect of the dependence b_m (T), shown in Fig. 6.18. According to the cluster model [4], the decay of instable nanoclusters occurs at temperature $T'_g \approx T_g - 50$ K, holding back loosely packed matrix in glassy state, owing to which this structural component is devitrificated within the temperature range of $T'_g \div T_g$. Such effect results to rapid reduction of polymer mechanical properties within the indicated temperature range [51]. As it follows from the data of Fig. 6.18, precisely in this temperature range the highest intercomponent adhesion level is observed and its value approaches to the corresponding characteristic for nanocomposites polypropylene/Na⁺-montmorillonite.

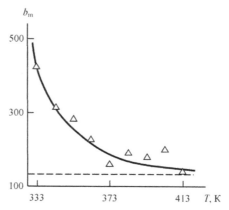

FIGURE 6.18   The dependence of parameter b_m on testing temperature T for PC. The horizontal shaded line shows the maximum value b_m for nanocomposites polypropylene/Na⁺-montmorillonite [50].

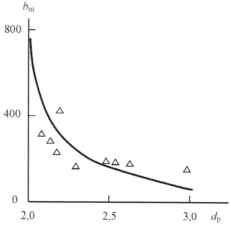

FIGURE 6.19   The dependence of parameter b_m on nanoclusters surface fractal dimension d_{surf} for PC [50].

It can be supposed with a high probability degree that adhesion level depends on the structure of nanoclusters surface, coming into contact with loosely packed matrix, which is characterized by the dimension d_{surf}. In Fig. 6.19, the dependence $b_m (d_{surf})$ for PC is adduced, from which rapid reduction b_m (or intercomponent adhesion level enhancement) follows at d_{surf} growth or, roughly speaking, at nanoclusters surface roughness enhancement.

The authors [48] showed that the interfacial adhesion level for composites polyhydroxyether/graphite was raised at the decrease of polymer matrix and filler particles surface fractal dimensions difference. The similar approach was used by the authors of Ref. [50], who calculated nanoclusters d_f^{cl} and loosely packed matrix $d_f^{l.m}$ fractal dimensions difference Δd_f:

$$\Delta d_f = d_f^{cl} - d_f^{l.m.}, \qquad (28)$$

where d_f^{cl} is accepted equal to real solids maximum dimension (d_f^{cl} = 2.95 [40]) in virtue of their dense packing and the value $d_f^{l.m}$ was calculated according to the mixtures rule (the equation from Chapter 5).

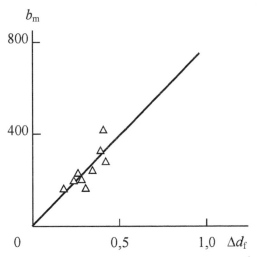

FIGURE 6.20 The dependence of parameter b_m on nanoclusters d_f^{cl} and loosely packed matrix $d_f^{l.m}$ structures fractal dimensions difference Δd_f for PC [50].

In Fig. 6.20, the dependence of b_m on the difference Δd_f is adduced, from which b_m decrease or intercomponent adhesion level enhancement at Δd_f reduction or values $d_f^{l.m}$ and $d_f^{l.m}$ growing similarity follows. This dependence demonstrates, that the greatest intercomponent adhesion level, corresponding to b_m = 0, is reached at Δd_f = 0.95 and is equal to ~ 780.

The data of Figs. 6.14 and 6.18 combination shows, that the value $b_m \approx 200$ corresponds to perfect adhesion by Kerner. In its turn, the Figs. 6.16 and 6.17 plots data demonstrated, that the value $b_m \approx 200$ could be obtained either at $d_{surf} > 2.5$ or at Δd_f < 0.3, accordingly. The obtained earlier results showed [24], that the condition $d_{surf} >$ 2.5 was reached at $r_{cl} < 7.5$Å or $T > 373$ K, that again corresponded well to the stated above results. And at last, the $\Delta d_f \approx 0.3$ or $d_f^{1.m} \approx 2,65$ according to the equation (in Chapter 5) was also obtained at $T \approx 373$K.

Hence, at the indicated above conditions fulfillment within the temperature range of $T < T_g'$ for PC perfect intercomponent adhesion can be obtained, corresponding to Kerner equation, and then the value E_p estimation should be carried out according to the Eq. (15). At $T = 293$ K ($\varphi_{cl} = 0.56$, $E_m = 0.85$GPa) the value E_p will be equal to 8.9 GPa, that approximately in 6 times larger, than the value E_p for serial industrial PC brands at the indicated temperature.

Let us note the practically important feature of the obtained above results. As it was shown, the perfect intercomponent adhesion corresponds to $b_m \approx 200$, but not $b_m = 0$. This means, that the real adhesion in natural nanocomposites can be higher than the perfect one by Kerner, that was shown experimentally on the example of particulate-filled polymer nanocomposites [17, 52]. This effect was named as nanoadhesion and its realization gives large possibilities for elasticity modulus increase of both natural and artificial nanocomposites. So, the introduction in aromatic polyamide (phenylone) of 0.3 mas.% aerosil only at nanoadhesion availability gives the same nanocomposite elasticity modulus enhancement effect, as the introduction of 3 mas.% of organoclay, which at present is assumed as one of the most effective nanofillers [9]. This assumes, that the value $E_p = 8.9$ GPa for PC is not a limiting one, at any rate, theoretically. Let us note in addition, that the indicated E_p values can be obtained at the natural nanocomposites nanofiller (nanoclusters) elasticity modulus magnitude $E_{cl} = 2.0$ GPa, that is, at the condition $E_{cl} < E_p$. Such result possibility follows from the polymer composites structure fractal concept [53], namely, the model [44], in which the Eqs. (15) ÷ (17) do not contain nanofiller elasticity modulus, and reinforcement percolation model [35].

The condition $d_{surf} < 2.5$, that is, $r_{cl} < 7.5$ Å or $N_{cl} < 5$, in practice can be realized by the nanosystems mechanosynthesis principles using, the grounds of which are stated in Ref. [54]. However, another more simple and, hence, more technological method of desirable structure attainment realization is possible, one from which will be considered in subsequent section.

Hence, the stated above results demonstrated, that the adhesion level between natural nanocomposite structural components depended on nanoclusters and loosely packed matrix structures closeness. This level change can result to polymer elasticity modulus significant increase. A number of these effect practical realization methods were considered [50].

The mentioned above dependence of intercomponent adhesion level on nanoclusters radius r_{cl} assumes more general dependence of parameter b_m on nanoclusters geometry. The authors [55] carried out calculation of accessible for contact sites of nanoclusters surface and loosely packed matrix number N_u according to the Eq. (23) for two cases. the nanocluster is simulated as a cylinder with diameter D_{cl} and length l_{st}, where l_{st} is statistical segment length, therefore, in the first case its butt-end is contacting with loosely packed matrix nanocluster surface and then $L = D_{cl}$ and in the second case with its side (cylindrical) surface and then $L = l_{st}$. In Fig. 6.21, the dependences of parameter b_m on value N_u, corresponding to the two considered above cases, are adduced. As one can see, in both cases, for the range of $T = 293 \div 363$ K l_{st}, where interactions nanoclusters – loosely packed matrix are character-ized by powerful friction between them, the value b_m does not depend on N_u, as it was expected. For the range of $T = 373 \div 413$ K, where between nanoclusters and loosely packed matrix perfect adhesion is observed, the linear dependences $b_m (N_u)$ are obtained. However, at using value D_{cl} as Lb_m reduction or intercomponent adhe-sion level enhancement at N_u decreasing is obtained and at $N_u = 0$ b_m value reaches its minimum magnitude $b_m = 0$. In other words, in this case the minimum level of in-tercomponent adhesion is reached at intercomponent bonds formation sites (nodes) absence that is physically incorrect [48]. And on the contrary at the condition $L = l_{st} b_m$ the reduction (intercomponent adhesion level enhancement) at the increase of contacts number N_u between nanoclusters and loosely packed matrix is observed, that is obvious from the physical point of view. Thus, the data of Fig. 6.21 indicate unequivocally, that the intercomponent adhesion is realized over side (cylindrical) nanoclusters surface and butt-end surfaces in this effect formation do not participate.

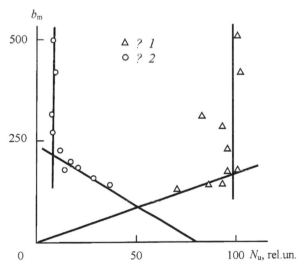

FIGURE 6.21 The dependences of parameter b_m on a number of accessible for intercomponent bonds formation sizes on nanocluster surface N_u at the condition $L = D_{cl}$ (1) and $L = l_{st}$ (2) for PC [55].

Let us consider geometrical aspects intercomponent interactions in natural nano-composites. In Fig. 6.22, the dependence of nanoclusters butt-end S_b and side (cy-lindrical) S_c surfaces areas on testing temperature T for PC are adduced. As one can see, the following criterion corresponds to the transition from strong friction to perfect adhesion at $T = 373K$ [55]:

$$S_b \approx S_c. \tag{29}$$

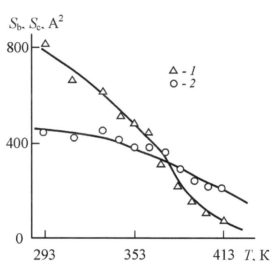

FIGURE 6.22 The dependences of nanoclusters butt-end S_b (1) and cylindrical S_c (2) surfaces areas on testing temperature T for PC [55].

Hence, the intercomponent interaction type transition from the large friction nanoclusters − loosely packed matrix to the perfect adhesion between them is de-fined by nanoclusters geometry: at $S_b > S_c$ the interactions of the first type is realized and at $S_b < S_c$ – the second one. Proceeding from this, it is expected that intercompo-nent interactions level is defined by the ratio S_b/S_c. Actually, the adduced in Fig. 6.23 data demonstrate b_m reduction at the indicated ratio decrease, but at the criterion (29) realization or $S_b/S_c \approx 1$ S_b/S_c decreasing does not result to b_m reduction and at $S_b/S_c < 1$ intercomponent adhesion level remains maximum high and constant [55].

Hence, the stated above results have demonstrated, that interactions nanoclu-sters-loosely packed matrix type (large friction or perfect adhesion) is defined by nanoclusters butt-end and side (cylindrical) surfaces areas ratio or their geometry if the first from the mentioned areas is larger that the second one then a large friction nanoclusters-loosely packed matrix is realized; if the second one exceeds the first one, then between the indicated structural components perfect adhesion is realized. In the second from the indicated cases intercomponent adhesion level does not depend on the mentioned areas ratio and remains maximum high and constant. In other words, the adhesion nanoclusters-loosely packed matrix is realized by nanoclusters cylindrical surface.

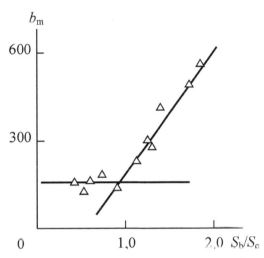

FIGURE 6.23 The dependence of parameter b_m on nanoclusters butt-end and cylindrical surfaces are ratio S_b/S_c value for PC [55].

The stated above results were experimentally confirmed by the EPR-spectroscopy method [56]. The Eqs. (1) and (6) comparison shows, that dipole–dipole interaction energy ΔH_{dd} has structural origin, namely [56]:

$$\Delta H_{dd} \approx \left(\frac{v_{cl}}{n_{cl}} \right) \qquad (30)$$

As estimations according to the Eq. (30) showed, within the temperature range of $T = 293 \div 413K$ for PC ΔH_{dd} increasing from 0.118 up to 0.328 Ersteds was observed.

Let us consider dipole-dipole interaction energy ΔH_{dd} intercommunication with nanoclusters geometry. In Fig. 6.24, the dependence of ΔH_{dd} on the ratio S_c/S_b for PC is adduced. As one can see, the linear growth ΔH_{dd} at ratio S_c/S_b increasing is observed, that is, either at S_c enhancement or at S_b reduction. Such character of the adduced in Fig. 6.24 dependence indicates unequivocally, that the contact nanoclusters-loosely packed matrix is realized on nanocluster cylindrical surface. Such effect was to be expected, since emerging from the butt-end surface statistically distributed polymer chains complicated the indicated contact realization unlike relatively smooth cylindrical surfaces. It is natural to suppose, that dipole-dipole interactions intensification or ΔH_{dd} increasing results to natural nanocomposites elasticity modulus E_p enhancement. The second as natural supposition at PC consideration as nano-composite is the influence on the value E_p of nanoclusters (nanofiller) relative fraction φ_{cl}, which is determined according to the percolation relationship (in Chapter 5).

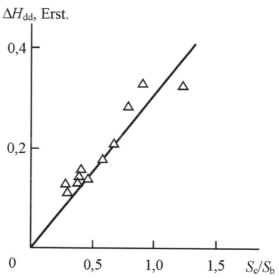

FIGURE 6.24   The dependence of dipole-dipole interaction energy ΔH_{dd} on nanoclusters cylindrical S_c and butt-end S_b surfaces areas ratio for PC [56].

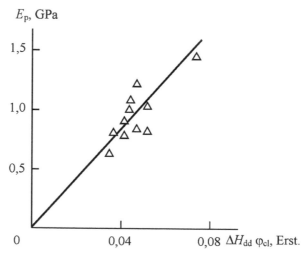

FIGURE 6.25   The dependence of elasticity modulus E_p on complex argument $(\Delta H_{dd}\varphi_{cl})$ for PC [56].

In Fig. 6.25, the dependence of elasticity modulus E_p on complex argument $(\Delta H_{dd}\varphi_{cl})$ for PC is presented. As one can see, this dependence is a linear one, passes through coordinates origin and is described analytically by the following empirical equation [56].

$$E_p = 21\,(\Delta H_{dd}\varphi_{cl}),\ GPa, \tag{31}$$

which with the appreciation of the Eq. (30) can be rewritten as follows [56]:

$$E_p = 21 \times 10^{-26} \left(\frac{\phi_{cl}V_{cl}}{n_{cl}} \right),\ GPa. \tag{32}$$

The Eq. (32) demonstrates clearly, that the value E_p and, hence polymer reinforcement degree is a function of its structural characteristics, described within the frameworks of the cluster model [3–5]. Let us note, that since parameters v_{cl} and φ_{cl} are a function of testing temperature, then the parameter n_{cl} is the most suitable factor for the value E_p regulation for practical purposes. In Fig. 6.26, the dependence E_p (n_{cl}) for PC at $T = 293$ K is adduced, calculated according to the Eq. (32), where the values v_{cl} and φ_{cl} were calculated according to the equations (in Chapter 5). As one can see, at small n_{cl} (<10) the sharp growth E_p is observed and at the smallest possible value $n_{cl} = 2$ the magnitude $E_p \approx 13.5$ GPa. Since for PC $E_{l.m.} = 0.85$ GPa, then it gives the greatest reinforcement degree $E_p/E_m \approx 15.9$. Let us note, that the greatest attainable reinforcement degree for artificial nanocomposites (polymers filled with inorganic nanofiller) cannot exceed 12 [9]. It is notable, that the shown in Fig. 6.26 dependence E_p (n_{cl}) for PC is identical completely by dependence shape to the dependence of elasticity modulus of nanofiller particles diameter for elastomeric nanocomposites [57].

Hence, the presented above results have shown that elasticity modulus of amorphous glassy polycarbonate, considered as natural nanocomposite, are defined completely by its suprasegmental structure state. This state can be described quantitatively within the frameworks of the cluster model of polymers amorphous state structure and characterized by local order level. Natural nanocomposites reinforcement degree can essentially exceed analogous parameter for artificial nanocomposites [56].

As it has been shown above (see the Eqs. (7) and (15)), the nanocluster relative fraction increasing results to polymers elasticity modulus enhancement similarly to nanofiller contents enhancement in artificial nanocomposites. Therefore, the necessity of quantitative description and subsequent comparison of reinforcement degree for the two indicated above nanocomposites classes appears. The authors [58, 59] fulfilled the comparative analysis of reinforcement degree by nanoclusters and by layered silicate (organoclay) for polyarylate and nanocomposite epoxy polymer/Na$^+$-montmorillonite [60], accordingly.

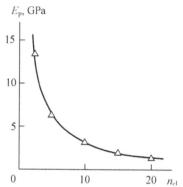

FIGURE 6.26 The dependence of elasticity modulus E_p on segments number n_{cl} per one nanocluster, calculated according to the Eq. (32) for PC at $T = 293K$ [56]

In Fig. 6.27, theoretical dependences of reinforcement degree E_n/E_m on nanofiller contents φ_n, calculated according to the Eqs. (15) ÷ (17), are adduced. Besides, in the same figure the experimental values (E_n/E_m) for nanocomposites epoxy polymer Na^+-montmorillonite (EP/MMT) at $T < T_g$ and $T > T_g$ (where T and T_g are testing and glass transition temperatures, respectively) are indicated by points. As one can see, for glassy epoxy matrix the experimental data correspond to the Eq. (16), that is, zero adhesional strength at a large friction coefficient and for devitrificated matrix – to the Eq. (15), that is, the perfect adhesion between nanofiller and polymer matrix, described by Kerner equation. Let us note that the authors [17] explained the distinction indicated above by a much larger length of epoxy polymer segment in the second case.

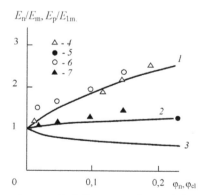

FIGURE 6.27 The dependences of reinforcement degree E_n/E_m and $E_p/E_{1m.}$ on the contents of nanofiller φ_n and nanoclusters φ_{cl}, accordingly. 1÷3 – theoretical dependences (E_n/E_m) (φ_n), corresponding to the Eqs. (15) ÷ (17); 4, 5 – the experimental data $(E_p/E_{1m.})$ for Par at $T = T_g$ ÷ T_g (4) and $T < T'_g$ (5); 6, 7 – the experimental data (E_n/E_m) (φ_n) for EP/MMT at $T > T_g$ (6) and $T < T_g$ (7) [59].

To obtain the similar comparison for natural nanocomposite (polymer) is impossible, since at $T \geq T_g$ nanoclusters are disintegrated and polymer ceases to be quasi-two-phase system [5]. However, within the frameworks of two-stage glass transition concept [11] it has been shown, that at temperature T'_g, which is approximately equal to $T_g - 50$ K, instable (small) nanoclusters decay occurs, that results to loosely packed matrix devitrification at the indicated temperature [5]. Thus, within the range of temperature $T'_g \div T_g$ natural nanocomposite (polymer) is an analog of nanocomposite with glassy matrix [58]. As one can see, for the temperatures within the range of $T = T'_g \div T_g$ ($\varphi_{cl} = 0.060.19$) the value $E_p/E_{l.m}$ corresponds to the Eq. (15), that is, perfect adhesion nanoclusters-loosely packed matrix and at $T < T'_g$ ($\varphi_{cl} > 0.24$) – to the Eq. (16), that is, to zero adhesional strength at a large friction coefficient. Hence, the data of Fig. 6.27 demonstrated clearly the complete similarity, both qualitative and quantitative, of natural (Par) and artificial (EP/MMT) nanocomposites reinforcement degree behavior. Another microcomposite model (e.g., accounting for the layered silicate particles strong anisotropy) application can change the picture quantitatively only. The data of Fig. 6.27 qualitatively give the correspondence of reinforcement degree of nanocomposites indicated classes at the identical initial conditions.

Hence, the analogy in behavior of reinforcement degree of polyarylate by nanoclusters and nanocomposite epoxy polymer/Na$^+$-montmorillonite by layered silicate gives another reason for the consideration of polymer as natural nanocomposite. Again strong influence of interfacial (intercomponent) adhesion level on nanocomposites of any class reinforcement degree is confirmed [17].

6.5   THE METHODS OF NATURAL NANOCOMPOSITES NANOSTRUCTURE REGULATION

As it has been noted above, at present it is generally acknowledged [2], that macromolecular formations and polymer systems are always natural nanostructural systems in virtue of their structure features. In this connection the question of using this feature for polymeric materials properties and operating characteristics improvement arises. It is obvious enough that for structure-properties relationships receiving the quantitative nanostructural model of the indicated materials is necessary. It is also obvious that if the dependence of specific property on material structure state is unequivocal, then there will be quite sufficient modes to achieve this state. The cluster model of such state [3–5] is the most suitable for polymers amorphous state structure description. It has been shown, that this model basic structural element (cluster) is nanoparticles (nanocluster) (see Section 6.15.1). The cluster model was used successfully for cross-linked polymers structure and properties description [61]. Therefore, the authors [62] fulfilled nanostructures regulation modes and the latter influence on rarely cross-linked epoxy polymer properties study within the frameworks of the indicated model.

In Ref. [62], the studied object was an epoxy polymer on the basis of resin UP5–181, cured by iso-methyltetrahydrophthalic anhydride in the ratio by mass 1:0.56. Testing specimens were obtained by the hydrostatic extrusion method. The indicated method choice is due to the fact, that high hydrostatic pressure imposition in deformation process prevents the defects formation and growth, resulting to the material failure [64]. The extrusion strain ε_e was calculated and makes up 0.14, 0.25, 0.36, 0.43 and 0.52. The obtained by hydrostatic extrusion specimens were annealed at maximum temperature 353 K during 15 min.

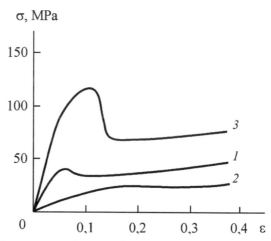

FIGURE 6.28   The stress – strain ($\sigma - \varepsilon$) diagrams for initial (1), extruded up to $\varepsilon_e = 0.52$ (2) and annealed (3) REP samples [62].

The hydrostatic extrusion and subsequent annealing of rarely cross-linked epoxy polymer (REP) result to very essential changes of its mechanical behavior and properties, in addition unexpected ones enough. The qualitative changes of REP mechanical behavior can be monitored according to the corresponding changes of the stress – strain ($\sigma - \varepsilon$) diagrams, shown in Fig. 6.28. The initial REP shows the expected enough behavior and both its elasticity modulus E and yield stress σ_Y are typical for such polymers at testing temperature T being distant from glass transition temperature T_g on about 40 K [51]. The small (≈ 3 MPa) stress drop beyond yield stress is observed, that is also typical for amorphous polymers [61]. However, REP extrusion up to $\varepsilon_e = 0.52$ results to stress drop $\Delta\sigma_Y$ ("yield tooth") disappearance and to the essential E and σ_Y reduction. Besides, the diagram $\sigma - \varepsilon$ itself is now more like the similar diagram for rubber, than for glassy polymer. This specimen annealing at maximum temperature $T_{an} = 353$ K gives no less strong, but diametrically opposite effect – yield stress and elasticity modulus increase sharply (the latter in about twice in comparison with the initial REP and more than one order in comparison with the extruded specimen). Besides, the strongly pronounced "yield tooth" appears. Let us note, that specimen shrinkage at annealing is small ($\approx$10%), that makes up about 20% of ε_e [62].

The common picture of parameters E and σ_Y change as a function of ε_e is presented in Figs. 6.29 and 6.30 accordingly. As one can see, both indicated parameters showed common tendencies at ε_e change: up to $\varepsilon_e \approx 0.36$ inclusive E and σ_Y weak increase at ε_e growth is observed, moreover their absolute values for extruded and annealed specimens are close, but at $\varepsilon_e > 0.36$ the strongly pronounced antibatness of these parameters for the indicated specimen types is displayed. The cluster model of polymers amorphous state structure and developed within its frameworks polymers yielding treatment allows to explain such behavior of the studied samples [35, 65].

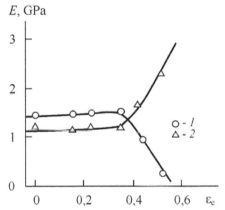

FIGURE 6.29   The dependences of elasticity modulus E_p on extrusion strain ε_e for extrudated (1) and annealed (2) REP [62].

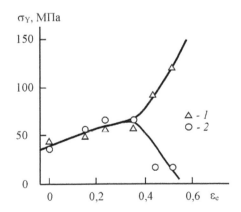

FIGURE 6.30   The dependences of yield stress σ_Y on extrusion strain ε_e for extrudated (1) and annealed (2) REP [62].

The cluster model supposes that polymers amorphous state structure represents the local order domains (nanoclusters), surrounded by loosely packed matrix. Nanoclusters consist of several collinear densely packed statistical segments of different macromolecules and in virtue of this they offer the analog of crystallite with stretched chains. There are two types of nanoclusters – stable, consisting of a relatively large segments number, and instable, consisting of a less number of such segments [65]. At temperature increase or mechanical stress application the instable nanoclusters disintegrate in the first place, that results to the two well-known effects. The first from them is known as two-stage glass transition process [11] and it supposes that at $T'_g = T_g - 50$ K disintegration of instable nanoclusters, restraining loosely packed matrix in glass state, occurs that defines devitrification of the latter [3, 5]. The well-known rapid polymers mechanical properties reduction at approaching to T_g [51] is the consequence of this. The second effect consists of instable nanoclusters decay at σ_Y under mechanical stress action, loosely packed matrix mechanical devitrification and, as consequence, glassy polymers rubber-like behavior on cold flow plateau [65]. The stress drop $\Delta\sigma_Y$ beyond yield stress is due to just instable nanoclusters decay and therefore $\Delta\sigma_Y$ value serves as characteristic of these nanoclusters fraction [5]. Proceeding from this brief description, the experimental results, adduced in Figs. 28 ÷ 30, can be interpreted.

The rarely cross-linked epoxy polymer on the basis of resin UP5–181 has low glass transition temperature T_g, which can be estimated according to shrinkage measurements data as equal ≈ 333K. This means, that the testing temperature $T = 293$ K and T'_g for it are close, that is confirmed by small $\Delta\sigma_Y$ value for the initial REP. It assumes nanocluster (nanostructures) small relative fraction φ_{cl} [3–5] and, since these nanoclusters have arbitrary orientation, ε_e increase results rapidly enough to their decay, that induces loosely packed matrix mechanical devitrification at $\varepsilon_e > 0.36$. Devitrificated loosely packed matrix gives insignificant contribution to E_p [66, 67], equal practically to zero, that results to sharp (discrete) elasticity modulus decrease. Besides, at $T > T'_g$ φ_{cl} rapid decay is observed, that is, segments number decrease in both stable and instable nanocluster [5]. Since just these parameters (E and φ_{cl}) check σ_Y value, then their decrease defines yield stress sharp lessening. Now extruded at $\varepsilon_e > 0.36$ REP presents as matter of fact rubber with high cross-linking degree, that is reflected by its diagram $\sigma - \varepsilon$ (Fig. 6.28, curve 2).

The polymer oriented chains shrinkage occurs at the extruded REP annealing at temperature higher than T_g. Since this process is realized within a narrow temperature range and during a small time interval, then a large number of instable nanoclusters is formed. This effect is intensified by available molecular orientation, that is, by preliminary favorable segments arrangement, and it is reflected by $\Delta\sigma_Y$ strong increase (Fig. 6.28, curve 3).

The φ_{cl} enhancement results to E_p growth (Fig. 6.29) and φ_{cl} and E_p combined increase – to σ_Y considerable growth (Fig. 6.30).

The considered structural changes can be described quantitatively within the frameworks of the cluster model. The nanoclusters relative fraction ϕ_{cl} can be calculated according to the method, stated in Ref. [68].

The shown in Fig. 6.31 dependences ϕ_{cl} (ε_e) have the character expected from the adduced above description and are its quantitative conformation. The adduced in Fig. 6.32 dependence of density ρ of REP extruded specimens on ε_e is similar to the dependence ϕ_{cl} (ε_e), that was to be expected, since densely packed segments fraction decrease must be reflected in ρ reduction.

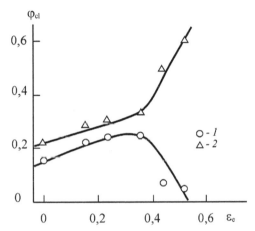

FIGURE 6.31   The dependences of nanoclusters relative fraction ϕ_{cl} on extrusion strain ε_e for extruded (1) and annealed (2) REP [62].

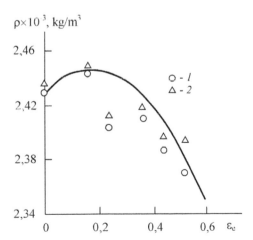

FIGURE 6.32   The dependence of specimens density ρ on extrusion strain ε_e for extruded (1) and annealed (2) REP [62].

In Ref. [69] the supposition was made that ρ change can be conditioned to microcracks network formation in specimen that results to ρ reduction at large ε_e (0.43 and 0.52), which are close to the limiting ones. The ρ relative change ($\Delta\rho$) can be estimated according to the equation

$$\Delta\rho = \frac{\rho^{max} - \rho^{min}}{\rho^{max}}, \tag{33}$$

where ρ^{max} and ρ^{min} are the greatest and the smallest density values. This estimation gives $\Delta\rho \approx 0.01$. This value can be reasonable for free volume increase, which is necessary for loosely matrix devitrification (accounting for closeness of T and T'_g), but it is obviously small if to assume as real microcracks formation. As the experiments have shown, REP extrusion at $\varepsilon_e > 0.52$ is impossible owing to specimen cracking during extrusion process. This allows to suppose that value $\varepsilon_e = 0.52$ is close to the critical one. Therefore, the critical dilatation $\Delta\delta_{cr}$ value, which is necessary for microcracks cluster formation, can be estimated as follows [40]:

$$\Delta\delta_{cr} = \frac{2(1+v)(2-3v)}{11-19v}, \tag{34}$$

where v is Poisson's ratio.

Accepting the average value $v \approx 0.35$, we obtain $\Delta\delta_{cr} = 0.60$, that is essentially higher than the estimation $\Delta\rho$ made earlier. These calculations assume that ρ decrease at $\varepsilon_e = 0.43$ and 0.52 is due to instable nanoclusters decay and to corresponding REP structure loosening.

The stated above data give a clear example of large possibilities of polymer properties operation through its structure change. From the plots of Fig. 6.29, it follows that annealing of REP extruded up to $\varepsilon_e = 0.52$ results to elasticity modulus increase in more than 8 times and from the data of Fig. 6.30 yield stress increase in 6 times follows. From the practical point of view the extrusion and subsequent annealing of rarely cross-linked epoxy polymers allow to obtain materials, which are just as good by stiffness and strength as densely cross-linked epoxy polymers, but exceeding the latter by plasticity degree. Let us note, that besides extrusion and annealing other modes of polymers nanostructure operation exist: plasticization [70], filling [26, 71], films obtaining from different solvents [72] and so on.

Hence, the stated above results demonstrated that neither cross-linking degree nor molecular orientation level defined cross-linked polymers final properties. The factor, controlling properties is a state of suprasegmental (nanocluster) structure, which, in its turn, can be goal-directly regulated by molecular orientation and thermal treatment application [62].

In the stated above treatment not only nanostructure integral characteristics (macromolecular entanglements cluster network density v_{cl} or nanocluster relative fraction φ_{cl}), but also separate nanocluster parameters are important (see Section

6.1). In this case of particulate-filled polymer nanocomposites (artificial nanocomposites) it is well known that their elasticity modulus sharply increases at nanofiller particles size decrease [17]. The similar effect was noted above for REP, subjected to different kinds of processing (see Fig. 6.28). Therefore, the authors [73] carried out the study of the dependence of elasticity modulus E on nanoclusters size for REP.

It has been shown earlier on the example of PC, that the value E_p is defined completely by natural nanocomposite (polymer) structure according to the Eq. (32) (see Fig. 6.26)

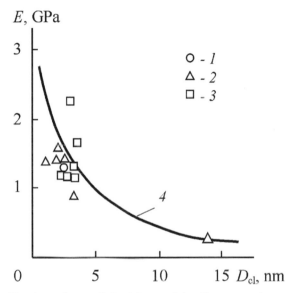

FIGURE 6.33   The dependence of elasticity modulus E_p on nanoclusters diameter D_{cl} for initial (1), extruded (2) and annealed (3) REP. 4 – calculation according to the Eq. (32) [73].

In Fig. 6.33, the dependence of E_p on nanoclusters diameter D_{cl}, determined according to the equation (in Chapter 5), for REP subjected to the indicated processing kinds at ε_e values within the range of $0.16 \div 0.52$ is adduced. As one can see, like in the case of artificial nanocomposites, for REP strong (approximately of order of magnitude) growth is observed at nanoclusters size decrease from 3 up to 0.9 nm. This fact confirms again, that REP elasticity modulus is defined by neither cross-linking degree nor molecular orientation level, but it depends only on epoxy polymer nanocluster structure state, simulated as natural nanocomposite [73].

Another method of the theoretical dependence $E_p (D_{cl})$ calculation for natural nanocomposites (polymers) is given in Ref. [74]. The authors [75] have shown, that the elasticity modulus E value for fractal objects, which are polymers [4], is given by the following percolation relationship:

$$K_T, G \sim (p - p_c)^\eta, \tag{35}$$

where K_T is bulk modulus, G is shear modulus, p is solid-state component volume fraction, p_c is percolation threshold, η is exponent.

The following equation for the exponent η was obtained at a fractal structure simulation as Serpinsky carpet [75]:

$$\frac{\eta}{v_p} = d - 1, \tag{36}$$

where v_p is correlation length index in percolation theory, d is dimension of Euclidean space, in which a fractal is considered.

As it is known [4], the polymers nanocluster structure represents itself the percolation system, for which $p = \varphi_{cl}$, $p_c = 0.34$ [35] and further it can be written:

$$\frac{R_{cl}}{l_{st}} \sim \left(\phi_{cl} - 0.34\right)^{i_p}, \tag{37}$$

where R_{cl} is the distance between nanoclusters, determined according to the Eq. (4.63), l_{st} is statistical segment length, v_p is correlation length index, accepted equal to 0.8 [77].

Since in the considered case the change E_p at n_{cl} variation is interesting first of all, then the authors [74] accepted $v_{cl} = \text{const} = 2.5 \times 10^{27} \, \text{m}^{-3}$, $l_{st} = \text{const} = 0.434$ nm. The value E_p calculation according to the Eqs. (35) and (37) allows to determine this parameter according to the formula [74]:

$$E_p = 28.9 \left(\phi_{cl} - 0.34\right)^{(d-1)\,p}, \, \text{GPa}. \tag{38}$$

In Fig. 6.34, the theoretical dependence (a solid line) of E_p on nanoclusters size (diameter) D_{cl}, calculated according to the Eq. (38) is adduced. As one can see, the strong growth E_p at D_{cl} decreasing is observed, which is identical to the shown one in Fig. 6.33. The adduced in Fig. 6.34 experimental data for REP, subjected to hydrostatic extrusion and subsequent annealing, correspond well enough to calculation according to the Eq. (38). The decrease D_{cl} from 3.2 up to 0.7 nm results again to E_p growth on order of magnitude [74].

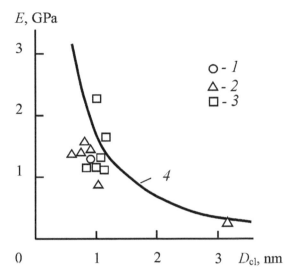

FIGURE 6.34 The dependence of elasticity modulus E_p on nanoclusters diameter D_{cl} for initial (1), extruded (2) and annealed (3) REP. 4 – calculation according to the Eq. (38) [74].

The similar effect can be obtained for linear amorphous polycarbonate (PC) as well. Calculation according to the Eq. (38) shows, n_{cl} reduction from 16 (the experimental value n_{cl} at $T = 293K$ for PC [5]) up to 2 results to E_p growth from 1.5 up to 5.8 GPa and making of structureless ($n_{cl} = 1$) PC will allow to obtain $E_p \approx 9.2$ GPa, that is, comparable with obtained one for composites on the basis of PC.

6.6 CONCLUSION

The results give purely practical aspect of such theoretical concepts as the cluster model of polymers amorphous state stricture and fractal analysis application for the description of structure and properties of polymers, treated as natural nanocomposites. The necessary nanostructure goal-directed making will allow to obtain polymers, not yielding (and even exceeding) by their properties to the composites, produced on their basis. Structureless (defect-free) polymers are imagined the most perspective in this respect. Such polymers can be natural replacement for a large number of elaborated at present polymer nanocomposites. The application of structureless polymers as artificial nanocomposites polymer matrix can give much larger effect. Such approach allows to obtain polymeric materials, comparable by their characteristics with metals (e.g., with aluminum).

KEYWORDS

- **components**
- **hybrid polymer systems**
- **mechanism**
- **models**
- **natural composites**
- **polymers**
- **properties**
- **semicrystalline**
- **structure**

REFERENCES

1. Kardos, I. L., & Raisoni, I. (1975). The Potential Mechanical Response of Macro Molecular Systems F Composite Analogy, Polymer Engineering Science, *15(3)*, 183–189.
2. Ivanches, S. S., & Ozerin, A. N. (2006). A Nanostructures in Polymeric Systems, Vysokomolek Soed B, *48(8)*, 1531–1544.
3. Kozlov, G. V., & Novikov, V. U. (2001). The Cluster Model of Polymers Amorphous State, Uspekhi Fizicheskikh Nauk, *171(7)*, 717–764.
4. Kozlov, G. V., & Zaikov, G. E. (2004). Structure of the Polymer Amorphous State, utrecht, Boston, Brill Academic Publishers, 465p.
5. Kozlov, G. V., Ovcharenko, E. N., & Mikitaev, A. K. (2009). Structures of the Polymer Amorphous State, Moscow, Publishers of the Mendeleev, D. I., RKhTU, 392p.
6. Kozlov, G. V., & Novikov, V. U. (1998). Synergetics and Fractal Analysis of Cross-Linked Polymers, Moscow, Klassika, 112p.
7. Burya, A. I., Kozlov, G. V., Novikov, V. U., & Ivanova, V. S. (2003). Synergetics of Super Segmental Structure of Amorphous Glassy Polymers Matter of 3rd International Conference "Research and Development in Mechanical Industry RaDMI-03", September 19–23, Herceg Novi, Serbia and Montenegro, 645–647.
8. Bashorov, M. T., Kozlov, G. V., & Mikitaev, A. K. (2010). Nano Structures and Properties of Amorphous Glassy Polymers, Moscow, Publishers of the Mendeleev, D. I. RKhTU, 269p.
9. Malamatov, Kh A., Kozlov, G. V., & Mikitaev, M. A. (2006). Reinforcements Mechanisms of Polymers Nano Composites, Moscow, Publishers of the Mendeleev, D. I. RKhTU, 240p.
10. Kozlov, G. V., Gazaev, M. A., Novikov, V. U., & Mikitaev, A. K. (1996). Simulations of Amorphous Polymers Structures As Percolations Cluster, Pis'ma v ZhTF, *22(16)*, 31–38.
11. Belousov, V. N., Kotsev, Kh. B., & Mikitaev, A. K. (1983). Two-step of Amorphous Polymers Glass Transition Doklady ANSSSR, *270(5)*, 1145–1147.
12. Ivanova, V. S., Kuzeev, I. R., & Zakirnichnaya, M. M. (1998). Synergetics and Fractals Universality of Metal Mechanical Behaviours, Ufa, Publishers of UGNTU, 366p.
13. Berstein, V. A., & Egorov, V. M. (1990). Differential Scanning Calorimetry in physics-Chemistry of the Polymers, Leningrad, Khimiya, 256p.
14. Bashorov, M. T., Kozlov, G. V., & Mikitaev, A. K. (2009). A Nano clusters Synergetics in Amorphous Glassy Polymers Structure, Inzhenernaya Fizika, *4*, 39–42.

15. Bashorov, M. T., Kozlov, G. V., & Mikitaev, A. K. (2009). A Nano structures in Polymers Formation synergetics, Regulation Methods and Influence on the properties, Material Ovedenie, 9, 39–51.
16. Shevchenko, Ya V., & Bal'makov, M. D. (2002). A Particles-Centravs as Nano world Objects, Fizika I Khimiya Stekla, 28(6), 631–636
17. Mikitaev, A. K., Kozlov, G. V., & Zaikov, G. E. (2008). Polymer Nano Composites, variety of Structural Forms and Applications, New York, Nova Science Publishers, Inclusive, 318p.
18. Buchachenko, A. L. (2003). The Nano Chemistry Direct Way to High Technologies of New Century, Uspekhi Khimii, 72(5), 419–437.
19. Formanis, G. E. (2003). Self-Assembly of Nanoparticles is Nano world Special Properties Spite, Proceedings of International Interdisciplinary Symposium "Fractals and Applied Synergetics", FiPS-03", Moscow, Publishers of MGOU, 303–308.
20. Bashorov, M. T., Kozlov, G. V., Shustov, G. B., & Mikitaev, A. K. (2009). The Estimation of Fractal Dimension of Nano clusters Surface in Polymers, Izvestiya Vuzov, Severo-Kavkazsk, Region, Estestv Nauki, 6, 44–46.
21. Magomedov, G. M., & Kozlov, G. V. (2010). Synthesis Structure and Properties of Cross-Linked Polymers and Nano Composites on its Basis Moscow, Publishers of Natural Sciences Academy, 464p.
22. Kozlov, G. V. (2011). Polymers as Natural Nano composites The Missing Opportunities, Recent Patents on Chemical Engineering, 4(1), 53–77.
23. Bovenko, V. N., & Startsev, V. M. (1994). The Discretely-Wave Nature of Amorphous Poliimide Supra Molecular Organization, Vysokomolek Soed B, 36(6), 1004–1008.
24. Bashorov, M. T., Kozlov, G. V., & Mikitaev, A. K. September (2009). Polymers as Natural Nano Composites, an Interfacial Regions Identifications, Proceedings of 12th International Symposium "Order, Disorder and Oxides Properties" Rostov-Na-Donu-Loo, 17–22, 280–282.
25. Magomedov, G. M., Kozlov, G. V., & Amirshikhova, Z. M. (2009). Cross-linked Polymers as Natural Nano Composites, an Interfacial Regions Identifications, Izvestiya DGPU, Estestv I Tochn Nauki, 4, 19–22.
26. Kozlov, G. V., Yanovskii, Yu G., & Zaikov, G. E. (2010). Structure and Properties of Particulate-Filled Polymer Composites, the Fractal Analysis, New York, Nova science Publishers, Inclusive, 282p.
27. Bashorov, M. T., Kozlov, G. V., Shustov, G. B., & Mikitaev, A. K. (2009). Polymers as Natural Nano Composites, the Filling Degree Estimations, Fundamental'nye Issledovaniya, 4, 15–18.
28. Vasserman, A. M., & Kovarskii, A. L. (1986). A Spin Probes and Labels in Physics-chemistry of Polymers, Moscow, Nauka, 246p.
29. Korst, N. N., & Antsiferova, L. I. (1978). A Slow Molecular Motions Study by Stable Radicals EPR Method, Uspekhi Fizicheskikh Nauka, 126(1), 67–99.
30. Yech, G. S. (1979). The General Notions on Amorphous Polymers Structure, Local Order and Chain Conformation Degrees, Vysokomolek Soed A, 21(11), 2433–2446.
31. Perepechko, I. I. (1978). Introduction in Physics of Polymer, Moscow, Khimiya, 312p.
32. Kozlov, G. V., & Zaikov, G. E. (2001). The Generalized Description of Local Order in Polymers, in Book Fractals and Local Order in Polymeric Materials, Kozlov, G. V., Zaikov, G. E. (Ed), New York, Nova Science Publishers, Inclusive, 55–63.
33. Tager, A. A. (1978). Physics-chemistry of Polymers Moscow, Khimiya, 416p.
34. Bashorov, M. T., Kozlov, G. V., Malamatov, Kh A., & Mikitaev, A. K. (2008). Amorphous Glassy Polymers Reinforcement Mechanisms by Nanostructures, Matter of IV International Science-Practical Conference "New Polymer Composite Materials" Nal'chik, KBSU, 47–51.
35. Bobryshev, A. N., Koromazov, V. N., Babin, L. O., & Solomatov, V. I. (1994). Synergetics of Composite Materials, lipetsk, NPO ORIUS, 154p.

36. Aphashagova Kh Z., Kozlov, G. V., Burya, A. T., & Mikitaev, A. K. (2007). The Prediction of Particulate-filled Polymer Nano Composites Reinforcement Degree, Material Ovedenie, *(9)*, 10–13.

37. Sheng N., Boyce, M. C., Parks, D. M., Rutledge, G. C., Ales, J. I., & Cohen, R. E. (2004). Multi scale Micromechanical Modeling of Polymer/Clay Nano composites and the Effective Clay Particle, Polymer, *45(2)*, 487–506.

38. Dickie, R. A. (1980). The Mechanical Properties (Small Strains) of Multiphase Polymer Blends, in Book Polymer Blends, Paul, D. R., Newman S. (Ed) New York, San Francisko, London, Academic Press, *1*, 397–437.

39. Ahmed, S., & Jones, F. R. (1990). A Review of Particulate Reinforcement Theories for Polymer Composites, Journal Matter Science, *25(12)*, 4933–4942.

40. Balankin, A. S. (1991). Synergetics of Deformable Body Moscow, Publishers of Ministry Defences SSSR, 404p.

41. Bashorov, M. T., Kozlov, G. V., & Mikitaev, A. K. (2010). Polymers as Natural Nano composites, Description of Elasticity Modulus within the Frameworks of Micromechanical Models, Plastics Massy, *11*, 41–43

42. Lipatov, Yu S. (1980). Interfacial Phenomena in Polymers, kiev, Naukova Dumka, 260p.

43. Yanovskii, Yu G., Bashorov, M. T., Kozlov, G. V., & Karnet, Yu N. (2012). Polymeric Mediums as Natural Nano Composites, Inter Component Interactions Geometry, Proceedings of All-Russian Conference "Mechanics and Nano Mechanics of Structurally-Complex and Heterogeneous Mediums Achievements, Problems, Perspectives" Moscow, IPROM, 110–117.

44. Tugov, I. I., & Shaulov, Yu A. (1990). A Particulate-Filled Composites Elasticity Modulus, Vysokomolek Soed B, *32(7)*, 527–529.

45. Piggott, M. R., & Leidner, Y. (1974). Micro Conceptions about Filled Polymers Y Applied Polymer Science, *18(7)*, 1619–1623.

46. Chen, Z. Y., Deutch, Y. M., & Meakin, P. (1984). Translational Friction Coefficient of Diffusion limited Aggregates Y Chemistry Physics, *80(6)*, 2982–2983.

47. Kozlov, G. V., Beloshenko, V. A., & Varyukhin, V. N. (1998). Simulation of Cross-Linked Polymers Structure as Diffusion-Limited Aggregate, Ukrainskii Fizicheskii Zhurnal, *43(3)*, 322–323.

48. Novikov, V. U., Kozlov, G. V., & Burlyan, O. Y. (2000). The Fractal Approach to Interfacial Layer in Filled Polymers, Mekhanika Kompozitnykh Materialov, *36(1)*, 3–32.

49. Stanley, E. H. (1986). A Fractal Surfaces and "Termite" Model for Two-Component Random Materials, In book Fractals in Physics Pietronero, L., Tosatti, E. (Ed) Amsterdam, Oxford, New York, Tokyo, North-Holland, 463–477.

50. Bashorov, M. T., Kozlov, G. V., Zaikov, G. E., & Mikitaev, A. K. (2009). Polymers as Natural Nano Composites, Adhesion between Structural Components, Khimicheskaya Fizika i Mezoskopiya, *11(2)*, 196–203.

51. Dibenedetto, A. T., & Trachte, K. L. (1970). The Brittle Fracture of Amorphous Thermoplastic Polymers Y Applied Polymer Science, *14(11)*, 2249–2262.

52. Burya, A. I., Lipatov, Yu S., Arlamova, N. T., & Kozlov, G. V. October 25 (2007). Patent by Useful Model N27 199, Polymer Composition, It is registered in Ukraine Patents State Resister.

53. Novikov, V. U., & Kozlov, G. V. (1999). Fractals Parameterizations of Filled Polymers Structure, Mekhanika Kompozitnykh Materialov, *35(3)*, 269–290.

54. Potapov, A. A. (2008). A Nano Systems Design Principles, Nano i Mikrosistemnaya Tekhnika, *3(4)*, 277–280.

55. Bashorov, M. T., Kozlov, G. V., Zaikov, G. E., & Mikitaev, A. K. (2009). Polymers as Natural Nanocomposites 3, the Geometry of Inter component Interactions, Chemistry and Chemical Technology, *3(4)*, 277–280.

56. Bashorov, M. T., Kozlov, G. V., Zaikov, G. E., & Mikitaev, A. K. (2009). Polymers as Natural Nano Composites 1, the Reinforcement Structural Model, Chemistry and Chemical Technology, *3(2)*, 107–110.

57. Edwards, D. C. (1990). Polymer-Filler Interactions in Rubber Reinforcement I Mater Science, *25(12)*, 4175–4185.

58. Bashorov, M. T., Kozlov, G. V., & Mikitaev, A. K. (2009). Polymers as Natural Nano Composites, the Comparative Analysis of Reinforcement Mechanism, Nanotekhnika, *4*, 43–45.

59. Bashorov, M. T., Kozlov, G. V., Zaikov, G. E., & Mikitaev, A. K. (2009). Polymers as Natural Nanocomposites 2 the Comparative Analysis of Reinforcement Mechanism, Chemistry and Chemical Technology, *3(3)*, 183–185.

60. Chen, Y. S., Poliks, M. D., Ober, C. K., Zhang Y., Wiesner U., & Giannelis E. (2002). Study of the Interlayer Expansion Mechanism and Thermal-Mechanical Properties of Surface-Initiated Epoxy Nanocomposites, Polymer, *43(17)*, 4895–4904.

61. Kozlov, G. V., Beloshenko, V. A., Varyukhin, V. N., & Lipatov, Yu S. (1999). Application of Cluster Model for the Description of Epoxy Polymers Structure and Properties, Polymer, *40(4)*, 1045–1051.

62. Bashorov, M. T., Kozlov, G. V., & Mikitaev, A. K. (2009). Nano structures in Cross-Linked Epoxy Polymers and their Influence on Mechanical Properties, Fizika I Khimiya Obrabotki Materialov, *(2)*, 76–80.

63. Beloshenko, V. A., Shustov, G. B., Slobodina, V. G., Kozlov, G. V., Varyukhin, V. N., Temiraev, K. B., & Gazaev, M. A., 13 June (1995). Patent on Invention "The Method of Rod-Like Articles Manufacture from Polymers" Claim for Invention Rights N95109832 Patent N2105670 Priority it is Registered in Inventions State Register of Russian Federation February 27 (1998).

64. Aloev, V. Z., & Kozlov, G. V., (2002). Physics of Orientational Phenomena in Polymeric Materials, nalchik, Polygraph-Service and T288p.

65. Kozlov, G. V., Beloshenko, V. A., Garaev, M. A., & Novikov, V. U. (1996). Mechanisms of Yielding and Forced High-Elasticity of Cross-Linked Polymers, Mekhanika Kompozitnykh Materialov, *32(2)*, 270–278.

66. Shogenov, V. N., Belousov, V. N., Potapov, V. V., Kozlov, G. V., & Prut, E. V. (1991). The Glassy Polyarylate surfone Curves Stress–strain Description within the Frameworks of High-Elasticity Concepts, Vysokomolek Soed F, *33(1)*, 155–160.

67. Kozlov, G. V., Beloshenko, V. A., & Shogenov, V. N. (1999). The Amorphous Polymers Structural Relaxation Description within the Frameworks of the Cluster Model, Fiziko-Khimicheskaya Mekhanika Materialov, *35(5)*, 105–108.

68. Kozlov, G. V., Burya, A. I. & Shustov, G. B. (2005). The Influence of Rotating Electromagnetic Field on Glass Transition and Structure of Carbon Plastics on the Basis of lhenylone Fizika I Khimiya Obrabotki Materialov, *5*, 81–84.

69. Pakter, M. K., Beloshenko, V. A., Beresnev, B. I., Zaika, T. R., Abdrakhmanova, L. A., & Berai, N. I. (1990). Influences of Hydrostatic Processing on Densely Cross-Linked epoxy Polymers Structural Organization Formation, Vysokomolek Soed F, *32(10)*, 2039–2046.

70. Kozlov, G. V., Sanditov, D. S., & Lipatov, Yu S. (2001). Structural and Mechanical Properties of Amorphous Polymers in Yielding Region, in book Fractals and Local Order in Polymeric Materials, Kozlov, G. V., Zaikov, G. E. (Ed). New York, Nova Science Publishers, Inclusive, 65–82.

71. Kozlov, G. V., Yanovskii, Yu G., & Zaikov, G. E. (2011). Synergetics and Fractal Analysis of Polymer Composites Filled with Short Fibers, New York, Nova Science Publishers, Inclusive, 223p.

72. Shogenov, V. N., & Kozlov, G. V. (2002). Fractal Clusters in Physics-Chemistry of Polymers, Nal'chik, Polygraphservice and T, 270p.

73. Kozlov, G. V., & Mikitaev, A. K. (2010) Polymers as Natural Nano composites Unrealized Potential Saarbrücken, Lambert Academic Publishing, 323p.
74. Magomedov, G. M., Kozlov, G. V., & Zaikov, G. E. (2011). Structure and Properties of Cross-Linked Polymers, Shawbury, a smithers Group Company, 492p.
75. Bergman, D. Y., & Kantor, Y. (1984). Critical Properties of an Elastic Fractal Physical Revision Letters, *53(6),* 511–514.
76. Malamatov, Kh A., & Kozlov, G. V. (2005). The Fractal model of Polymer-Polymeric Nanocomposites Elasticity, Proceedings of Fourth International Interdisciplinary Symposium "Fractals and Applied Synergetics FaAS-05", Moscow, Interkontakt Nauka, 119–122.
77. Sokolov, I. M. (1986). Dimensions and Other Geometrical Critical Exponents in Percolation Theory, Uspekhi Fizicheskikh Nauka, *151(2),* 221–248.

CHAPTER 7

A CLUSTER MODEL OF POLYMERS AMORPHOUS: AN ENGINEERING INSIGHT

G. V. KOZLOV[1], I. V. DOLBIN[1], JOZEF RICHERT[2],
O. V. STOYANOV[3], and G. E. ZAIKOV[4]

[1]Kabardino-Balkarian State University, Nal'chik – 360004, Chernyshevsky st., 173, Russian Federation; E-mail: I_dolbin@mail.ru

[2]Institut Inzynierii Materialow Polimerowych I Barwnikow, 55 M. Sklodowskiej-Curie str., 87-100 Torun, Poland; E-mail: j.richert@impib.pl

[3]Kazan National Research Technological University, Kazan, Tatarstan, Russia; F-mail: OV_Stoyanov@mail.ru

[4]N. M. Emanuel Institute of Biochemical Physics of Russian Academy of Sciences, Moscow 119334, Kosygin st., 4, Russian Federation; E-mail: Chembio@sky.chph.ras.ru

CONTENTS

7.1   Introduction 210
7.2   The Model 210
7.3   The Approach 210
Keywords 248
References 248

7.1   INTRODUCTION

A cluster model of polymers amorphous state structure allows to introduce principally new treatment of structure defect (in the full sense of this term) for the indicated state [1, 2]. As it is known [3], real solids structure contains a considerable number of defects. The given concept is the basis of dislocations theory, widely applied for crystalline solids behavior description. Achieved in this field successes predetermine the attempts of authors number [4–11] to use the indicated concept in reference to amorphous polymers. Additionally used for crystalline lattices notions are often transposed to the structure of amorphous polymers. As a rule, the basis for this transposition serves formal resemblance of stress – strain (s – e) curves for crystalline and amorphous solids.

7.2   THE MODEL

In relation to the structure of amorphous polymers for a long time the most ambiguous point [12–14] was the presence or the absence of the local (short-range) order in this connection points of view of various authors on this problem were significantly different. The availability of the local order can significantly affect the definition of the structure defect in amorphous polymers, if in the general case the order-disorder transition or vice versa is taken for the defect. For example, any violation (interruption) of the long-range order in crystalline solids represents a defect (dislocation, vacancy, etc.), and a monocrystal with the perfect long-range order is the ideal defect-free structure with the perfect long-range order. It is known [15], that sufficiently bulky samples of 100% crystalline polymer cannot be obtained, and all characteristics of such hypothetical polymers are determined by the extrapolation method. That is why the Flory "felt" model [16, 17] can be suggested as the ideal defect-free structure for amorphous state of polymers. This model assumes that amorphous polymers consist of interpenetrating macromolecular coils personifying the full disorder (chaos). Proceeding from this, as the defect in polymers amorphous state a violation (interruption) of full disorder must be accepted, that is, formation of the local (or long-range) order [1]. It should also be noted that the formal resemblance of the curves s – e for crystalline solids and amorphous polymers appears far incomplete and the behavior of these classes of materials displays principal differences, which will be discussed in detail below.

7.3   THE APPROACH

Turning back to the suggested concept of amorphous polymer structural defect, let us note that a segment including in the cluster can be considered as the linear defect – the analog of dislocation in crystalline solids. Since in the cluster model the length of such segment is accepted equal to the length of statistical segment, l_{st}, and their amount per volume unit is equal to the density of entanglements cluster network, v_{cl}, then the density of linear defects, ρ_d, per volume unit of the polymer can be expressed as follows [1]:

$$\rho_d = v_{cl} \cdot l_{st}.$$

(1)

The offered treatment allows application of well-developed mathematical apparatus of the dislocation theory for the description of amorphous polymers properties. Its confirmation by the X-raying methods was stated in Ref. [18].

Further on, the rightfulness of application of the structural defect concept to polymers yielding process description will be considered. As a rule, previously assumed concepts of defects in polymers were primarily used for the description of this process or even exclusively for this purpose [4–11]. Theoretical shear strength of crystals was first calculated by Frenkel, basing on a simple model of two atoms series, displaced in relation to one another by the shear stress (Fig. 7.1a) [3]. According to this model, critical shear stress τ_0 is expressed as follows [3]:

$$\tau_0 = \frac{G}{2\pi},$$

(2)

where G is the shear modulus.

Slightly changed, this model was used in the case of polymers yielding [6], from where the following equation was obtained:

$$\tau_{0Y} = \frac{G}{\pi\sqrt{3}},$$

(3)

where τ_{0Y} is a theoretical value of the shear stress at yielding.

Special attention should be paid to the fact that characterizes principally different behavior of crystalline metals compared with polymers. As it is known [3, 19], τ_{0Y}/τ_Y ratio (where τ_Y is experimentally determined shear stress at yielding) is much higher for metals than for polymers. For five metals possessing the face-centered cubic or hexagonal lattices the following ratios were obtained: $\tau_{0Y}/\tau_Y = 37,400$, 22,720 (according to the data of Ref. [3]), whereas for five polymers this ratio makes 2.9 , 6.3 [6]. In essence, sufficient closeness of τ_{0Y}/τ_Y ratio values to one may already be the proof for the possibility of realizing of Frenkel mechanism in polymers (in contrast with metals), but it will be shown below that for polymers a small modification of the law of shear stress t periodic change used commonly gives τ_{0Y}/τ_Y values very close to one [20].

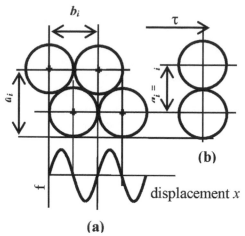

FIGURE 7.1   Schematic representation of deformation of two sequences of atoms according to the Frenkel model. Positions before (a) and after (b) deformation [3].

As it has been shown in Ref. [21], dislocation analogies are also true for amorphous metals. In essence, the authors [21] consider the atoms construction distortion (which induces appearance of elastic stress fields) as a linear defect (dislocation) being practically immovable. It is clear that such approach correlates completely with the offered above structural defect concept. Within the frameworks of this concept, Fig. 1a may be considered as a cross-section of a cluster (crystallite) and, hence, the shear of segments in the latter according to the Frenkel mechanism – as a mechanism limiting yielding process in polymers. This is proved by the experimental data [32], which shown that glassy polymers yielding process is realized namely in densely packed regions. Other data [23] indicate that these densely packed regions are clusters. In other words, one can state that yielding process is associated with clusters (crystallites) stability loss in the shear stresses field [24].

In Ref. [25], the asymmetrical periodic function is adduced, showing the dependence of shear stress t on shear strain γ_{sh} (Fig. 7.2). As it has been shown before [19], asymmetry of this function and corresponding decrease of the energetic barrier height overcome by macromolecules segments in the elementary yielding act are due to the formation of fluctuation free volume voids during deformation (that is the specific feature of polymers [26]). The data in Fig. 7.2 indicate that in the initial part of periodic curve from zero up to the maximum dependence of t on displacement x can be simulated by a sine-shaped function with a period shorter than in Fig. 7.1. In this case, the function t (x) can be presented as follows:

$$\tau(x) = k \sin\left(\frac{6\pi x}{b_i}\right), \qquad (4)$$

that is fully corresponds to Frenkel conclusion, except for arbitrarily chosen numerical coefficient in brackets (6 instead of 2).

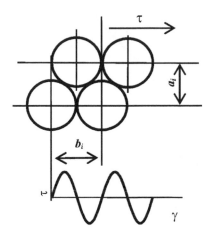

FIGURE 7.2 Schematic picture of shear deformation and corresponding stress – strain (t – γ_{ah}) function [25].

Further calculation of τ_o by method, described in Ref. [3], and its comparison with the experimental values τ_x indicated their close correspondence for nine amorphous and semicrystalline polymers (Fig. 7.3), which proves the possibility of realization of the above-offered yielding mechanism at the segmental level [18].

Inconsistency of τ_{oY} and τ_Y values for metals results to a search for another mechanism of yielding realization. At present, it is commonly accepted that this mechanism is the motion of dislocations by sliding planes of the crystal [3]. This implies that interatomic interaction forces, directed transversely to the crystal sliding plane, can be overcome in case ofthe presence of local displacements number, determined by stresses periodic field in the lattice. This is strictly different from macroscopic shear process, during which all bonds are broken simultaneously (the Frenkel model). It seems obvious that with the help of dislocations total shear strain will be realized at the applying much lower external stress than for the process including simultaneous breakage of all atomic bonds by the sliding plane [3].

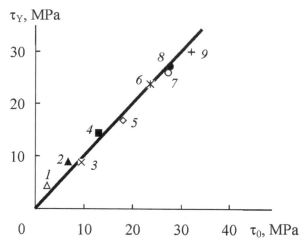

FIGURE 7.3 The relationship between theoretical τ_o and experimental τ_Y shear stresses at yielding for polytetrafluoroethylene (1), HDPE (2), polypropelene (3), polyamide-6 (4), poly (vinil chloride) (5), polyhydroxiester (6), PC (7), polysulfone (8) and PAr (9) [20].

Payerls and Nabarro [3] were the first who calculated the shear stress necessary for the dislocations motion, τ_{dm}. They used a sinusoidal approximation and deduced the expression for τ_{dm} as follows:

$$\tau_{dm} = \frac{2G}{1-v}e^{-2\pi a_i/b_i(1-v)}, \qquad (5)$$

where n is the Poisson's ratio and parameters a_i and b_i are of the same meaning as in Fig. 7.2.

By substituting reasonable n value, for example, 0.35 [27], and assuming $a_i = b_i$, the following value for τ_{dm} is obtained: $\tau_{dm} = 2 \cdot 10^{-4}G$. Though for metals this value is higher than the observed τ_Y, it is much closer to them than the stress calculated using simple shear model (the Frenkel model, Fig. 7.1).

However, for polymers the situation is opposite: analogous calculation indicates that their τ_{dm} does not exceed 0.2 MPa, which is by two orders of magnitude, approximately, lower than the observed τ_Y values.

Let us consider further the free path length of dislocations, λ_d. As it is known for metals [3], in which the main role in plastic deformation belongs to the mobile dislocations, λ_d assesses as $\sim 10^4$ Å. For polymers, this parameter can be estimated as follows [28]:

$$\lambda_Y = \frac{\varepsilon_Y}{b\rho_d}, \qquad (6)$$

where ε_Y is the yield strain, b is Burgers vector, ρ_d is the density if linear defects, determined according to the Eq. (1).

The value ε_Y assesses as ~0.10 [29] and the value of Burgers vector b can be estimated according to the equation [30]:

$$b = \left(\frac{60,7}{C_\infty}\right)^{1/2}, \text{Å.} \tag{7}$$

The values for different polymers, λ_d assessed by the Eq. (6) is about 2.5 Å. The same distance, which a segment passes at shearing, when it occupies the position, shown in Fig. 7.1b, that can be simply calculated from purely geometrical considerations. Hence, this assessment also indicates no reasons for assuming any sufficient free path length of dislocations in polymers rather than transition of a segment (or several segments) of macromolecule from one quasiequilibrium state to another [31].

It is commonly known [3, 25] that for crystalline materials Baily Hirsh relationship between shear stress, τ_Y, and dislocation density, ρ_d, is fulfilled:

$$\tau_Y = \tau_{in} + \alpha G b \rho_d^{1/2}, \tag{8}$$

where τ_{in} is the initial internal stress, a is the efficiency constant.

The Eq. (8) is also true for amorphous metals [21]. In Ref. [20] it was used for describing mechanical behavior of polymers on the example of these materials main classes representatives. For this purpose, the data for amorphous glassy PAr [32], semicrystalline HDPE [33] and cross-linked epoxy polymers of amine and anhydride curing types (EP) were used [34]. Different loading schemes were used: uniaxial tension of film samples [32], high-speed bending [33] and uniaxial compression [34]. In Fig. 7.4, the relations between calculated and experimental values τ_Y for the indicated polymers are adduced, which correspond to the Eq. (8). As one can see, they are linear and pass through the coordinates origin (i.e., $\tau_{in} = 0$), but a values for linear and cross-linked polymers are different. Thus in the frameworks of the offered defect concept the Baily Hirsh relationship is also true for polymers. This means that dislocation analogs are true for any linear defect, distorting the material ideal structure and creating the elastic stresses field [20]. From this point of view high defectness degree of polymers will be noted: $\rho_d \gg 10^9$, 10^{14} cm^{-2} for amorphous metals [21], and $\rho_d \gg 10^{14}$ cm^{-2} for polymers [28].

Hence, the stated above results indicate that in contrast with metals, for polymers realization of the Frenkel mechanism during yielding is much more probable rather than the defects motion (Fig. 7.1). This is due to the above-discussed (even diametrically opposed) differences in the structure of crystalline metals and polymers [1].

As it has been shown above using position spectroscopy methods [22], the yielding in polymers is realized in densely packed regions of their structure. Theoretical analysis within the frameworks of the plasticity fractal concept [35] demonstrates that the Poisson's ratio value in the yielding point, v_Y, can be estimated as follows:

$$v_Y = vc + 0.5\,(1 - c), \tag{9}$$

where n is Poisson's ratio value in elastic strains field, c is the relative fraction of elastically deformed polymer.

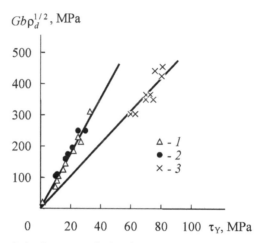

FIGURE 7.4 The relation between calculated and shears stress at yielding τ_Y, corresponding to the Eq. (8), for PAr (1), HDPE (2) and EP (3) [20].

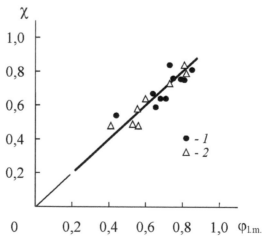

FIGURE 7.5 The relation between relative fraction of loosely packed matrix $\varphi_{l.m}$ and probability of elastic state realization c for PC (1) and PAr (2) [23].

In Fig. 7.5, the comparison of values c and $\varphi_{1.m}$ for PC and PAr is adduced, which has shown their good correspondence. The data of Fig. 7.5 assume that the loosely packed matrix can be identified as the elastic deformation region and clusters are identified as the region of inelastic (plastic) deformation [23]. These results prove the conclusion made in Ref. [22] about proceeding of inelastic deformation processes in dense packing regions of amorphous glassy polymers and indicate correctness of the plasticity fractal theory at their description.

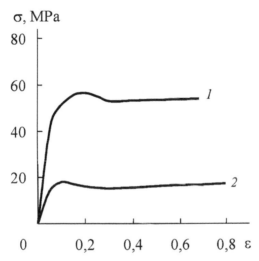

FIGURE 7.6 The stress-strain (s – e) diagrams for PAr at testing temperatures 293 (1) and 453 K (2) [24].

The yielding process of amorphous glassy polymers is often considered as their mechanical devitrification [36]. However, if typical stress – strain (s – e) plot for such polymers is considered (Fig. 7.6), then one can see, that behind the yield stress σ_Y the forced elasticity (cold flow) plateau begins and its stress σ_p is practically equal to σ_Y, that is, σ_p has the value of order of several tens MPa, whereas fordevitrificated polymer this value is, at least, on the order of magnitude lower. Furthermore, σ_p is a function of the temperature of tests T, whereas for devitrificated polymer such dependence must be much weaker what is important, possess the opposite tendency (σ_p enhancement at T increase). This disparity is solved easily within the frameworks of the cluster model, where cold flow of polymers is associated with devitrivicated loosely packed matrix deformation, in which clusters "are floating."However, thermal devitrification of the loosely packed matrix occurs at the temperature T_g' – which is approximately 50 K lower than T_g. That is why it should be expected that amorphous polymer in the temperature range $T_g' \div T_g$ will be subjected to yielding under the application of even extremely low stress (of about 1

MPa). Nevertheless, as the plots in Fig. 7.6 show, this does not occur and s – e curve for PAr in $T_g' \div T_g$ range is qualitatively similar to the plot s – e at $T<T_g\phi$ (Curve 1). Thus is should be assumed that devitrification of the loosely packed matrix is the consequence of the yielding process realization, but not its criterion. Taking into account realization of inelastic deformation process in the clusters (Fig. 7.5) one can suggest that the sufficient condition of yield in the polymer is the loss of stability by the local order regions in the external mechanical stress field, after which the deformation process proceeds without increasing the stress s (at least, nominal one), contrary to deformation below the yield stress, where a monotonous increase of s is observed (Fig. 7.6).

Now using the model suggested by the authors [37] one can demonstrate that the clusters lose their stability, when stress in the polymer reaches the macroscopic yield stress, σ_Y. Since the clusters are postulated as the set of densely packed collinear segments, and arbitrary orientation of cluster axes in relation to the applied tensile stress s should be expected, then they can be simulated as "inclined plates" (IP) [37], for which the following expression is true [37]:

$$\tau_Y < \tau_{IP} = 24G_{cl}\varepsilon_0\left(1+v_2\right)/\left(2-v_2\right), \tag{10}$$

where τ_Y is the shear stress in the yielding point, t_{IP} is the shear stress in IP (cluster), G_{cl} is the shear modulus, which is due to the clusters availability and determined from the plots.

Since the Eq. (10) characterizes inelastic deformation of clusters, the following can be accepted: $v_2 = 0.5$. Further on, under the assumption that $\tau_Y = t_{IP}$, the expression for the minimal (with regard to inequality in the left part of the Eq. (10)) proper strain ε_0^{min} is obtained [24]:

$$\varepsilon_0^{min} = \frac{\tau_Y}{\sqrt{24G_{cl}}} \tag{11}$$

The condition for IP (clusters) stability looks as follows [37]:

$$q = \sqrt{\frac{3}{2}} \cdot \frac{\varepsilon_0}{\tau_Y} \left\{ \left| 1 + \frac{\tilde{\varepsilon}_0}{\varepsilon_0} \right| - \sqrt{\frac{3}{8}} \frac{\tau_Y}{G_{cl}\varepsilon_0\left(1+v_2\right)} \right\}, \tag{12}$$

where q is the parameter, characterizing plastic deformation, $\tilde{\varepsilon}_0$ is the proper strain of the loosely packed matrix.

The cluster stability violation condition is fulfillment of the following inequality [37]:

$$q £ 0. \tag{13}$$

Comparison of the Eqs. (12) and (13) gives the following criterion of stability loss for IP (clusters) [24]:

$$\left| 1 + \frac{\tilde{\varepsilon}_0}{\varepsilon_0} \right| = \sqrt{\frac{3}{8}} \frac{\tau_Y^T}{G_{cl} \varepsilon_0 (1 + v_2)} \sqrt{b^2 - 4ac} \, , \qquad (14)$$

from which theoretical stress τ_Y (τ_Y^T) can be determined, after reaching of which the criterion (13) is fulfilled.

To perform quantitative estimations, one should make two simplifying assumptions [24]. Firstly, for IP the following condition is fulfilled [37]:

$$0 \le \sin^2 \theta_{IP} \left(\tilde{\varepsilon}_0 / \varepsilon_0 \right) \le 1, \qquad (15)$$

where q_{IP} is the angle between the normal to IP and the main axis of proper strain.

Since for arbitrarily oriented IP (clusters) $\sin^2 q_{IP} = 0.5$, then for fulfillment of the condition (15) the assumption is enough. Secondly, the Eq. (11) gives the minimal value of ε_o, and for the sake of convenience of calculations parameters τ_Y and G_{cl} were replaced by σ_Y and E, respectively. E value is greater than the elasticity modulus E_{cl} due to the availability of clusters. That is why to compensate two mentioned effects the strain ε_o, estimated according to the Eq. (11), was twice increased. The final equation looks as follows [24]:

$$\varepsilon_0 \approx 0,64 \frac{\sigma_Y}{E} = 0,64\varepsilon_{cl} \, , \qquad (16)$$

where ε_{cl} is the elastic component of macroscopic yield strain [38], which corresponds to strains ε_o and $\tilde{\varepsilon}_0$ by the physical significance [37].

Combination of the Eqs. (14) and (16) together with the plots similar to the ones shown before (see Chapter 1), where from E_{cl} (G_{cl}) can be determined, allows to estimate theoretical yield stress σ_Y^T and compare it with experimental values σ_Y. Such comparison is adduced in Fig. 7.7, whichdemonstrates satisfactory conformity between σ_Y^T and σ_Y that proves the suggestion made in Ref. [24] and justifies the above-made assumptions.

Hence, realization of the yielding process in amorphous glassy polymers requires clusters stability loss in the mechanical stress field, after which mechanical devitrification of the loosely packed matrix proceeds. Similar criterion was obtained for semicrystalline polymers [24].

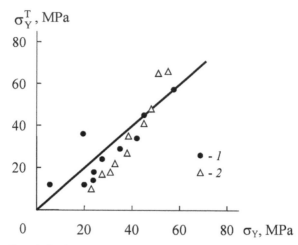

FIGURE 7.7   The relation between experimental σ_Y and calculated according to the Eq. (14) yield stress values for PAr (1) and PC (2) [24].

As results obtained in Refs. [34, 39] have shown, the behavior of cross-linked polymers is just slightly different from the above-described one for linear PC and PAr. However, further progress in this field is quite difficult due to, at least, two reasons: excessive overestimation of the chemical cross-links role and the quantitative structural model absence. In Ref. [39], the yielding mechanism of cross-linked polymer has been offered, based on the application of the cluster model and the latest developments in the deformable solid body synergetics field [40] on the example of two already above-mentioned epoxy polymers of amine (EP-1) and anhydrazide (EP-2) curing type.

Figure 7.8 shows the plots s – e for EP-2 under uniaxial compression of the sample up to failure (curve 1) and at successive loading up to strain e exceeding the yield strain ε_Y (curves 2–4). Comparison of these plots indicates consecutive lowering of the "yield tooth" under constant cold flow stress, σ_p. High values of σ_p assume corresponding values of stable clusters network density v_{cl}^{st}, which is much higher than the chemical cross-links network density v_c [34]. Thus though the behavior of a cross-linked polymer on the cold flow plateau is described within the frameworks of the rubber high-elasticity theory, the stable clusters network in this part of s – e plots is preserved. The only process proceeding is the decay of instable clusters, determining the loosely packed matrix devitrification. This process begins at the stress equal to proportionality limit that correlates with the data from Ref. [41], where the action of this stress and temperature $T_2 = T_g\phi$ is assumed analogous. The analogy between cold flow and glass transition processes is partial only: the only one component, the loosely packed matrix, is devitrivicated. Besides, complete decay of instable clusters occurs not in the point of yielding reaching at σ_Y, but at the beginning of cold flow plateau at σ_p. This can be observed from s – e diagrams shown in Fig. 7.8. As a consequence, the yielding is regulated not by the loosely packed matrix devitrification, but by other mechanism. As it is shown above, as such mechanism the stability loss by clusters in the mechanical stress field can be assumed, which also follows from the well-known fact of derivative ds/de turning to zero in the yield point [42]. According to the Ref. [40] critical shear strain g_* leading to the loss of shear stability by a solid is equal to:

$$\gamma_* = \frac{1}{mn},$$
(17)

where m and n are exponents in the Mie equation [27] setting the interconnection between the interaction energy and distance between particles. The value of parameter $1/mn$ can be expressed via the Poisson's ratio, n [27]:

$$\frac{1}{mn} = \frac{1-2v}{6(1+v)}$$
(18)

From the Eq. (18) it follows [18]:

$$\frac{1}{mn} = \gamma_* = \frac{\sigma_Y}{E}$$
(19)

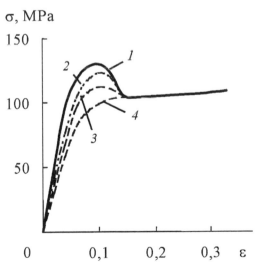

FIGURE 7.8   The stress – strain (s – e) diagrams at loading up to failure (1) and at cyclic load exertion (2–4) for EP-2: 2 – the first loading cycle; 3 – the second loading cycle; 4 – the third loading cycle [39].

The Eq. (19) gives the strain value with no regard to viscoelastic effect, that is, diagram s – e deviation from linearity behind the proportionality limit. Taking into account that tensile strain is twice greater, approximately, than the corresponding shear strain [42], theoretical yield strain ε_Y^T, corresponding to the stability loss by a solid, can be calculated. In Fig. 9 the comparison of experimental ε_Y and ε_Y^T yield strain magnitudes is fulfilled. Approximate equality of these parameters is observed that assumes association of the yielding with the stability loss by polymers. More precisely, we are dealing with the stability loss by clusters, because parameter n depends upon the cluster network density v_{cl} and ε_Y value is proportional to v_{cl} [32].

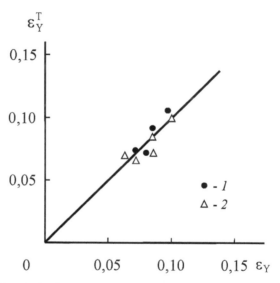

FIGURE 7.9   The relation between experimental ε_Y and theoretical ε_Y^T yield strain values for epoxy polymers EP-1 (1) and EP-2 (2) [39].

The authors [43] consider the possibility of intercommunication of polymers yield strain and these materials suprasegmental structure evolution, which is constituent part of hierarchical systems behavior [44, 45]. It is supposed that within the frameworks of this general concept polymer suprasegmental structures occupy their temporal and energetic "niches" in general hierarchy of real world structures [46]. As a structure quantitative model the authors [43] use a cluster model of polymers amorphous state structure [18, 24]. Ten groups of polymers, belonging to different deformation schemes in wide range of strain rates and temperatures, were used for obtaining possible greater results community. The yield strain ε_Y was chosen as the parameter, characterizing suprasegmental structures stability in the mechanical stresses field.

The Gibbs function of suprasegmental (cluster) structure self-assembly at temperature $T = T_g - \Delta T$ was calculated as follows [45]:

$$\Delta \tilde{G}^{im} = \Delta S \Delta T \, , \tag{20}$$

where ΔS is entropy change in this process course, which can be estimated as follows [47]:

$$\Delta S = (3 \, , 5) \cdot k \cdot f_g \cdot \ln f_g \tag{21}$$

In the Eq. (21), the coefficient $(3 \, , 5)$ takes into account conformational molecular changes contribution to ΔS, k is Boltzmann constant, f_g is a polymer relative fluctuation free volume.

Let us consider now the results of the concept [44, 45] application to polymers yielding process description. The yielding can be considered as polymer structure loss of its stability in the mechanical stresses field and the yield strain ε_Y is measure of this process resistance. In Ref. [44], it is indicated that specific lifetime of suprasegmental structures t^{im} is connected with $\Delta \tilde{G}^{im}$ as follows:

$$t^{im} \sim \exp\left(-\Delta \tilde{G}^{im}/RT\right) \tag{22}$$

where R is the universal gas constant.

Assuming that $t^{im} \sim t_Y$ (t_Y is time, necessary for yield stress σ_Y achievement) and taking into account, that $\varepsilon_Y \sim t_Y$, it can be written [43]:

$$\varepsilon_Y \sim \exp\left(-\Delta \tilde{G}^{im}/RT\right) \tag{23}$$

Let us note, that the Eq. (23) correctness in reference to polymers means, that yielding process in them is controlled by supra segmental structures thermodynamical stability.

In Fig. 7.10, the dependence of ε_Y on $\exp\left(-\Delta \tilde{G}^{im}/RT\right)$, corresponding to the Eq. (23), for all groups of the considered in Ref. [43] polymers. Despite definite (and expected) data scattering it is obvious, that all data break down intotwo branches, the one of which is approximated by a straight line. Such division reasons are quite obvious: the data with negative value of $\Delta \tilde{G}^{im}$ cover one (right) branch and with positive ones – the other (left) one. The last group consists of semicrystalline polymers with devitrificated at testing temperature loosely packed matrix (polytetrafluoroethylene, polyethylenes, polypropylene). The cluster model [18, 24] postulates thermofluctuation character of clusters formation and their decay at $T \geq T_g$. Therefore, such clusters availability in the indicated semicrystalline polymers devitrificatedamorphous phase has quite another origin, namely, it is due to amorphous chains tightness in crystallization process [48]. In practice this effect results to the condition $\Delta \tilde{G}^{im}$

>0 realization, which in the low-molecular substances case belongs to hypothetical nonexistent in reality transitions "an overheated liquid → solid body" [45].

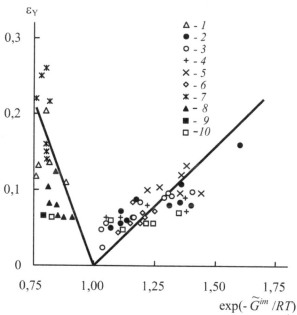

FIGURE 7.10 The correlation of yield strain ε_Y and parameter $\exp\left(-\Delta\tilde{G}^{im}/RT\right)$ for polymers with devitrificated (1, 7, 8, 9) and glassy (2, 3, 4, 5, 6, 10) loosely packed matrix [43].

Let us note one more feature, confirming existence reality of Fig. 7.10 plot left branch. At present ε_Y increase at T growth for polyethylenenes and ε_Y decreaseat the same conditions – for amorphous glassy polymers are well known [18]. One can see easily, that this experimental fact is explained completely by two branches availability in the plot of Fig. 7.10 that confirms again existence reality of suprasegmental structures, which are in quasiequilibrium with "free" segments. This quasiequilibrium is characterized by $\Delta\tilde{G}^{im}$ so, that in this case $\Delta\tilde{G}^{im} > 0$.

As it was to be expected the value $\varepsilon_Y = 0$ at $\exp\left(-\Delta\tilde{G}^{im}/RT\right)$ or $\Delta\tilde{G}^{im} = 0$. The last condition is achieved at $T = T_g$, where the yield strain is always equal to zero. Let us also note, that the data for semicrystalline polymers with vitrificated amorphous phase (polyamide-6, poly (ethylene terephthalate)) cover the right branch of the Fig. 7.10 plot. This means, that suprasegmental structures existence ($\Delta\tilde{G}^{im} > 0$) is due just to devitrificated amorphous phase availability, but no crystallinity. At $T<T_g$ amorphous phase viscosity, increases sharply and amorphous chains tightness cannot be exercised its action, displacing macromolecules parts, owing to that local

order formation has thermofluctuation character. The attention is paid to the obtained plot community, if to remember, that the values t_Y for impact and quasistatic tests are differed by five orders. This circumstance is explained simplyenough, since the value ε_Y can be written as follows [43]:

$$\varepsilon_Y = t_Y \dot{\varepsilon}, \tag{24}$$

where $\dot{\varepsilon}$ is strain rate and substitution of the Eq. (24) in the Eq. (23) shows that in the right part of the latter the factor $\dot{\varepsilon}^{-1}$ appears, which for the considered loading schemes changes by about five orders.

The Gibbs specific function notion for nonequilibrium phase transition "over-cooled liquid → solid body" is connected closely to local order notion (and, hence, fractality notion, see Chapter 1), since within the frameworks of the cluster model the indicated transition is equivalent to cluster formation start. The dependence of clusters relative fraction φ_{cl} on the value $\left|\Delta\tilde{G}^{im}\right|$ for PC and PAr is adduced. As one can see, this dependence is linear, φ_{cl} growth at $\left|\Delta\tilde{G}^{im}\right|$ increasing is observed and at $\left|\Delta\tilde{G}^{im}\right| = 0$ (i.e., for the selected standard temperature $T = T_g$) the cluster structure complete decay ($\varphi_{cl} = 0$) occurs.

The adduced above results can give one more, at least partial, explanation of "cell's effect." As it has been shown in Ref. [18], the following approximate relationship exists between ε_Y and Grüneisen parameter g_L:

$$\varepsilon_Y = \frac{1}{2\gamma_L} \tag{25}$$

Using this relationship and the plots of Fig. 7.10, it is easy to show, that the decrease and, hence, φ_{cl} reduction results to g_L growth, characterizing intermolecular bonds anharmonicity level. This parameter shows, how fast intermolecular interaction weakens at external (e.g., mechanical one [49]) force on polymer and the higher g_L the faster intermolecular interaction weakening occurs at other equal conditions. In other words, the greater $\Delta\tilde{G}^{im}$ and φ_{cl}, the smaller g_L and the higher polymer resistance to external influence. In Fig. 7.11, the dependence of g_L on mean number of statistical segments per one cluster n_{cl}, which demonstrates clearly the said above.

Hence, the stated above results shown that polymer yielding process can be described within the frameworks of the macrothermodynamical model. This is confirmed by the made in Ref. [44] conclusion about thermodynamical factor significance in those cases, when quasiequilibrium achievement is reached by mechanical stresses action. The existence possibility of structures with $\Delta\tilde{G}^{im} > 0$ (connected with transition "overheated liquid → solid body" [45] was shown and, at last, one more possible treatment of "cell's effect" was given within the frameworks of intermolecular bonds anharmonicity theory for polymers [49].

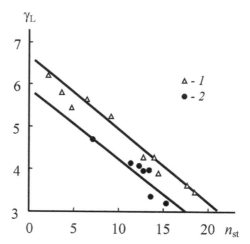

FIGURE 7.11 The dependence of Grüneisen parameter g_L on mean number of statistical segments per one cluster n_{cl} for PC (1) and PAr (2) [43].

If to consider the yielding process as polymer mechanical devitrification [36], then the same increment of fluctuation free volume f_g is required for the strain ε_Y achievement. This increment Δf_g can be connected with ε_Y as follows [50]:

$$\Delta f_g = \varepsilon_Y (1-2n) \tag{26}$$

Therefore, f_g decrease results to Δf_g growth and respectively, ε_Y enhancement.

Let us consider, which processes result to necessary for yielding realization fluctuation free volume increasing. Theoretically (within the frameworks of polymers plasticity fractal concept [35] and experimentally (by positrons annihilation method [22]) it has been shown, that the yielding process is realized in densely packed regions of polymer. It is obvious, that the clusters will be such regions in amorphous glassy polymer and in semicrystalline one – the clusters and crystallites [18]. The Eq. (9) allows to estimate relative fraction c of polymer, which remains in the elastic state.

In Fig. 7.12, the temperature dependence of c for HDPE is shown, from which one can see its decrease at T growth. The absolute values c change within the limits of 0.516 , 0.364 and the determined by polymer density crystallinity degree K for the considered HDPE is equal to 0.687 [51]. In other words, the value c in all cases exceeds amorphous regions fraction and this means the necessity ofsome part crystallites melting for yielding process realization. Thus, the conclusion should be made, that a semicrystalline HDPE yielding process includes its crystallites partial mechanical melting (disordering). For the first time Kargin and Sogolova [52] made such conclusion and it remains up to now prevalent in polymers mechanics [53].

The concept [35] allows to obtain quantitative estimation of crystallites fraction χ_{cr}, subjecting to partial melting − recrystallization process, subtracting amorphous phase fraction in HDPE from c. The temperature dependence of χ_{cr} was also shown in Fig. 7.12, from which one can see, that χ_{cr} value is changed within the limits of 0.203 , 0.051, decreasing at T growth.

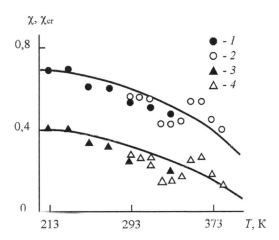

FIGURE 7.12 The dependences of elastically deformed regions fraction c (1, 2) and crystallites fraction, subjected to partial melting, χ_{cr} (3, 4) on testing temperature T in impact (1, 3) and quasistatic (2, 4) tests for HDPE [51].

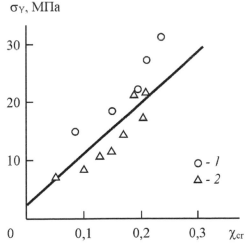

FIGURE 7.13 The dependence of yield stress σ_Y on crystallites fraction, subjecting to partial melting χ_{cr} in impact (1) and quasistatic (2) tests for HDPE [51].

It is natural to assume, that the yield stress is connected with parameter χ_{cr} as follows: the greater χ_{cr}, the larger the energy, consumed for melting and the higher σ_Y. The data of Fig. 7.13 confirm this assumption and the dependence $\sigma_Y(\chi_{cr})$ is extrapolated to finite σ_Y value, since not only crystallites, but also clusters participate in yielding process. As it was to be expected [54], the crystalline regions role in yielding process realization is much larger, than the amorphous ones.

Within the frameworks of the cluster model [55] it has been assumed, that the segment joined to cluster means fluctuation free volume microvoid "shrinkage" and vice versa. In this case the microvoids number ΔN_h, forming in polymer dilation process, should be approximately equal to segments number ΔN_f, subjected to partial melting process. These parameters can be estimated by following methods. The value ΔN_h is equal to [51]:

$$\Delta N_h = \frac{\Delta f_g}{V_h},\qquad(27)$$

where V_h is free volume microvoid volume, which can be estimated according to the kinetic theory of free volume [27].

Macromolecules total length L per polymer volume unit is estimated as follows [56]:

$$L = \frac{1}{S},\qquad(28)$$

where S is macromolecule cross-sectional area.

The length of macromolecules L_{cr}, subjected to partial melting process, per polymer volume unit is equal to [51]:

$$L_{cr} = L\chi_{cr}\qquad(29)$$

Further the parameter ΔN_c can be calculated [51]:

$$\Delta N_\chi = \frac{L_{cr}}{l_{st}} = \frac{L\chi_{cr}}{l_0 C_\infty}\qquad(30)$$

The comparison of ΔN_h and ΔN_Y values is adduced in Fig. 7.14, from which their satisfactorycorrespondence follows. This is confirmed by the conclusion, that crystallites partial mechanical melting (disordering) is necessary for fluctuation free volume f_g growth up to the value, required for polymers mechanical devitrification realization [51].

Whenever work is done on a solid, there is also a flow of heat necessitated by the deformation. The first law of thermodynamics:

$$dU = dQ + dW \qquad (31)$$

where Eq. (31) states that the internal energy change dU of a sample is equal to the sum of the work dW performed on the sample and the heat flow dQ into the sample. This relation is valid for any deformation, whether reversible or irreversible. There are twothermodynamically irreversible cases for which dQ and dW are equal by absolute value and opposite by sign: uniaxial deformation of a Newtonian fluid and ideal elastic-plastic deformation. For amorphous glassy polymers deformation has essentially different character: the ratio dQ/dW 1 and changes within the limits of 0.36 , 0.75 depending on testing condition [57]. In other words, the thermodynamically ideal plasticity is not realized for these materials. The reason of this effect is thermodynamical nonequilibrium of polymers structure. Within the frameworks of the fractal analysis it has been shown that it results to polymers yielding process realization not only in entire samplevolume, but also in its part [see Eq. (9)] [35]. Besides, it has been demonstrated experimentally and theoretically that amorphous glassy polymer structural component, in which the yielding is realized, is densely packed local order regions (clusters) [22, 23].

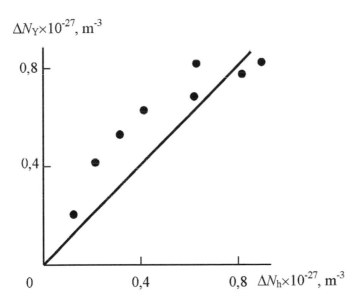

FIGURE 7.14 The relation between increment of segments number ΔN_Y in crystallites, subjecting to partial melting, and increment of free volume microvoids number ΔN_h, which is necessary for yielding process realization, for HDPE [51].

Lately the mathematical apparatus of fractional integration and differentiation [58, 59] was used for fractal objects description, which is amorphous glassy polymers structure. It has been shown [60] that Kantor's set fractal dimension coincides with an integral fractional exponent, which indicates system states fraction, remaining during its entire evolution (in our case deformation). As it is known [61], Kantor's set ("dust") is considered in one-dimensional Euclidean space ($d = 1$) and therefore its fractal dimension obey the condition $d_f \pounds 1$. This means, that for fractals, which are considered in Euclidean spaces with $d > 2$ ($d = 2, 3,...$) the fractional part of fractal dimension should be taken as fractional exponent v_{fr} [62, 63]:

$$v_{fr} = d_f - (d - 1) \qquad (32)$$

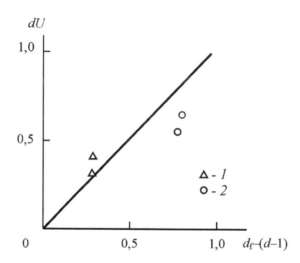

FIGURE 7.15 The dependence of relative fraction of latent energy dU on fractional exponent $v_{fr} = d_f - (d - 1)$ value for PC (1) and PMMA (2) [64].

The value v_{fr} characterizes that states (structure) part of system (polymer), which remains during its entire evolution (deformation). In Fig. 7.15, the dependence of latent energy fraction dU at PC and poly (methyl methacrylate) (PMMA) deformation on $v_{fr} = d_f - (d - 1)$ [64] is shown. The value dU was estimated as $(W - Q)/W$. In Fig. 7.15, the theoretical dependence dU (v_{fr}) is adducted, plotted according to the conditions $dU = 0$ at $v_{fr} = 0$ or $d_f = 2$ and $dU = 1$ at $v_{fr} = 1$ or $d_f = 3$ ($d = 3$), that is, at v_{fr} or d_f limiting values [64]. The experimental data correspond well to this theoretical dependence, from which it follows [64]:

$$dU = v_{fr} = d_f - (d - 1). \qquad (33)$$

Let us consider two limiting cases of the adduced in Fig. 7.15 dependence at $v_{fr} = 0$ and 1.0, both at $d = 3$. In the first case ($d_f = 2$) the value $dU = 0$ or, as it follows from dU definition (the Eq. (31)), $dW = dQ$ and polymer possesses an ideal elastic-plastic deformation. Within the frameworks of the fractal analysis $d_f = 2$ means, that $\varphi_{cl} = 1.0$, that is, amorphous glassy polymer structure represents itself one gigantic cluster. However, as it has been shown above, the condition $d_f = 2$ achievement for polymers is impossible in virtue of entropic tightness of chains, joining clusters, and therefore $d_f > 2$ for real amorphous glassy polymers. This explains the experimental observation for the indicated polymers: $dU^1 \, 0$ or $|dW| \, ^1| dQê$ [57]. At $v_{fr} = 1.0$ or $d_f = d = 3$ polymers structure loses its fractal properties and becomes Euclidean object (true rubber). In this case from the plot of Fig. 7.15 $dU = 1.0$ follows. However, it has been shown experimentally, that for true rubbers, which are deformed by thermodynamically reversible deformation $dU = 0$. This apparent discrepancy is explained as follows [64]. Fig. 7.15 was plotted for the conditions of inelastic deformation, whereas at $d_f = d$ only elastic deformation is possible. Hence, at $v_{fr} = 1.0$ or $d_f = d = 3$ deformation typediscrete (jump – like) change occurs from $dU®$ 1 up to $dU = 0$. This point becomes an initial one for fractal object deformation in Euclidean space with the next according to the custom dimension $d = 4$, where 3 $\leq d_f £ 4$ and all said above can be repeated in reference to this space: at $d = 4$ and $d_f = 3$ the value $v_{fr} = 0$ and $dU = 0$. Let us note in conclusion that exactly the exponent v_{fr} controls the value of deformation (fracture) energy of fractal objects as a function of processlength scale. Let us note that the equality $dU = \varphi_{1.m.}$ was shown, from which structural sense of fractional exponent in polymers inelastic deformation process follows: $v_{fr} = \varphi_{1.m.}$ [64].

Mittag-Lefelvre function [59] usage is one more method of a diagrams s – e description within the frameworks of the fractional derivatives mathematical calculus. A nonlinear dependences, similar to a diagrams s – e for polymers, are described with the aid of the following equation [65]:

$$\sigma(\varepsilon) = \sigma_0 \left[1 - E_{v_{fr},1} \left(-\varepsilon^{v_{fr}} \right) \right], \tag{34}$$

where σ_0 is the greatest stress for polymer in case of linear dependence s(e) (of ideal plasticity), is the Mittag-Lefelvre function [65]:

$$E_{v_{fr},1} \left(-\varepsilon^{v_{fr}} \right) = \sum_{k=0}^{\infty} \frac{\varepsilon^{v_{fr}k}}{\tilde{A}\left(v_{fr}k + \beta \right)}, \; v_{fr} > 0, \, b > 0, \tag{35}$$

where Γ is Eiler gamma-function.

As it follows from the Eq. (34), in the considered case $b = 1$ and gamma-function is calculated as follows [40]:

$$\tilde{A}\left(v_{fr}k + 1 \right) = \sqrt{\frac{\pi}{2}} \left(v_{fr}k - \varepsilon^{v_{fr}k} \right)^{v_{fr}k - \varepsilon^{-v_{fr}}} e^{-\left(v_{fr}k - \varepsilon^{v_{fr}} \right)} \tag{36}$$

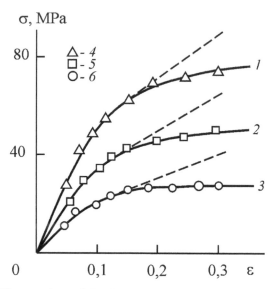

FIGURE 7.16   The experimental (1 ÷ 3) and calculated according to the equations (4.34) , (4.36) (4 ÷ 6) diagrams s – e for PAr at T = 293 (1, 4), 353 (2, 5) and 433 K (3, 6). The shaded lines indicate calculated diagrams s – e for forced high-elasticity part without v_{fr} change [66].

In Fig. 7.16, the comparison of experimental and calculated according to the equations (34) , (36) diagrams s – e for PAr at three testing temperatures is adduced. The values σ_0 were determined as the product $E\varepsilon_Y$ [66]. As it follows from the data of Fig. 7.16, the diagrams s – e on the part from proportionality limit up to yield stress are well described well within the frameworks of the Mittag-Lefelvre function. Let us note that two necessary for these parameters (σ_0 and v_{fr}) are the function of polymers structural state, but not filled parameters. This is a principal question, since the usage in this case of empirical fitted constants, as, for example, in Ref. [67], reduces significantly using method value [60, 65].

In the initial linear part (elastic deformation) calculation according to the Eq. (34) was not fulfilled, since in it deformation is submitted to Hooke law and, hence, is not nonlinear. At stresses greater than yield stress (high-elasticity part) calculation according to the Eq. (34) gives stronger stress growth (stronger strain hardening), than experimentally observed (that it has been shown by shaded lines in Fig. 7.16). The experimental and theoretical dependences s – e matching on cold flow part within the frameworks of the Eq. (34) can be obtained at supposition v_{fr} = 0.88 for T = 293 K and v_{fr} = 1.0 for the two remaining testing temperatures. This effect explanation was given within the frameworks of the cluster model of polymers amorphous state structure [39], where it has been shown that in a yielding point small (instable) clusters, restraining loosely packed matrix in glassy state, break down. As a result of such mechanical devitrification glassy polymers behavior on the forced high-elasticity (cold flow) plateau is submitted to rubber high-elasticity laws and, hence, $d_f \rightarrow d = 3$ [68]. The stress decay behind yield stress, so-called "yield tooth", can be described similarly [66]. An instable clusters decay in yielding point results to clusters relative fraction φ_{cl} reduction, corresponding to d_f growthand v_{fr} enhancement (the Eq. (33)) and, as it follows from the Eq. (34), to stress reduction. Let us note in conclusion that the offered in Ref. [69] techniques allow to predict parameters, which are necessary for diagrams s – e description within the frameworks of the considered method, that is, E, ε_y and d_f.

In Ref. [70], it has been shown that rigid-chain polymers can be have a several sub-states within the limits of glassy state. For polypiromellithimide three such sub-states are observed on the dependences of modulus ds/de, determined according to the slope of tangents to diagram s – e, on strain e [70]. However, such dependences of ds/de on e have much more general character: in Fig. 7.17 three similar dependences for PC are adduced, which were plotted according to the data of Ref. [49]. If in Ref. [70] the transition from part I to part II corresponds to polypiromellithimide structure phase change with axial crystalline structure formation [71], then for the adduced in Fig. 7.17 similar dependences for PC this transition corresponds to deformation type change from elastic to inelastic one (proportionality limit [72]). The dependences (ds/de) (e) community deserves more intent consideration that was fulfilled by the authors [73] within the frameworks of fractal analysis.

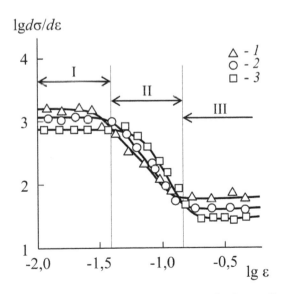

FIGURE 7.17 The dependences of modulus ds/de, determined according to the slope of tangents to diagram s – e, on strain e in double logarithmic coordinates for PC at $T = 293$ (1), 343 (2) and 373 K (3) [73].

The dependence of physical fractal density r on measurement scale L in double logarithmic coordinates was shown. For $L<L_{min}$ and $L>L_{max}$ Euclidean behavior is observed and within the range of $L = L_{min}, L_{max}$ – fractal one [40]. Let us pay attention to the complete analogy of the plots of Fig. 7.17.

There is one more theoretical model allowing nonstandard treatment of the shown in Fig. 7.17 dependences ds/de (e). Kopelman was offered the fractal descriptions of chemical reactions kinetics, using the following simple relationship [74]:

$$k \sim t^{-h}, \tag{37}$$

where k is reaction rate, t is its duration, h is reactionary medium nonhomogeneity (heterogeneity) exponent ($0<h<1$), which turns into zero only for Euclidean (homogeneous) mediums and the Eq. (37) becomes classical one: k = const.

From the Eq. (37) it follows, that in case h^1 0, that is, for heterogeneous (fractal) mediums the reaction rate k reduces at reaction proceeding. One should attention to qualitative analogy of curves s – e and the dependences of conversion degree on reaction duration Q (t) for a large number of polymers synthesis reactions [75]. Still greater interest for subsequent theoretical developments presents complete qualitative analogy of diagrams s – e and strange attractor trajectories, which can be have "yield tooth," strain hardening and so on [76].

If to consider deformation process as polymer structure reaction with supplied, from outside mechanical energy to consider, then the modulus ds/de will be $k$analog (Fig 7.17). The said above allows to assume, that deformation on parts I and III (elasticity and cold flow) proceeds in Euclidean space and on part II (yielding) – in fractal one.

The fractal analysis main rules usage for polymers structure and properties description [68, 77] allows to make quantitative estimation of measurement scale L change at polymer deformation. There are a several methods of such estimation and the authors [73] use the simplest from them as ensuring the greatest clearness. As it was noted in Chapter 1, the self-similarity (fractality) range of amorphous glassy polymers structure coincides with cluster structure existence range: the lower scale of self-similarity corresponds to statistical segment length l_{st} and the upper one – to distance between clusters R_{cl}. The simplest method of measurement scale L estimation is the usage of well-known Richardson equation.

In the case of affine deformation the value R_{cl} will be changed proportionally to drawing ratio l [70]. This change value can be estimated from the equation [79]:

$$R_{cl}\lambda = l_{st}\frac{2(1-v)}{(1-2v)} \tag{38}$$

From the Eq. (38) n increase follows – the d_f increase at drawing ratio l growth. In its turn, the value l is connected with the strain e by a simple relationship (in the case of affine deformation) [80]:

$$1 = 1 + e \qquad (39)$$

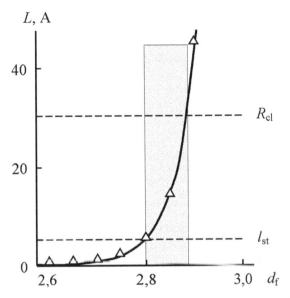

FIGURE 7.18 The dependence of measurement scale L on structure fractal dimension d_f for PC at $T = 293$ K. horizontal shaded lines indicate nondeformed PC structure self-similarity boundaries (l_{st} and R_{cl}) and the shaded region – deformation fractal behavior range [73].

In Fig. 7.18, the dependence of L on d_f is adduced, which is calculated according to the Eqs. (38) and (39) combination and at the condition, that the parameters $C_\infty d_f$ are connected with each other as follows [18]:

$$C_\infty = \frac{2d_f}{d(d-1)(d-d_f)} + \frac{4}{3} \qquad (40)$$

As it follows from the data of Fig. 7.18, L growth at d_f increase is observed and within the range of $d_f \approx 2.80$, 2.89 polymer deformation proceeds in fractal space. At $d_f > 2.90$ the deformation space transition from fractal to Euclidean one is observed (PC yielding is achieved at $d_f = 2.85$ [23]) and structure PC approaching to true rubber state ($d_f = d = 3$) induces very fast L growth. Let us consider the conditions of transition from part II to part III (Fig. 7.17). The value R_{cl} by $R_{cl}l$ according to the indicated above reasons it assumption $D_{ch} = 1.0$, that is, part of chain, stretched completely between clusters, let us obtain [73]:

$$\frac{L_{cl}}{l_{st}} = \frac{R_{cl}\lambda}{l_{st}}, \qquad (41)$$

That is, drawing ratio critical value λ_{cr}, corresponding to the transition from part II to part III (from fractal behavior to Euclidean one) or the transition from yielding to cold flow, is equal to [73]:

$$\lambda_{cr} = \frac{L_{cl}}{R_{cl} l_{st}}, \tag{42}$$

that corresponds to the greatest attainable molecular draw [81].

For PC at $T = 293$ K and the indicated above values L_{cl} and $R_{cl}\lambda_{cr}$ will be equal to 2.54. Taking into account, that drawing ratio at uniaxial deformation $\lambda''_{cr} = \lambda_{cr}^{1/3}$, let us obtain $\lambda''_{cr} = 1.364$ or critical value of strain of transition. Let us note that within the frameworks of the cluster model of polymers amorphous state structure [18] chains deformation in loosely packed matrix only is assumed and since the Eq. (42) gives molecular drawing ratio, which is determined in the experiment according to the relationship [81]:

$$\varepsilon_{cr} = \varepsilon''_{cr}\left(1 - \phi_{cl}\right) \tag{43}$$

Let us obtain $\varepsilon_{cr} = 0.117$ according to the Eq. (43). The experimental value of yield strain ε_Y for PC at $T = 293$ K is equal to 0.106. This means, that the transition to PC cold flow begins immediately beyond yield stress, which is observed experimentally [23].

Therefore, the stated above results show, that the assumed earlier substates within the limits of glassy state are due to transitions from deformation in Euclidean space to deformation in fractal space and vice versa. These transitions are controlled by deformation scale change, induced by external load (mechanical energy) application. From the physical point of view this postulate has very simple explanation: if size of structural element, deforming deformation proceeding, hits in the range of sizes $L_{min} - L_{max}$ (Fig. 7.2), then deformation proceeds in fractal space, if it does not hit – in Euclidean one. In part I intermolecular bonds are deformed elastically on scales of 3 , 4Å ($L<L_{min}$), in part II – cluster structure elements with sizes of order of 6 , 30Å [18, 24] ($L_{min}<L<L_{max}$) and in part III chains fragments with length of L_{cl} or of order of several tens of Ånströms ($L>L_{max}$). In Euclidean space the dependence s – e will be linear ($ds/de = $ const) and in fractal one – curvilinear, since fractal space requires deformation deceleration with time. The yielding process realization is possible only in fractal space. The stated model of deformation mechanisms is correct only in the case of polymers structure presentation as physical fractal.

In the general case polymers structure is multifractal, for behavior description of which in deformation process in principle its three dimensions knowledge is enough: fractal (Hausdorff) dimension d_f, informational one d_i and correlation one d_c [82]. Each from the indicated dimension describes multifractal definite properties change and these dimensions combined application allows to obtain more or less complete picture of yielding process [73].

As it is known [83], a glassy polymers behavior on cold flow plateau (part III in Fig. 7.17) is well described within the frameworks of the rubber high-elasticity theory. In Ref. [39] it has been shown that this is due to mechanical devitrification of an amorphous polymers loosely packed matrix. Besides, it has been shown [82, 84] that behavior of polymers in rubber-like state is described correctly under assumption, that their structure is a regular fractal, for which the identity is valid:

$$d_1 = d_c = d_f. \tag{44}$$

A glassy polymers structure in the general case is multifractal [85], for which the inequality is true [82, 84]:

$$d_c < d_1 < d_f. \tag{45}$$

Proceeding from the said above and also with appreciation of the known fact, that rubbers do not have to some extent clearly expressed yielding point the authors [73] proposed hypothesis, that glassy polymer structural state changed from multifractal up to regular fractal, that is, criterion (44) fulfillment, was the condition of its yielding state achievement. In other words, yielding in polymers is realized only in the case, if their structure is multifractal, that is, if it submits to the inequality (45).

Let us consider now this hypothesis experimental confirmations and dimensions d_1 and d_c estimation methods in reference to amorphous glassy polymers multifractal structure. As it is known [82], the informational dimension d_1 characterizes behavior Shennone informational entropy I (e):

$$I(\varepsilon) = \sum_{i}^{M} P_i \ln P_i, \tag{46}$$

where M is the minimum number of d-dimensional cubes with side e, necessary for all elements of structure coverage, P_i is the event probability, that structure point belongs to i-th element of coverage with volume ε^d.

In its turn, polymer structure entropy change ΔS, which is due to fluctuation free volume f_g, can be determined according to the Eq. (21). Comparison of the Eqs. (21) and (46) shows that entropy change in the first from them is due to f_g change probability and to an approximation of constant the values I (e) and ΔS correspond to each other. Further polymer behavior at deformation can be described by the following relationship [84]:

$$\Delta S = -c \left(\sum_{j=1}^{d} \lambda_i^{d_1} - 1 \right), \tag{47}$$

where c is constant, l_1 is drawing ratio.

Hence, the comparison of the Eqs. (21), (46) and (47) shows that polymer behavior at deformation is defined by change f_g exactly, if this parameter is considered as probabilistic measure. Let us remind, that f_c such definition exists actually within the frameworks of lattice models, where this parameter is connected with the ratio of free volume microvoids number N_h and lattice nodes number N (N_h/N) [49]. The similar definition is given and for P_i in the Eq. (46) [86].

The value d_I can be determined according to the following equation [82]:

$$d_I = \lim_{\varepsilon \to 0} \left[\frac{\sum_{i=1}^{M} P_i(\varepsilon) \ln P_i(\varepsilon)}{\ln \varepsilon} \right] \qquad (48)$$

Since polymer structure is a physical fractal (multifractal), then for it fractal behavior is observed only in a some finite range of scalesand the statistical segment length l_{st} is accepted as lower scale. Hence, assuming $P_i(e) = f_g$ and$\varepsilon \to l_{st}$ the Eq. (48) can be transformed into the following one [73]:

$$d_I = c_1 \frac{Rf_g \ln f_g}{\ln l_{st}}, \qquad (49)$$

where c_1 is constant.

As it is known [23], in yielding point for PC $d_f = 2.82$ (or n = 0.41), that allows to determine the value f_g according to the previous equationsand constant c_1 – from the condition (44). Then, using the known equality $l_1 = 1 + e_1$, similar to the Eq. (39), the yield strain ε_Y theoretical value can be calculated, determining constant c by matching method and the value Δf_g for ΔS calculation – as values f_g difference for nondeformed polymer and polymer in yielding point. Since in Ref. [73] thin films are subjected to deformation, then as the first approximation $d = 1$ is assumed in the Eq. (47) sum sign. In Fig. 7.19,comparison of the temperature dependences of experimental and calculated by the indicated method values ε_Y for PC is adduced. As one can see, the good correspondence of theory and experiment is obtained by both the dependences $\varepsilon_Y(T)$ course and ε_Y absolute values and discrepancy of experimental and theoretical values ε_Y at large T is due to the fact, that for calculation simplicity the authors [73] did not take into consideration transverse strains influence onvalue ε_Y in the Eq. (47).

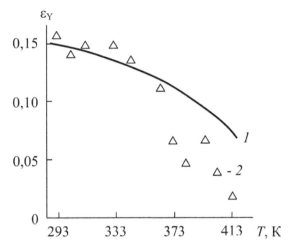

FIGURE 7.19   The comparison of experimental (1) and calculated according to the equation (47) temperature dependences of yield strain ε_Y for PC [73].

Williford [87] proposed, that the value d_1 of multifractal corresponded to its surface dimension (the first subfractal) – either sample surface or fracture surface. For this supposition checking the authors [73] calculate a fracture surface fractal dimension for brittle (d_{fr}^{br}) and ductile (d_{fr}^{duc}) failure types according to the equations [40]:

$$d_{fr}^{br} = \frac{10(1+v)}{7-3v} \tag{50}$$

and

$$d_{fr}^{duc} = \frac{2(1+4v)}{1+2v} \tag{51}$$

In Fig. 7.20, comparison of the temperature dependences of d_1, d_{fr}^{br} and d_{fr}^{duc} for PC is adduced. As one can see, at low T ($T<373$ K) the value d_1 corresponds to d_{fr}^{br} well enough and at higher temperatures ($T>383$ K) it is close to d_{fr}^{duc} . In Fig. 7.20, the shaded region shows the temperature range corresponding to the brittle-ductile transition for PC [73]. It is significant that this interval beginning ($T = 373$ K) coincides with loosely packed matrix devitrification temperature T'_g, which is approximately 50 K lower than polymer glass transition temperature T_g [18], for PC equal to ~ 423 K [88].

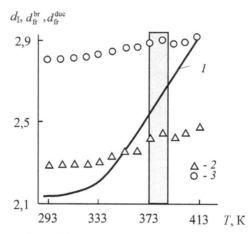

FIGURE 7.20 Comparison of the temperature dependences of informational dimension d_1 (1), fractal dimensions of fracture surface at brittle d_{fr}^{br} (2) and ductile d_{fr}^{duc} (3) failure for PC. The temperature range of brittle-ductile transition is shown by shaded region [73].

The correlation dimension d_c is connected with multifractal structure internal energy U [61] and it can be estimated according to the equation [82]:

$$\Delta U = -c_2 \left(\lambda_F^{d-d_c} \right), \tag{52}$$

where ΔU is internal energy change in deformation process, c_2 is constant, l_F is macroscopic drawing ratio, which in the case of uniaxial deformation is equal to λ_1. The value l_F is determined as follows [82]:

$$l_F = \lambda_1 \lambda_2 \lambda_3, \tag{53}$$

where λ_2 and λ_3 are transverse drawing rations, connected with λ_1 by the simple relationships [82]:

$$\lambda_2 = 1 + \varepsilon_2, \tag{54}$$

$$\lambda_3 = 1 + \varepsilon_3, \tag{55}$$

$$\varepsilon_2 = \varepsilon_3 = \nu \varepsilon_1. \tag{56}$$

The temperature dependence of d_c can be calculated, as and earlier, assuming from the condition (44) that at yielding $d_c = d_f = 2,82$, estimating ΔU as one half of product $\sigma_Y \varepsilon_Y$ (with appreciation of practically triangular form of curve s – e up to yield stress) and determining the constant c_2 by the indicated above mode. Comparison of the temperature dependences of multifractal three characteristic dimensions d_c, d_1 and d_f, calculated according to the Eqs. (47) and (52), is adduced in Fig. 7.21.

As it follows from the plots of this figure, for PC the inequality (45) is confirmed, which as a matter of fact is multifractal definition. The dependences d_c (T) and d_1 (T) are similar and their absolute values are close, that is explained by the indicated above intercommunication of f_g and U change [89]. Let us note, that dimension d_1 controls only yield strain ε_Y and dimension d_c – both ε_Y and yield stress σ_Y. At approaching to glass transition temperature, that is, at $T \rightarrow T_g$, the values d_c, d_1 and d_f become approximately equal, that is, rubber is a regular fractal. Thus, with multifractal formalism positions the glass transition can be considered as the transition of structure from multifractal to the regular fractal. Additionally it is easy to show the fulfillment of the structure thermodynamical stability condition [90]:

$$d_f (d - d_1) = d_c (d - d_c). \tag{57}$$

Hence, the authors [73] considered the multifractal concept of amorphous glassy polymers yielding process. It is based on the hypothesis, that the yielding process presents itself structural transition from multifractal to regular fractal. In a amorphous glassy polymers Shennone informational entropy to an approximation of constant coincides with entropy, which is due to polymer fluctuation free volume change. The approximate quantitative estimations confirm the offered hypothesis correctness. The postulate about yielding process realization possibility only for polymers fractal structure, presented as a physical fractal within the definite range of linear scales, is the main key conclusion from the stated above results. What is more, the fulfilled estimations strengthen this definition – yielding is possible for polymer multifractal structure only and represents itself the structural transition multifractal – regular fractal [73].

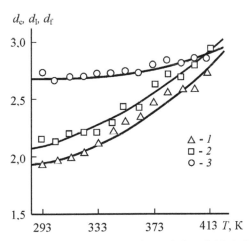

FIGURE 7.21   The temperature dependences of correlation d_c (1), informational d_1 (2) and Hausdorff d_f (3) dimensions of PC multifractal structure [73].

Let us consider in the present chapter conclusion the treatment of dependences of yield stress on strain rate and crystalline phase structure for semicrystalline polymers [77]. As it known [91], the clusters relative fraction φ_{cl} is an order parameter of polymers structure in strict physical significance of this term and since the local order was postulated as having thermofluctuation origin, then φ_{cl} should be a function of testing temporal scale in virtue of super-position temperature-time.

Having determined value t as duration of linear part of diagram load – time (P – t) in impact tests and accepting is equal to clusters relative fraction in quasistatic tensile tests [92], the values φ_{cl} (t) can be estimated. In Fig. 7.22, the dependences φ_{cl} (t) on strain rate for HDPE and polypropylene (PP) are shown, whichdemonstrate φ_{cl} increase at strain rate growth, that is, tests temporal scale decrease.

The elasticity modulus E value decreases at $\dot{\varepsilon}$ growth (Fig. 7.23). This effect cause (intermolecular bonds strong anharmonicity) is described in detail in Refs. [49, 94, 95]. In Fig. 7.24, the dependences of elasticity modulus E on structure fractal dimension d_f are adduced for HDPE and PP, which turned out to be linear and passing through coordinates origin. As it is known [84], the relation between E and d_f is given by the equation:

$$E = Gd_f, \tag{58}$$

where G is a shear modulus.

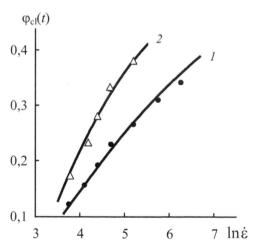

FIGURE 7.22 The dependences of clusters relative fraction φ_{cl} (t) on strain rate $\dot{\varepsilon}$ in logarithmic coordinates for HDPE (1) and PP (2) [93].

From the Eq. (58) and Fig. 7.24 it follows, that for both HDPE and PP G is const. This fact is very important for further interpretation [77].

Let us estimate now determination methods of crystalline σ_Y^{cr} and noncrystalline σ_Y^{nc} regions contribution to yield stress σ_Y of semicrystalline polymers. The value σ_Y^{cr} can be determined as follows [96]:

$$\sigma_Y^{cr} = \frac{Gb}{2\pi}\left(\frac{K}{S}\right)^{1/2},$$

(59)

where b is Burgers vector, determined according to the Eq. (7), K is crystallinity degree, S is cross-sectional area, which is equal to 14.35 and 26.90 Å² for HDPE and PP, accordingly [97].

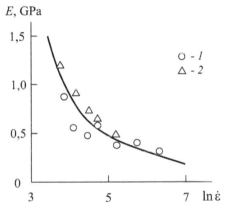

FIGURE 7.23 The dependence of elasticity modulus E on strain rate $\dot{\varepsilon}$ in logarithmic coordinates for HDPE (1) and PP (2) [93].

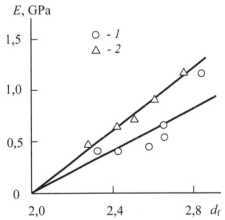

FIGURE 7.24 The dependences of elasticity modulus E on fractal dimension of structure d_f for HDPE (1) and PP (2) [93].

As one can see, all included in the Eq. (59) parameters for each polymer are constant, from which it follows, that σ_Y^{cr} =const. The value σ_Y^{cr} = 16.3 MPa for HDPE and 17.6 MPa – for PP. Thus, the change, namely, increase σ_Y at $\dot{\varepsilon}$ growth is defined by polymers noncrystalline regions contribution σ_Y^{nc} [77].

The Grist dislocation model [98] assumes the formation in polymers crystalline regions of screw dislocation (or such dislocation pair) with Burgers vector b and the yield process is realized at the formation of critical nucleus domain with size u^*:

$$u^* = \frac{Bb}{2\pi\tau_Y},$$

(60)

where B is an elastic constant, τ_Y shear yield stress.

In its turn, the domain with size U^* is formed at energetic barrier DG^* overcoming [98]:

$$\Delta G^* = \frac{Bl^2 l_d}{2\pi}\left[\ln\left(\frac{u^*}{r_0}\right) - 1\right],$$

(61)

where l_d is dislocation length, which is equal to crystalline thickness, r_0 is dislocation core radius.

In the model [98], it has been assumed, that nucleus domain with size u^* is formed in defect-free part of semicrystalline polymer, that is, in crystallite. Within the frameworks of model [1] and in respect to these polymers amorphous phase structure such region is loosely packed matrix, surrounding a local order region (cluster), whose structure is close enough to defect-free polymer structure, postulated by the Flory "felt" model [16, 17]. In such treatment the value u^* can be determined as follows [43]:

$$u^* = R_{cl} - r_{cl},$$

(62)

where R_{cl} is one half of distance between neighboring clusters centers, r_{cl} is actually cluster radius.

The value R_{cl} is determined according to the equation [18]:

$$R_{cl} = 18\left(\frac{2\nu_e}{F}\right)^{-1/3}, \text{Å},$$

(63)

where ν_e is macromolecular binary hooking network density, values of which for HDPE and PP are adduced in Ref. [99], $F = 4$.

The value r_{cl} can be determined as follows [77]:

$$r_{cl} = \left(\frac{n_{st} S}{\pi \eta_{pac}} \right), \tag{64}$$

where n_{st} is statistical segments number per one cluster (n_{st} = 12 for HDPE and 15 – for PP [77]), η_{pac} is packing coefficient, for the case of dense packing equal to 0.868 [100].

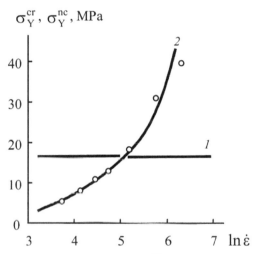

$\sigma_Y^{cr}, \sigma_Y^{nc}$, MPa

FIGURE 7.25 The dependences of crystalline σ_Y^{cr} (1) and noncrystalline σ_Y^{nc} (2) regions contributions in yield stress value σ_Y on strain rate in logarithmic coordinates for HDPE [93].

The shear modulus G, which, has been noted above, is independent on $\dot{\varepsilon}$, is accepted as B. Now the value σ_Y^{nc} can be determined according to the Eq. (60), assuming, that n_{st} changes proportionally to φ_{cl} (Fig. 7.22). In Fig. 25, the dependences of σ_Y^{cr} and σ_Y^{nc} on strain rate $\dot{\varepsilon}$ are adduced for HDPE (the similar picture was obtained for PP). As it follows from the data of this figure, at the condition σ_Y^{cr} = const the value σ_Y^{nc} grows at $\dot{\varepsilon}$ increase and at $\dot{\varepsilon} \gg 150$ s^{-1} $\sigma_Y^{nc} > \sigma_Y^{cr}$, that is, a noncrystalline regions contribution in σ_Y begins to prevail [93].

In Fig. 7.26, the comparison of the experimental σ_Y and calculated theoretically σ_Y^T as sum ($\sigma_Y^{cr} + \sigma_Y^{nc}$) dependences of yield stress on $\dot{\varepsilon}$ for HDPE and PP are adduced. As one can see, a good correspondence of theory and experiment is obtained, confirming the offered above model of yielding process for semicrystalline polymers.

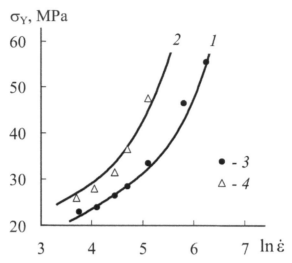

FIGURE 7.26   Comparison of the experimental (1, 2) and theoretical (3, 4) dependences of yield stress σ_Y on strain rate $\dot\varepsilon$ in logarithmic coordinates for HDPE (1, 3) and PP (2, 4) [93].

Let us note in conclusion the following. As it follows from the comparison of the data of Figs. 7.23, 7.24 and 7.26, σ_Y increase occurs at E reduction and G constancy, that contradicts to the assumed earlier σ_Y and E proportionality [101]. The postulated in Ref. [101] σ_Y and E proportionality is only an individual case, which is valid either at invariable structural state or at the indicated state, changing by definite monotonous mode [77].

Hence, the stated above results demonstrate nonzero contribution of noncrystalline regions in yield stress even for such semicrystalline polymers, which have devitrificated amorphous phase in testing conditions. At definite conditions noncrystalline regions contribution can be prevailed. Polymers yield stress and elastic constants proportionality is not a general rule and is fulfilled only at definite conditions.

The authors [102] use the considered above model [93] for branched polyethylenes (BPE) yielding process description. As it is known [103], the crystallinity degree K, determined by samples density, can be expressed as follows:

$$K = \alpha_c + \alpha_{if},$$ (65)

Where α_c and α_{if} are chains units fractions in perfect crystallites and anisotropic interfacial regions, accordingly.

The Eq. (59) with replacement of K by α_c was used for the value σ_Y^{cr} estimation. Such estimations have shown that the value σ_Y^{cr} is always smaller than macroscopic yield stress σ_Y. In Fig. 7.27, the dependence of crystalline phase relative contribution in yield stress σ_Y^{cr}/σ_Y on α_c is adduced. This dependence is linear and at $\alpha_c =$ 0 the trivial result $\sigma_Y^{cr} = 0$ is obtained. Let us note that this extrapolation assumes $\sigma_Y{}^1$ 0 at $\alpha_c = 0$. At large α_c the crystalline phase contribution is prevailed and at $\alpha_c =$ 0.75 $\sigma_Y^{cr}/\sigma_Y = 1$ (Fig. 7.27).

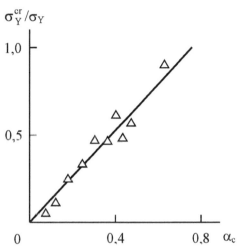

FIGURE 7.27   The dependence of crystalline regions relative contribution σ_Y^{cr}/σ_Y in yield stress on perfect crystallites fraction α_c for series of BPE [102].

The dependence $\alpha_{if}(\alpha_c)$ showed α_{if} linear reduction at α_c growth. Such α_{if} change and simultaneous σ_Y^{cr}/σ_Y increasing (Fig. 7.27) at α_c growth assume local order degree reduction, determining noncrystalline regions contribution in σ_Y, at crystallinity degree enhancement. Besides, the correlation $\alpha_{if}(\alpha_c)$ shows that local order regions of BPE noncrystalline regions are concentrated mainly in anisotropic interfacial regions, as it has been assumed earlier [18]. At $\alpha_c = 0.82$ $\alpha_{if} = 0$, that corresponds to the data of Fig. 7.27.

The common fraction of the ordered regions (clusters and crystallites) φ_{ord} can be determined according to the percolation relationship [91]:

$$\varphi_{ord} = 0.03 \, (T_m - T)^{0.55}, \tag{66}$$

where T_m and T are melting and testing temperatures, respectively.

Then clusters relative fraction φ_{cl} is estimated from the obvious relationship [102]:

$$\varphi_{cl} = \varphi_{ord} - \alpha_c. \tag{67}$$

The comparison of the experimental and estimated as a sum of crystalline and noncrystalline regions contributions in yield stress theoretical values α_Y shows their good correspondence for considered BPE.

It has been noted earlier [104, 105], that stress decay beyond stress ("yield tooth") for polyethylenes is expressed the stronger the greatervalue α_c is. For amorphous polymers it has been shown that the indicated "yield tooth" is due to instable clusters decay in yielding process and this decay is expressed the clearer the higher instable clusters relative fraction [39]. By analogy with the indicated mechanism the authors [102] assume that "yield tooth" will be the stronger the larger crystallites fraction is subjected to mechanical disordering (partial melting) in yielding process. The indicated fraction of crystallites χ_{cr} is determined as difference [102]:

$$\chi_{cr} = c - \alpha_{am}, \tag{68}$$

where c is polymer fraction, subjecting to elastic deformation, α_{am} is fraction of amorphous phase.

The value c can be determined within the frameworks of polymers plasticity fractal concept [35] according to the Eq. (9). The dependence of χ_{cr} on α_c shows χ_{cr} increasing at α_c growth [102]. At small α_c values all crystallites are subjected to disordering owing to that "yield tooth" in BPE curves stress–strain is absent and these curves are acquired the form, which is typical for rubbers. Hence, stress decay beyond yield stress intensification is due to χ_{cr} growth [103].

KEYWORDS

- amorphous state
- cluster model
- crystalline solids
- defects
- mechanism
- polymers yielding
- solid structure
- structure

REFERENCES

1. Kozlov, G. V., Belousov, V. N., Serdyuk, V. D., & Kuznetsov, E. N. (1995). Defect of Polymers Amorphous State Structure, Fizika i Technika Yysokikh Davlenii, *5(3)*, 59–64.
2. Kozlov, G. V., Beloshenko, V. A., & Varyukhin, V. N. (1996). Evolution of Dissipative Structures in Yielding Process of Cross-Linked Polymers, Prikladnaya Mekhanika i Tekhnicheskaya Fizika, *37(3)*, 115–119.

3. Honeycombe, R. W. K. (1968). The Plastic deformation of Metals London, Edward Arnold Publishers Ltd, 398p.
4. Argon, A. S. (1988). A Theory for the Low-Temperature Plastic Deformation of Glassy Polymers, Philosophy Magnetic, 1974, 29(1), 149–167.
5. Argon, A. S. (1973). Physical Basis of Distortional and Dilatational Plastic Flow in Glassy Polymers, Journal Macromolecule Science-Physics, 88(3–4), 573–596.
6. Bowden, P. B., & Raha, S. A. (1974). Molecular model for Yield and Flow in Amorphous Glassy Polymers Making use of a Dislocation Analogue Philosophy Magnetic, 29(1), 149–165.
7. Escaig, B. (1978). The Physics of Plastic Behavior of Crystalline and Amorphous Solids, Annual Physics, 3(2), 207–220.
8. Pechhold, W. R., & Stoll, B. (1982). Motion of Segment Dislocations as a Model for Glass Relaxation, Polymer Bull, 7(4), 413–416.
9. Sinani, A. B., & Stepanov, V. A. (1981). Prediction of Glassy Polymer Deformation Properties with the Aid of Dislocation Analogues, Mekhanika Kompozitnykh Materialov, 17(1), 109–115.
10. Oleinik, E. F., Rudnev, S. M., Salamatina, O. B., Nazarenko, S. I., & Grigoryan, G. A. (1986). Two Modes of Glassy Polymers Plastic Deformation, Doklady AN SSSR, 286(1), 135–138.
11. Melot, D., Escaig, B., Lefebvre, J. M., Eustache, R. R., & Laupretre, F. (1994). Mechanical Properties of Unsaturated Polyester Resins in Relation to their Chemical Structure 2, Plastic Deformation Behavior Journal Polymer Science. Part B. Polymer Physics, 32(11), (1805)–(1811).
12. Boyer, R. F. (1976). General Reflections on the Symposium on Physical Structure of the Amorphous State Journal Macromolecule Science, Physics B, 12(2), 253–301.
13. Fischer, E. W., & Dettenmaier, M. (1978). Structure of Polymer Glasses and Melts, J Non-Crystals Solids, 31(1–2), 181–205.
14. Wendorff, J. H. (1982). The Structure of Amorphous Polymers, Polymer, 23 (4), 543–557.
15. Nikol'skii, V. G., Plate, I. V., Fazlyev, F. A., Fedorova, E. A., Filippov, V. V., & Yudaeva, L. V. (1983). Structure of Polyole fins Thin Films, Obtained by Quenching of Melt up to 77K, Vysokomolek Soed A, 25(11), 2366–2371.
16. Flory, P. J. (1984). Conformations of Macro Molecules in Condensed Phases, Pure Applied Chemistry, 56(3), 305–312.
17. Flory, P. J. (1976). Spatial Configuration of Macromolecular Chains, British Polymer J, 8,1–10.
18. Kozlov, G. V., & Zaikov, G. E. (2004). Structure of the Polymer Amorphous State, utrecht, Boston, Brill Academic Publishers, 465p.
19. Kozlov, G. V., Shogenov, V. N., & Mikitaev, A. K. (1988). A Free Volume Role in Amorphous Polymers Forced Elasticity Process, Doklady AN SSSR, 298(1), 142–144.
20. Kozlov, G. V., Afaunova, Z. I., & Zaikov, G. E. (2002). The Theoretical Estimation of Polymers Yield Stress, Electronic Zhurnal "Issledovano v Rossii", 98p 1071–1080 http://zhurnal ape relarn ru/articles/ (2002)/098.PDF
21. Liu, R. S., & Li, J. Y. (1989). On the Structural Defects and Microscopic Mechanism of the High Strength of Amorphous Alloys, Mater Science Engineering A, 114(1), 127–132.
22. Alexanyan, G. G., Berlin, A. A., Gol'danskii, A. V., Grineva, N. S., Onitshuk, V. A., Shantarovich, V. P., & Safonov, G. P. (1986). Study by the Positrons Annihilation Method of Annealing and Plastic Deformation Influence on Polyarylate Microstructure. Khimicheskaya Fizika, 5(9), 1225–1234.
23. Balankin, A. S., Bugrimov, A. L., Kozlov, G. V., Mikitaev, A. K., & Sanditov, D. S. (1992). The Fractal Structure and Physical-Mechanical Properties of Amorphous Glassy Polymers, Doklady, AN, 326(3), 463–466.

24. Kozlov, G. V., & Novikov, V. U. (2001). The Cluster Model of Polymers Amorphous State, Uspekhi Fizicheskikh Nauka, *171(7)*, 717–764.
25. McClintock, F. A., & Argon, A. S. (1966). Mechanical Behavior of Materials, Massachusetts, Addison-Wesley Publishing Company, Inclusive, 432p.
26. Wu S. (1992). Secondary Relaxation, Brittle-Ductile Transition Temperature, and Chain Structure Journal Applied Polymer Science, *46(4)*, 619–624.
27. Sanditov, D. S., & Bartenev, G. M. (1982). Physical Properties of Disordered Structures, Novosibirsk, Nauka, 256p.
28. Mil'man, L. D., & Kozlov, G. V. (1986). Polycarbonate Non elastic Deformation Simulation in Impact Loading Conditions with the Aid of Dislocations Analogues, in Book Poly condensation Processes and Polymers, Nal'chik, KBSUp 130–141.
29. Shogenov, V. N., Kozlov, G. V., & Mikitaev, A. K. (1989). Prediction of Forced Elasticity of Rigid-Chain Polymers, Vysokomolek Soed A, *31(8)*, 1766–1770.
30. Sanditov, D. S., & Kozlov, G. V. (1993). About Correlation Nature between Elastic Moduli and Glass Transition Temperature of Amorphous Polymers, Fizika i Khimiya Stekla, *19(4)*, 593–601.
31. Peschanskaya, N. N., Bershtein, V. A., & Stepanov, V. A. (1978). The Connection of Glassy Polymers Creep Activation Energy with Cohesion Energy, Fizika Tverdogo Tela, *20(11)*, 3371–3374.
32. Shogenov, V. N., Belousov, V. N., Potapov, V. V., Kozlov, G. V., & Prut, E. V. (1991). The Glassy Polyarylate Sulfone Curves Stress-Strain Description within the Frameworks of High-Elasticity Concepts, Vysokomolek Soed A, *33(1)*, 155–160.
33. Mashukov, N. I., Gladyshev, G. P., & Kozlov, G. V. (1991). Structure and Properties of High Density Polyethylene Modified by High-Disperse Mixture Fe and FeO, Vysokomolek Soed A, *33(12)*, 2538–2546.
34. Beloshenko, V. A. & Kozlov, G. V. (1994). The Cluster Models Applications for Epoxy Polymers Yielding Process, Mechanika Kompozitnykh Material, *30(4)*, 451–454.
35. Balankin, A. S., & Bugrimov, A. L. (1992). The Fractal Theory of Polymers Plasticity, Vysokomolek Soed A, *34(10)*, 135–139.
36. Andrianova, G. P., & Kargin, V. A. (1970). To Necking theory at Polymers Tension, Vysokomolek Soed A, *12(1)*, 3–8.
37. Kachanova, I. M., & Roitburd, A. L. (1989). Plastic Deformation Effect on New Phase Inclusion Equilibrium Form and Thermo dynamical Hysteresis, Fizika Tverdogo Tela, *31(4)*, 1–9.
38. Hartmann, B., Lee, G. F., & Cole, R. F. (1986). Tensile Yield in Polyethylene, Polymer Engineering Science, *26(8)*, 554–559.
39. Kozlov, G. V., Beloshenko, V. A., Gazaev, M. A., & Novikov, V. U. (1996). Mechanisms of Yielding and Forced High-Elasticity of Cross-Linked Polymers, Mekhanika Kompozitnykh Materialov, *32(2)*, 270–278.
40. Balankin, A. S. (1991). Synergetics of Deformable body Moscow, Publishers of Ministry of Defence SSSR, 404p.
41. Filyanov, E. M. (1987). The Connection of Activation Parameters of Cross-Linked Polymers Deformation in Transient Region with Glassy State Properties Vysokomolek Soed A, *29(5)*, 975–981.
42. Kozlov, G. V., & Sanditov, D. S. (1992). The Activation Parameters of Glassy Polymers Deformation in Impact Loading Conditions Vysokomolek Soed B, *34(11)*, 67–72.
43. Kozlov, G. V., Shustov, G. B., Zaikov, G. E., Burmistr, M. V., & Korenyako, V. A. (2003). Polymers Yielding Description within the Frame works of Thermo dynamical Hierarchical Model, Voprosy Khimii I Khimicheskoi Tekhnologii, *1*, 68–72.
44. Gladyshev, G. P. (1988). Thermodynamic and Macro Kinetics of Natural hierarchical Processes, Moscow, Nauka, 290p.

45. Gladyshev, G. P., & Gladyshev, D. P. (1994). The Approximate Thermo Dynamical Equation for None Equilibrium Phase Transitions, Zhurnal Fizicheskoi Khimii, *68(5)*, 790–792.
46. Kozlov, G. V. & Zaikov, G. E. (2003). Thermodynamics of Polymer Structure Formation in an Amorphous State, In Book Fractal Analysis of Polymers from Synthesis to Composites Kozlov, G. V., Zaikov, G. E., &Novikov, V. U., (Ed) New York, Nova Science Publishers, Inc, 89–97.
47. Matsuoka S. & Bair, H. E. (1977). The Temperature Drop in Glassy Polymers during Deformation, Journal Applied Physics, *48(10)*, 4058–4062.
48. Kozlov, G. V., & Zaikov, G. E. (2003). Formation Mechanisms of Local Order in Polymers Amorphous State Structure, Izvestiya KBNC RAN, *1(9)*, 54–57.
49. Kozlov, G. V., & Sanditov, D. S. (1994). An Harmonic Effects and Physical-Mechanical Properties of Polymers, Novosibirsk, Nauka, 261p.
50. Matsuoka S., Aloisio, C. Y., & Bair, H. E. (1973). Interpretation of Shift of Relaxation Time with Deformation in Glassy Polymers in Terms of Excess Enthalpy Journal Applied Physics, *44(10)*, 4265–4268.
51. Kosa, P. N., Serdyuk, V. D., Kozlov, G. V., & Sanditov, D. S. (1995). The Comparative Analysis of Forced Elasticity Process for Amorphous and Semi crystalline Polymers, Fizika I Tekhnika Vysokikh Davlenii, *5(4)*, 70–81.
52. Kargin, V. A., & Sogolova, I. I. (1953). The Study of Crystalline Polymers Mechanical Properties I Polyamides Zhurnal Fizicheskoi Khimii, *27(7)*, 1039–1049.
53. Gent, A. N., & Madan, S. (1989). Plastic yielding of Partially Crystalline Polymers Journal Polymer Sciience" Part B Polymer Physics, *27(7)*, 1529–1542.
54. Mashukov, N. I., Belousov, V. N., Kozlov, G. V., Ovcharenko, E. N., & Gladychev, G. P. (1990). The Connection of Forced Elasticity Stress and Structure for Semi Crystalline Polymers, Izvestiya AN SSSR, seriya khimicheskaya, *9*, 2143–2146.
55. Sanditov, D. S., Kozlov, G. V., Belousov, V. N., & Lipatov Yu S. (1994). The Cluster Model and Fluctuation Free Volume Model of Polymeric Glasses, Fizika i Khimiya Stekla, *20(1)*, 3–13.
56. Graessley, W. W., & Edwards, S. F. (1981). Entanglement Interactions in Polymers and the Chain Contour Concentration, Polymer, *22(10)*, 1329–1334.
57. Adams, G. W., & Farris, R. Y. (1989). Latent Energy of Deformation of Amorphous Polymers 1 Deformation Calorimetry Polymer, *30(9)*, 1824–1828.
58. Oldham, K., & Spanier, J. (1973). Fractional Calculus London, New York, Academic Press, 412p.
59. Samko, S. G., Kilbas, A. A., & Marishev, O. I. (1987). Integrals and Derivatives of Fractional Order and their Some Applications, Minsk, Nauka i Tekhnika, 688p.
60. Nigmatullin, R. R. (1992). Fractional Integral and its Physical Interpretation, Teoreticheskaya i Matematicheskaya Fizika, *90(3)*, 354–367.
61. Feder, F. (1989). Fractals, New York, Plenum Press, 248p.
62. Kozlov, G. V., Shustov, G. B., & Zaikov, G. E. (2002). Polymer Melt Structure Role in Hetero Chain Polyeters Thermo Oxidative Degradation Process, Zhurnal Prikladnoi Khimii, *75(3)*, 485–487.
63. Kozlov, G. V., Batyrova, H. M., & Zaikov, G. E. (2003). The Structural Treatment of a number of Effective Centers of Polymeric Chains in the Process of the Thermo Oxidative Degradation, J Applied Polymer Science, *89(7)*, 1764–1767.
64. Kozlov, G. V., Sanditov, D. S., & Ovcharenko, E. N. November (2001). Plastic deformation Energy and Structure of Amorphous Glassy Polymers, Preceding of International Interdisciplinary Seminar "Fractals and Applied Synergetics–FiAS-01", 26–30, Moscow, 81–83.
65. Meilanov, R. P., Sveshnikova, D. A., & Shabanov, O. M. (2001). Sorption Kinetics in Systems with Fractal Structure, Izvestiya VUZov, Severo-Kavkazsk, Region, Estestv Nauka, *1*, 63–66.

66. Kozlov, G. V., & Mikitaev, A. K. (2007). The Fractal Analysis of Yielding and Forced High-Elasticity Processes of Amorphous Glassy Polymers, Mater I-th All-Russian Science-Technology Conference "Nanostructures in Polymers and Polymer Nano Composites", Nal'chik, KBSU, 81–86.

67. Kekharsaeva, E. R., Mikitaev, A. K., & Aleroev, T. S. (2001). Models of Stress-Strain Characteristics of Chloro-Containing Polyesters On the Basis of Derivatives of Fractional order, Plastics Massy, *3*, 35.

68. Novikov, V. U., & Kozlov, G. V. (2000). Structure and Properties of the Polymers within the Frameworks of Fractal Approach, Uspekhi Khimii, *69(6)*, 572–599.

69. Shogenov, V. N., Kozlov, G. V., & Mikitaev, A. K. (2007). Prediction of Mechanical Behavior, Structure and Properties of Film polymer Samples at Quasistatic Tension, in Book Poly condensation Reactions and Polymers, Selected Works Nal'chik, KBSU, 252–270.

70. Lur'e, E. G., & Kovriga, V. V. (1977). To Question about Unity of Deformation Mechanism and Spontaneous Lengthening of Rigid-Chain Polymers, Mekhanika Polimerov, *4*, 587–593.

71. Lur'e, E. G., Kazaryan, L. G., Kovriga, V. V., Uchastkina, E. L., Lebedinskaya, M. L., Dobrokhotova, M. L., & Emel'yanova, L. M. (1970). The Features of Crystallization and Deformation of Polyimide Film PM Plastics Massy, *8*, 59–63.

72. Shogenov, V. N., Kozlov, G. V., & Mikitaev, A. K. (1989). Prediction of Rigid-Chain Polymers Mechanical Properties in Elasticity Region, Vysokomolek Soed B, *31(7)*, 553–557.

73. Kozlov, G. V., Yanovskii, Yu G., & Karnet, Yu N. (2008). The Generalized Fractal Model of Amorphous Glassy Polymers yielding Process, Mekhanika Kompozitsionnykh Materialov i Konstruktsii, *14(2)*, 174–187.

74. Kopelman, R. (1986). Exciton Dynamics Mentioned Fractal one Geometrical and Energetic Disorder, in Book Fractals in Physics Pietronero, L., Tosatti, E. (Ed) Amsterdam, Oxford, New York, Tokyo, North-Holland, 524–527.

75. Korshak, V. V., & Vinogradova, S. V. (1972). Non Equilibrium Poly Condensation, Moscow, Nauka, 696p.

76. Vatrushin, V. E., Dubinov, A. E., Selemir, V. D., & Stepanov, N. V. (1995). The Analysis of SHF Apparatus Complexity with Virtual Cathode as Dynamical Objects, in Book Fractals in Applied Physics Dubinov, A. E. (Ed) Arzamas-16, VNIIEF, 47–58.

77. Mikitaev, A. K., & Kozlov, G. V. (2008). Fractal Mechanics of Polymer Materials, Nal'chik, Publishers KBSU, 312p.

78. Kozlov, G. V., Belousov, V. N., & Mikitaev, A. K. (1998). Description of Solid Polymers as Quasitwophase Bodies, Fizika i Tekhika Vysokikh Davlenii, *8(1)*, 101–107.

79. Aloev, V. Z., Kozlov, G. V., & Beloshenko, V. A. (2001). Description of Extruded Componors Structure and Properties within the Frameworks of Fractal Analysis, Izvestiya VUZov, Severo-Kavkazsk Region, estesv Nauki, *1*, 53–56.

80. Argon, A. S., & Bessonov, M. I. (1977). Plastic Deformation in Poly imides, with New Implications on the Theory of Plastic Deformation of Glassy Polymers, Philosophy Magnetic, *35(4)*, 917–933.

81. Beloshenko, V. A., Kozlov, G. V., Slobodina, V. G., Prut, E. U., & Grinev, V. G. (1995). Thermal Shrinkage of Extrudates of Ultra-High-Molecular Polyethylene and Polymerization-Filled Compositions on its Basis, Vysokomolek Soed B, *37(6)*, 1089–1092.

82. Balankin, A. S., Izotov, A. D., & Lazarev, V. B. (1993). Synergetics and Fractal Thermo Mechanics of Inorganic Materials I Thermo Mechanics of Multi Fractals, Neorganicheskie Materialy, *29(4)*, 451–457.

83. Haward, R. N. (1993). Strain Hardening of Thermoplastics. Macromolecules, *26(22)*, 5860–5869.

84. Balankin, A. S. (1992). Elastic Properties of Fractals and Dynamics of Solids Brittle Fracture, Fizika Tverdogo Tela, *34(3b)*, 1245–1258.

85. Kozlov, G. V., Afaunova, Z. I., & Zaikov, G. E. (2005). Experimental Estimation of Multi Fractal Characteristics of Free Volume for Poly (Vinyl Acetate), Oxidation Communication, 28(4), 856–862.
86. Hentschel, H. G. E., Procaccia, I. (1983). The Infinite Numbers of Generalized Dimensions of Fractals and Strange Attractors Physics D, 8(3), 435–445.
87. Williford, R. E. (1988). Multi Fractal Fracture, Scripta Metallurgica, 22(11), 1749–1754.
88. Kalinchev, E. L., & Sakovtseva, M. B. (1983). Properties and Processing of Thermo Plastics Leningrad, Khimiya, 288p.
89. Sanditov, D. S., & Sangadiev, Sh S. (1988). About Internal Pressure and Micro Hardness of Inorganic Glasses, Fizika i Khimiya Stekla, 24(6), 741–751.
90. Balankin, A. S. (1991). The Theory of Elasticity and Entropic High-Elasticity of Fractals, Pis'ma v ZhTF, 17(17), 68–72.
91. Kozlov, G. V., Gazaev, M. A., Novikov, V. U., & Mikitaev, A. K. (1996). Simulations of Amorphous Polymers Structure as Percolation Cluster, Ris'ma v ZhTF, 22(16), 31–38.
92. Serdyuk, V. D., Kosa, P. N., & Kozlov, G. V. (1995). Simulation of Polymers Forced Elasticity Process with the Aid of Dislocation Analogues, Fizika i Tekhnika Vysokikh Davlenii, 5 (3), 37–42.
93. Kozlov, G. V. November 1, (2002). The Dependence of Yield Stress on Strain Rate for Semi Crystalline Polymers, Manuscript disposed to V I N I I I RAN, Moscow, (1884–B2002).
94. Kozlov, G. V., Shetov, R. A., & Mikitaev, A. K. (1987). Methods of Elasticity Modulus Measurement in Polymers Impact Tests, Vysokomolek Soed Av, 29(5), 1109 1110.
95. Sanditov, D. S., & Kozlov, G. V. (1995). Anharmonicity of Inter Atomic and Intermolecular Bonds and Physical-Mechanical Properties of Polymers, Fizika i Khimiya Stekla, 21(6), 547–576.
96. Belousov, V. N., Kozlov, G. V., Mashukov, N. I., & Lipatov, Yu S. (1993). The Application of Dislocation Analogues for Yield in Process Description in Crystallizable Polymers, Doklady, AN, 328(6), 706–708.
97. Aharoni, S. M. (1985). Correlations between Chain Parameters and Failure Characteristics of Polymers below their Glass Transition Temperature, Macro Molecules, 18(12), 2624–2630.
98. Crist, B. (1989). Yielding of Semi Crystalline Poly Ethylene a Quantatives Dislocation Model, Polymer Communication, 30(3), 69–71.
99. Wu, S. (1989). Chain Structure and Entanglement J Polymer Science Part B Polymer Physics, 27 (4), 723–741.
100. Kozlov, G. V., Beloshenko, V. A., & Varyukhin, V. N. (1988). Simulation of Cross-Linked Polymers Structure as Diffusion-Limited Aggregate, Ukrainskii Fizicheskii Zhurnal, 43(3), 322–323.
101. Brown, N. (1971). The Relationship between Yield Point and Modulus for Glassy Polymers, Mater Science Engineering, 8(1), 69–73.
102. Kozlov, G. V., Shustov, G. B., Aloev, V. Z., & Ovcharenko, E. N. (2002). Yielding Process of Branched Poly Ethylenes, Proceedings of II All-Russian Science-Practical Conference "innovations in Mechanical Engineering", Penza, PSU, 21–23.
103. Mandelkern, L. (1985). The Relation between Structure and Properties of Crystalline Polymers, Polymer J, 17(1), 337–350.
104. Peacock, A. J., & Mandelkern, L. (1990). The Mechanical Properties of Random Copolymers of Ethylene Force-Elongation Relations Journal Polymer Science Part B Polymer Physics, 28(11), 1917–1941.
105. Kennedy, M. A., Reacock, A. J., & Mandelkern, L. (1994). Tensile Properties of Crystalline Polymers Linear Polyethylene Macromolecules, 27(19), 5297–5310.

CHAPTER 8

A NOTE ON MODIFICATION OF EPOXY RESINS BY POLYISOCYANATES

N. R. PROKOPCHUK[1], E. T. KRUTS'KO[2], and F. V. MOREV[3]

[1]Corresponding Member of National Academy of Sciences of Belarus, Doctor of Chemical Sciences, Professor, Head of Department (BSTU);

[2]Doctor of Technical Sciences, Professor (BSTU);

[3]Postgraduate Belarusian State Technological University Sverdlova Str.13a , Minsk, Republic of Belarus; E-mail: prok_nr@mail.by

CONTENTS

Abstract 256
8.1   Introduction 256
8.2   Experimental Part...257
8.3   Conclusion 261
Keywords 262
References...262

ABSTRACT

Epoxy resin compositions modified by compounds containing isocyanate fragments are recently of considerable scientific and practical interest. However, only some of them are widely used. Urethane fragments in the molecular structure of epoxy units of oligomeric molecules cause the improvement of the deformation-strength properties of polymer materials. The aim of this work is the development and study of film-forming composites with improved physical, mechanical and chemical properties by modifying the epoxy oligomers by the polyisocyanates.

8.1   INTRODUCTION

Epoxy and polyurethane materials have superior performance properties, so they are used for the production of high quality coatings [1]. Each of the mentioned types of polymers has its advantages and disadvantages: epoxies have low shrinkage during hardening, high chemical resistance, hardness, adhesion to polar surfaces and high dielectric performance, but they are inferior to polyurethane materials in resistance to aromatized fuel, abrasion resistance, adhesion to aluminum and nonferrous metals. On the other hand, polyurethanes have limited resistance to alkalis and acids and they are inferior to epoxy resin in strength and hardness.

There are different ways of mutual modification of both types of polymers, which are nowadays presented in numerous publications, mostly in the patents, the number of which is progressively increasing. Analyzing the available references on the subject, we can identify the following ways of obtaining epoxy-urethane coatings: coatings obtained with the use of epoxy-urethane oligomers; coatings produced by hydroxyl-containing epoxy oligomers, isocyanates and their adducts, coatings produced by oligo-epoxides, polyisocyanates and other reactivity compounds, coatings produced without the use of isocyanates. In order to improve chemical resistance as well as the elasticity of epoxy resin composition blocked isocyanates were used on the basis of the epoxy resin ED-20 as a modifier [2]. As shown by the authors, the reaction of an epoxy resin with an isocyanate at a high temperature results in forming heat resistant poly oxazolidones. This gives hardness to the coating, improves its mechanical and electrical properties, and urethane linkage of the polymer coating, formed alongside with poly oxazolidones, makes it elastic.

It is known that by varying the chemical structure of hydroxyl component and polyisocyanate, we can alter significantly the hardening process conditions (temperature, rate of drying) as well as performance characteristics of the coatings. Due to the diversity of the chemical structure of the raw materials, the properties of anticorrosion coatings of paints and lacquers can vary widely. In order to properties of the resulting composite film-forming systems on the basis of epoxides and polyisocyanates it is necessary to study the effect of the quantitative and qualitative structure of the film-forming epoxy-isocyanate compositions on the physical and mechanical properties of coatings.

8.2 EXPERIMENTAL PART

This chapter presents the results of complex study of performance characteristics of coatings on the basis of epoxy resin modified by aliphatic polyisocyanate promising to protect steel constructions from corrosion in a humid atmosphere. It is known that an intense destruction of the polymer coating and corrosion of protected metal occur during the use under various atmospheric conditions.

In this connection the evaluation of the coatings resistance under atmospheric conditions and the prediction of their service life is an important task.

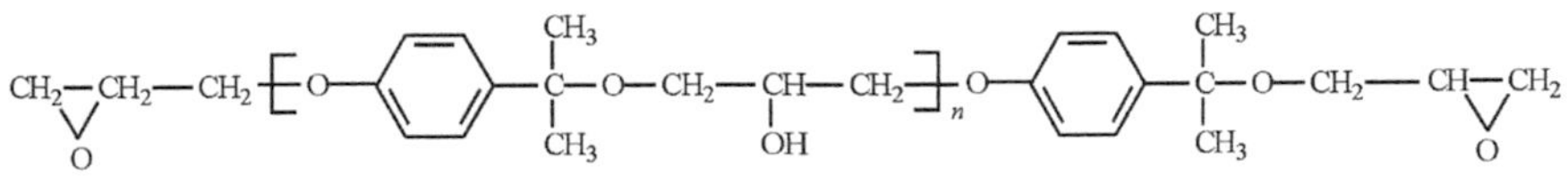

SCHEME 8.1

Epoxy brand ED-20 was used as the object of study (Scheme 8.1), the characteristics of which are presented in Table 8.1.

TABLE 8.1 Properties of Epoxy Resin ED-20

Index	Value
The average molecular weight	390–430
Epoxy group content, %	19.9–22.0
Hydroxyl content, %, not more	1.7
Density at 20 ° C, kg/m³	1166
Viscosity at 20 °C, Pa • s	13–28
Volatile substances content, %, not more	1.0

It is known that aliphatic isocyanates are less prone to yellowing under UV light, that's why they are preferred in varnish compositions. In this context, commercially produced aliphatic polyisocyanate 2K 100 was chosen as a modifier. Polyethylene polyamine was used as a hardener (TU 6–02–594–85).

Film-forming composites were obtained by adding a modifier in the amount of 1–5% by weight of dry matter to epoxy ED-20 resin and hardener (PEPA), followed by stirring the mixture to obtain a homogeneous mass. Films on the metal (copper, steel) and glass substrates were cast from the obtained lacquer solutions. The coatings were made with the help of hydraulic spraying. Plates with formed coatings were stored for 7 days at a temperature of 20–26°C before testing.

Impact resistance of coating samples was evaluated using the device U1 according to the standard ISO 6272 and State Standard 4765–73. The method of determining the strength of films being hit is based on the instantaneous deformation of the metal plate with a varnished coating during the load free fall on the specimen.

Hardness of a varnished coating was determined with the use of a pendulum device (ISO 1522). The essence of the method is to determine the decay time (the number of oscillations) of the pendulum in contact with its varnished coating.

Adhesive strength of the formed coating was determined in accordance with the standard method as well as ISO 2409 and State Standard 15140–78 using the cross-cut method. The essence of the method is to apply cross-cut of the finished coating, followed by a visual assessment of the state of the lattice coating.

Flexural strength of the coatings was determined using the device SHG1 (ISO 1519, State Standard 6806–73). To perform the determination a coated sample is bending slowly around the test cylinder, starting with a larger diameter at an angle of 180°. At one of the diameters of the cylinder a coating either cracks or breaks. In this case, we assume that the coating has an elasticity of the previous diameter of the test cylinder device in which it is not destroyed.

In epoxy-isocyanate catalyst systems without a catalyst, the main reaction at temperatures below 60 °C is the reaction of urethane formation through the interaction of isocyanate and secondary hydroxyl groups of epoxy-oligomers (Scheme 8.2).

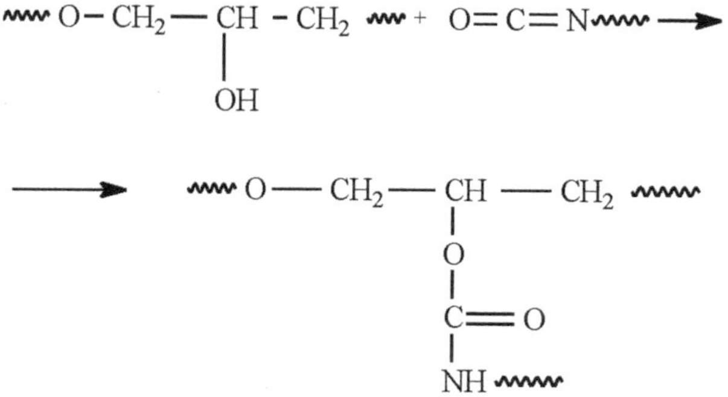

$$\text{\raise2pt\hbox{$\sim\!\!\sim$}}\ O-CH_2-CH-CH_2\ \sim\!\!\sim\ +\ O{=}C{=}N\!\!\sim\!\!\sim \longrightarrow$$
$$\underset{OH}{|}$$

$$\longrightarrow \quad \sim\!\!\sim O - CH_2 - CH - CH_2 \sim\!\!\sim$$
$$| $$
$$O$$
$$|$$
$$C{=}O$$
$$|$$
$$NH \sim\!\!\sim$$

SCHEME 8.2

In cases when the coating formation occurs in vivo (without heat), the formation of cross-linked polymer is complicated by the fact that the reactivity system can go into the glassy state, so that the hardening reaction practically stops. Thus, low temperature hardening of epoxy composites does not enable us to produce coatings with high performance. Another significant drawback of such hardening is its duration. Therefore, the formation of coatings was performed not only under natural conditions, but also at elevated temperatures. To do this, we have chosen two heat settings: 7 days without heat (temperature I) and 2 h at 100 °C (temperature II).

To determine the optimal ratio of epoxy oligomer – modifier to perform a complex physical and mechanical tests of coatings with different contents of the modifier a number of physical and mechanical tests of coatings with different contents of the modifier was performed. The results are given in Table 8.2. As seen from Table 8.2, the bending strength reaches a maximum when the content of the modifier is 2–3%. Impact strength also increases with the increasing of the modifier content in the composition of the selected concentration range modifier. In this case, the hardness of the coating remained at an acceptable level for the operating conditions used in lacquer coatings.

Physical and mechanical properties of the coatings formed at different temperatures of hardening differ significantly. Increasing hardening temperature affects the concentration dependences of physical and mechanical characteristics of the coatings over the whole range of the investigated ratios of components. Thus, the coating containing the same amount of modifier and cured at an elevated temperature, have higher relative hardness, impact strength and elasticity of less than coatings formed without heat. Thus, the coatings containing the same amount of modifier and hardened at an elevated temperature, possess higher relative hardness, impact strength and less elasticity than those formed without heat. Such changes in the behavior of the physical and mechanical properties of the composites can be explained by a significant change in the level of molecular mobility and packing density in their transition to more "hard" conditions of hardening. However, increasing the hardening temperature does not change the general character of the modifying effect, but only determines its value.

According to the adsorption theory, the adhesive strength of the coatings is caused by the formation of physical and chemical bonds between the macromolecules and the active sites of the solid surface. The observed increase in the adhesive strength of the film-forming compositions developed by us, apparently is connected with the flexibility of the emerging polymer grid, which contributes to a more favorable arrangement of the polymer chains with respect to adhesion-active sites of the substrate. We should also note the contribution to the improvement of the adhesive strength of urethane groups capable of forming bonds of the coordination type with a metal surface.

Internal strains occurring in the process of coatings formation and resulting in appearance of local connections between structural elements and adsorption interaction of a film-forming substrate with the service, affect the adhesion strength of the film to the metal. With the introduction of the modifier the flow of relaxation processes is facilitated in a grid formed by reducing the cross-linking density of the polymer due to its lateral flexible urethane branches.

The increased content of the modifier in the epoxy resin leads to higher impact resistance of coatings. This effect is probably associated with an increase in molecular mobility due to the introduction of flexible urethane units in the spatial structure of the polymer matrix, which contributes to the dissipation of mechanical energy input stroke [3].

It should be noted that the content of the modifier in the epoxy resin over 5 wt. % led to a marked gas release in the mixture and, consequently, to the defectiveness of the produced film.

TABLE 8.2   Physical and Mechanical Properties of the Coatings, Modified by Polyisocyanate

Temperature range	Modifier content, %	Hardness, rel. u.	Bending strength, mm	Impact strength, cm	Adhesion, score
I range	0.0	0.95	20	30	1
	1.0	0.74	12	30	1
	2.0	0.59	3	35	1
	3.0	0.49	1	45	1
	4.0	0.42	1	50	1
	5.0	0.37	1	50	1
II range	0.0	0.96	15	35	1
	1.0	0.89	3	45	1
	2.0	0.72	1	50	1
	3.0	0.60	1	50	1
	4.0	0.52	1	50	1
	5.0	0.48	1	50	1

It was also determined that the increase in the polyisocyanate to 5 wt.% causes significant deterioration of the protective characteristics of the coating due to their high water absorption, that can be explained by the increased porosity and lower density of the protective film, probably due to a significant plasticizing effect of the modifier. Due to the fact that the steel constructions and devices are used not only indoors, but also outdoors, in the water environment often aggressively affecting the surface of the metal, it seemed reasonable to evaluate the water absorption of protective layers of the formed coatings from the developed film-forming epoxy compositions.

Water absorption was determined by assessing the sorption capacity of varnish to water (see Fig. 8.1).

It is known that epoxy film formers possess high chemical resistance to concentrated acids and alkalis, and are widely used in chemical industry, for this reason it is useful to study the effect of the modifier on the chemical resistance of the coatings.

The effect of the modifier on the protective properties of the coatings formed at different temperatures has been studied. Relative evaluation of protective characteristics after exposure of samples in hostile environments for a month, performed with the use of the method [4], is shown in Table 8.3.

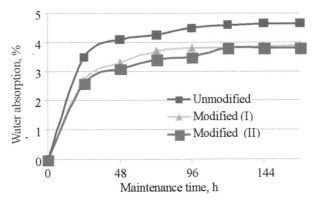

FIGURE 8.1 Water absorption of modified epoxy coatings.

As seen from Table 8.3, the obtained composites exhibit improved protective properties with respect to water, 3% NaCl, 10% NaOH, 25% H_2SO_4. These coatings can be recommended for use in the chemical industry to protect tanks, reactors, pipelines working in direct contact with hostile environments from corrosion.

TABLE 8.3 The Results of Testing Coatings Exposed in Various Environments For a Month

Modifier content, %	Relative value, %				
	H_2O	3% NaCl	10% NaOH	25% H_2SO_4	Petroleum solvent C2
0	60	40	50	20	100
3 (I drying mode)	80	70	70	50	100
3 (II drying mode)	80	70	70	50	100

Quantitative testing of the modifier content of more than 5 wt. % of isocyanate in epoxy compositions were not conducted, for the films are defective with a large number of pores due to gas release, however, a visual tendency of films to turbidity was seen, which may indicate a significant increase in water absorption with increasing modifier content in the film-forming composition.

8.3 CONCLUSION

New film-forming compositions on the basis of epoxy resin ED-20 and industrially produced isocyanate were synthesized.

Protective coatings with improved adhesion, high impact resistance, moisture- and water-resistance, and high resistance to other hostiler environments were obtained on their basis.

Optimal modifier content at which lacquer composition possesses the highest strain-strength and protective characteristics has been determined.

It has been determined that the polyisocyanate modifier catalyzes the hardening process of epoxy oligomers and increases their resistance to corrosion.

KEYWORDS

- **chemical modification**
- **cross-linked polymer**
- **epoxy materials**
- **epoxy resin**
- **film-forming composites**
- **mechanical and chemical properties**
- **polyurethane materials**

REFERENCES

1. Prokopchuk, N. R. (2004). Chemistry and Technology of Film Forming Materials, Prokopchuk, N. R., & Kruts'ko, E. T., Minsk BSTU, 423.
2. Electrically Insulating Varnish for Enameled Wires (1998) Pattern 2111994 RF C09D5/25, Fedoseyev, M. S. et.al, Institute of Technical Chemistry, the Urals, RAS–95122362/04, and Published 27.05.98, 6.
3. Zharin, D. E. (2002). Polyiso Cyanate Effect on Physical and Mechanical Properties of Epoxy Composites, Zharin, D. E., Plastics, 7, 38–41.
4. Karyakina, M. I. (1977). Laboratory Practical Tutorial on Testing Paint-work Materials and Coatings, Karyakina, M. I., Chemistry M, 240.

CHAPTER 9

TRENDS IN APPLICATION OF HYPERBRANCHED POLYMERS (HBPS) IN THE TEXTILE INDUSTRY

MAHDI HASANZADEH

Department of Textile Engineering, University of Guilan, Rasht, Iran

CONTENTS

Abstract 264
9.1   Introduction 264
9.2   Experimental Part...265
9.3   Results and Discussion 266
9.4   Conclusions...269
Keywords 269
References...269

ABSTRACT

Recently, the application of hyperbranched polymers (HBPs) in textile industry has been developed. This study focuses on the effect of hyperbranched polymer (HBP) treatment parameters such as solution concentration (wt.%), treatment temperature (°C) and time (min) on dye uptake of poly (ethylene terephthalate) (PET) fabric by acid dye. The experiments were conducted based on central composite design and mathematical model was developed.

9.1 INTRODUCTION

Over the past two decades, modification of fibers with hyperbranched polymers (HBPs) have attracted considerable attention due to their remarkable properties. These polymers are highly branched, polydisperse and three-dimensional macromolecules, which makes them a great choice to be used as modifying agent especially in textile fibers [1–3]. Many attempts have been made to develop textile fabrics with enhanced properties. For instance, in the study on applying HBP to cotton fabric [4, 5], it was demonstrated that HBP treatment on cotton fabrics has no undesirable effect on mechanical properties of fabrics. Furthermore, application of HBP to cotton fabrics reduced UV transmission and has good antibacterial activities. The study on dyeability of polypropylene (PP) fibers modified by HBP showed that the incorporation of HBP prior to fiber spinning considerably improved the color strength of polypropylene fiber with C.I. Disperse Blue 56 and has no significant effect on physical properties of the PP fibers [6]. In the most recent study in this field, fiber grade PET was compounded with polyester amide HBP and dyeability of resulted samples with disperse dyes was studied [7]. The results showed that the dyeability of dyed modified samples comprised of fiber grade PET films and a HBP were better than the neat PET and this was increased by increasing amount of HBP in presence or absence of a carrier. The dyeability of the samples was attributed to decrease in glass transition temperature for blended PET/HBP in comparison with neat PET [7]. However, there is no published report regarding the treatment of amine terminated HBPs on poly (ethylene terephthalate) (PET) fabric and study of its dyeability with acid dyes so far.

Response surface methodology (RSM) is a collection of mathematical and statistical techniques uses quantitative data from appropriate experiments to determine multiple regression equations between a response variable and a set of design variables. The main goal of RSM is to optimize the response, which is influenced by several independent variables, with minimum number of experiments [8]. Therefore, the application of RSM in HBP treatment process will be helpful in effort to find and optimize the treatment conditions.

In this paper, the effect of HBP treatment parameters such as solution concentration (wt.%), treatment temperature (°C) and time (min) on dye uptake (K/S value) of

PET fabric were investigated using central composite design (CCD). The RSM was used to evaluate main and combined effects of these parameters.

9.2   EXPERIMENTAL PART

An amine terminated HBP (Fig. 9.1) used in this study was synthesized and characterized as described in our previous research [9]. More details about experimental procedure in modification of PET fabrics with HBP can be found in it.

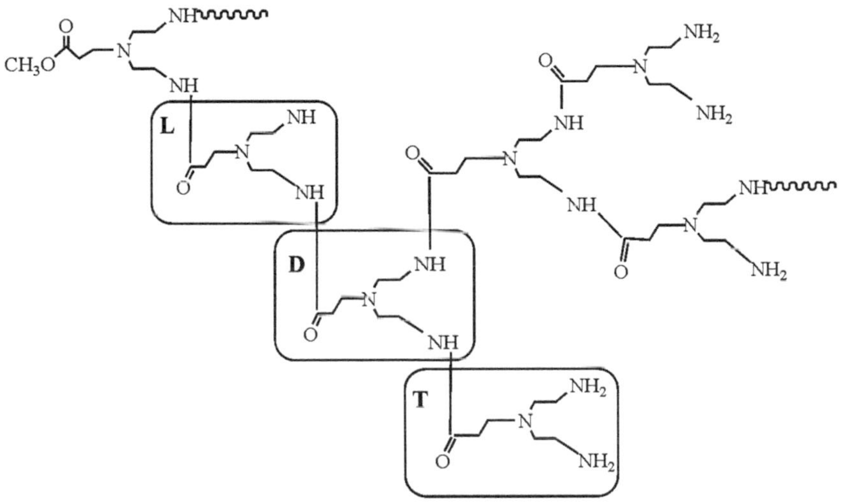

FIGURE 9.1   Chemical structure of an amine terminated hyperbranched polymer.

In this study, the dyeability of HBP treated PET fabrics was optimized with RSM by Design-Expert software including analysis of variance (ANOVA). The experiment was performed for at least three levels of each factor to fit a quadratic model. Based on preliminary experiments, HBP solution concentration (X_1), treatment temperature (X_2), and time (X_3) were determined as critical factors with significance effect on treated PET fabrics. The HBP treatment condition and fabrics K/S value are shown in Table 9.1. To evaluate the quadratic approximation of the CCD response surface model, an analysis of variance for the experimental response was carried out.

TABLE 9.1   Design of Experiment (Coded and Actual Values)

No.	Coded values			Actual values			Response
	X_1	X_2	X_3	HBP concentration (wt.%)	Temperature (°C)	Time (min)	K/S value
1	−1	−1	1	2	90	75	20.38
2	−1	1	1	2	130	75	21.57
3	−1	0	0	2	110	60	19.86
4	−1	1	−1	2	130	45	21.09
5	−1	−1	−1	2	90	45	19.78
6	0	0	0	6	110	60	22.68
7	0	0	0	6	110	60	22.18
8	0	0	0	6	110	60	22.67
9	0	0	0	6	110	60	22.97
10	0	0	0	6	110	60	22.64
11	0	0	0	6	110	60	22.26
12	0	0	−1	6	110	45	22.59
13	0	0	1	6	110	75	24.47
14	0	−1	0	6	90	60	22.58
15	0	1	0	6	130	60	26.07
16	1	−1	−1	10	90	45	23.35
17	1	−1	1	10	90	75	24.97
18	1	1	1	10	130	75	29.92
19	1	1	−1	10	130	45	26.01
20	1	0	0	10	110	60	23.23

9.3   RESULTS AND DISCUSSION

The results of analysis of variance for the response at 95% confidence level are obtained (Table 9.2) and showed that the p-value of terms X_1, X_2, X_3, X_1X_2, X_1X_3, X_1^2 and X_2^2 were significant ($p < 0.05$) and X_3^2, X_2X_3 were not significant ($p > 0.05$). Accordingly, the fitted equation in coded unit was obtained in the form of Eq. (1).

TABLE 9.2 Analysis of Variance for Response Surface

p-value	F-value	Source
<0.0001	25.98	Model
<0.0001	137.49	X_1-Concentration
<0.0001	41.33	X_2- Temperature
0.0025	16.05	X_3-Time
0.0223	7.30	X_1X_2
0.0406	5.52	X_1X_3
0.2794	1.31	X_2X_3
0.0095	10.24	X_1^2
0.0042	13.62	X_2^2
0.1177	2.93	X_3^2
Not significant	31.17	Lack of Fit
$R_{adj}^2 = 0.92$	$R^2 = 0.96$	Std.dev = 0.67

$$K/S=+22.65+2.48X_1+1.36X_2+0.85X_3$$
$$+0.64X_1X_2+0.56X_1X_3 \qquad (1)$$
$$-1.03X_1^2+1.75X_2^2$$

The high value of determination coefficient (R^2=0.96) indicated good agreements between the experimental and predicted K/S value of HBP treated PET fabrics.

Figure 9.2(a) shows the K/S values of dyed samples at different concentrations of HBP as well as at different temperatures for the constant time of the treatment. The surface plot showed that at any given temperature, the K/S values of samples increase with increasing of HBP concentration. The maximum value of K/S was observed at high HBP concentration and temperature. Figure 9.2(b) shows the response surface plot of interactions between varying HBP concentration and treatment time. For any given time, as concentration increased, the values of K/S also increased. As shown in Fig. 9.2(c), the impact of treatment temperature on K/S value of fabrics will change at different treatment time.

The plot of actual and predicted K/S values of samples with correlation coefficient of 0.96 is shown in Fig. 9.3. Actual values are the measured response data for a particular run, while the predicted values are the estimated values using the model. As shown in Fig. 9.3, there was a strong correlation between experimental and predicted values of K/S.

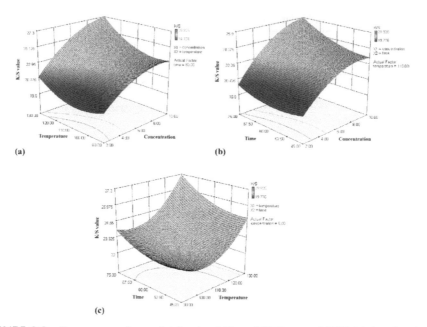

FIGURE 9.2   Response surfaces plot for dyeability of HBP treated PET fabrics showing the effect of: (a) HBP concentration and temperature, (b) HBP concentration and time, (c) temperature and time.

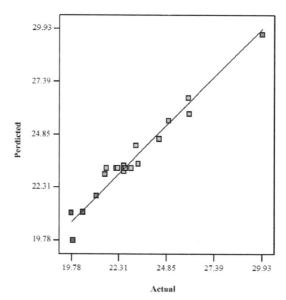

FIGURE 9.3   The actual and predicted plot for K/S value of dyed samples.

The optimal values of the HBP treatment conditions were established from the quadratic form of the RSM. Independent variables (HBP concentration, temperature, and time) were set in range and dependent variable (K/S value of dyed samples) was fixed at maximum. The optimum conditions in the tested range for the maximum K/S value of HBP treated samples were HBP concentration $(X_1) = 10$ wt.%, temperature $(X_2) = 130$ °C, and time $(X_3) = 75$ min. This optimum condition was a predicted value, thus to confirm the predictive ability of the RSM model for response, a further HBP treatment and color measurement was carried out according to the optimized conditions and the agreement between predicted and measured responses was verified.

9.4 CONCLUSIONS

The effects of hyperbranched polymer treatment conditions comprising HBP solution concentrations, treatment temperature and treatment time on dyeability (K/S value) of treated PET fabrics were investigated using response surface methodology. Central composite design was applied to determine the optimum level of the factors mentioned above. For maximum K/S value of HBP treated samples, the optimal condition with 10 wt.% of HBP, temperature 130 °C and the time of 75 min was found.

KEYWORDS

- **Dyeability**
- **Hyperbranched polymer**
- **PET fabric**
- **Response surface methodology**

REFERENCES

1. Jikei, M., & Kakimoto, M. (2001). Hyper Branched Polymers, a Promising New Class of Materials, *Progress in Polymer Science*, 26, 1233–1285.
2. Gao, C., & Yan, D. (2004). Hyper Branched Polymers, from Synthesis to Applications, *Progress in Polymer Science*, 29,183–275
3. Voit, B. I., & Lederer, A. (2009). Hyper Branched and Highly Branched Polymer Architectures Synthetic Strategies and Major Characterization Aspects, *Chemical Reviews*, 109, 5924–5973.
4. Zhang, F., Chen, Y., Lin, H., & Lu, Y. (2007). Synthesis of an Amino-Terminated Hyper Branched Polymer and its Application in Reactive Dyeing on Cotton as a Salt-Free Dyeing Auxiliary, *Coloration Technology*, 123, 351–357.
5. Zhang, F., Chen, Y., Lin, H., Wang, H., & Zhao, B. (2008). HBP-NH$_2$ Grafted Cotton Fiber, Preparation and Salt-Free Dyeing Properties, *Carbohydrate Polymer*, 74, 250–256.

6. Burkinshaw, S. M., Froehling, P. E., & Mignanelli, M. (2002). The Effect of Hyper Branched Polymers on the Dyeing of Poly propylene Fibers, *Dyes and Pigments, 53,* 229–235.
7. Khatibzadeh, M., Mohseni, M., & Moradian, S. (2010). Compounding Fiber Grade Polyethylene Terephthalate with a Hyper Branched Additive and Studying Its Dye Ability with a Disperse Dye, *Coloration Technology, 126,* 269–274.
8. Myers, R. H., Montgomery, D. C., & Anderson-Cook, C. M. (2009). Response Surface Methodology, Process and Product Optimization Using Designed Experiments, 3rd ed., John Wiley and Sons, New York.
9. Hasanzadeh, M., Moieni, T., & Hadavi Moghadam, B. (2013). Synthesis and Characterization of an Amine Terminated AB_2-type Hyper Branched Polymer and its Application in Dyeing of Poly(ethylene terephthalate) Fabric with Acid Dye, *Advances in Polymer Technology, 32,* 792–799.

CHAPTER 10

A COMPREHENSIVE REVIEW ON CHARACTERIZATION AND MODELING OF NONWOVEN STRUCTURES

M. KANAFCHIAN

University of Guilan, Rasht, Iran

CONTENTS

Abstract 272
10.1 Introduction 272
10.2 Structural Properties of Nonwoven Fabrics 291
10.3 Characterization of Nonwoven Fabrics 298
10.4 Modeling 313
10.5 Conclusion 321
Keywords 322
References 322

ABSTRACT

Nonwovens are a distinct class of textile materials made directly from fibers, thus avoiding the intermediate step of yarn production. The nonwoven industry as we know it today has grown from developments in the textile, paper and polymer processing industries. Nonwoven manufacturers operate in a highly competitive environment, where the reduction of production costs is of strategic imperative to stay in business. While the understanding and estimation of the production costs of nonwovens are very important, the literature review reveals a lack of research in this field, in the public domain. To fill the gap, this chapter is designed to lay the foundation for modeling and analysis of nonwoven structures.

10.1   INTRODUCTION

10.1.1   DEFINITION OF NONWOVEN

An average person is unlikely to be familiar with the term 'Nonwoven' and a few decades back there were no experts in this field. When the consumer hears the term Nonwovens it makes him think of something, which is not like traditional woven fabrics, something modern, advanced, hygienic, but he is not aware of any specific types of materials among those which carry the same name. The term nonwoven arises from more than half a century ago when nonwovens were often regarded as low-price substitutes for traditional textiles and were generally made from drylaid carded webs using converted textile processing machinery.

Nonwoven is a term used to describe a type of material made from textile fibers, which are not produced on conventional looms or knitting machines. They combine features from the textile, paper and plastic industries and an early description was 'web textiles' because this term reflects the essential nature of many of these products. The yarn spinning stage is omitted in the nonwoven processing of staple fibers, while bonding of the web by various methods, chemical, mechanical or thermal, replaces the weaving (or knitting) of yarns in traditional textiles. However, even in the early days of the industry, the process of stitch bonding, which originated in Eastern Europe in the 1950s, employed both layered and consolidating yarns, and the parallel developments in the paper and synthetic polymer fields, which have been crucial in shaping today's multibillion dollar nonwovens industry, had only tenuous links with textiles in the first place. But now the precise meaning of the term is somewhat clearer to the experts. According to the experts, Nonwovens is a class of textiles/sheet products, unique in industry, which is defined in the negative; that is, they are defined in what they are not.

Nonwovens are different than paper in that nonwovens usually consist entirely or at least contain a sizeable proportion of long fibers and/or they are bonded intermittently along the length of the fibers. Although paper consists of fiber webs, the fibers are bonded to each other so completely that the entire sheet comprises one unit. In nonwovens we have webs of fibers where fibers are not as rigidly bonded and to a large degree act as individuals [1].

The definitions of the nonwovens most commonly used nowadays are those by the Association of the Nonwovens Fabrics Industry (INDA) and the European Disposables and Nonwovens Association (EDANA).

INDA DEFINITION

Nonwovens are a sheet, web, or bat of natural and/or man-made fibers or filaments, excluding paper, that have not been converted into yarns, and that are bonded to each other by any of several means [2].

EDANA DEFINITION

Nonwovens are a manufactured sheet, web or bat of directionally or randomly oriented fibers, bonded by friction, and/or cohesion and/or adhesion, excluding paper or products which are woven, knitted, tufted stitch bonded incorporating binding yarns or filaments, or felted by wet milling, whether or not additionally needled. The fibers may be of natural or man-made origin. They may be staple or continuous or be formed in situ [3].

A definition of nonwoven is provided by ISO 9092:1988 which details the fibrous content and other conditions. However, felts, needled fabrics, tufted and stitch-bonded materials are usually grouped under this general heading for convenience, even though they may not strictly be described by the definition. They are made from all types of fiber, natural, regenerated man-made or synthetic or from fiber blends. One of their most significant features is their speed of manufacture, which is usually much faster than all other forms of fabric production; for example, spun bonding can be 2000 times faster than weaving. Therefore, they are very economical but also very versatile materials, which offer the opportunity to blend different fibers or fiber types with different binders in a variety of different physical forms, to produce a wide range of different properties. Nonwoven materials typically lack strength unless reinforced by a backing. They are broadly defined as sheet or web structures bonded together by entangling fiber or filaments mechanically, thermally or chemically. Nonwoven materials are flat or tufted porous sheets that are made directly from separate fibers, molten plastic or plastic film [4].

There upon, a nonwoven structure is different from some other textile structures because:

- it principally consists of individual fibers or layers of fibrous webs rather than yarns.
- it is anisotropic both in terms of its structure and properties due to both fiber alignment and the arrangement of the bonding points in its structure.
- it is usually not uniform in fabric weight and/or fabric thickness, or both.
- it is highly porous and permeable.

10.1.2 APPLICATION OF NONWOVEN

Nonwovens may be a limited-life, single-use fabric or a very durable fabric. They provide specific functions such as absorbency, liquid repellency, resilience, stretch, softness, strength, flame retardancy, wash ability, cushioning, filtering, bacterial barriers and sterility. These properties are often combined to create fabrics suited for specific jobs while achieving a good balance between product use-life and cost. They can mimic the appearance, texture and strength of a woven fabric, and can be as bulky as the thickest padding. Although it is not possible to list all the applications of nonwovens, some of the important applications are listed in Table 11.1 [5].

TABLE 10.1   Products That Use Nonwovens

Agriculture and Landscaping	Crop Covers, Turf protection products, Nursery overwintering, Weed control fabrics, Root bags, Containers, Capillary matting
Automotive	Trunk applications, Floor covers, Side liners, Front and back liners, Wheelhouse covers, Rear shelf trim panel covers, Seat applications, Listings, Cover slip sheets, Foam reinforcements, Transmission oil filters, Door trim panel carpets, Door trim panel padding, Vinyl and landau cover backings, Molded headliner substrates, Hood silencer pads, Dash insulators, Carpet tufting fabric and under, Padding
Clothing	Interlinings, Clothing and glove insulation, Bra and shoulder padding, Handbag components, Shoe components
Construction	Roofing and tile underlayment, Acoustical ceilings, Insulation, House wrap, Pipe wrap
Geotextiles	Asphalt overlay, Road and railroad beds, Soil stabilization, Drainage, Dam and stream embankments, Golf and tennis courts, Artificial turf, Sedimentation and erosion, Pond liners
Home Furnishings	Furniture construction sheeting, Insulators, arms, Cushion ticking, Dust covers, Decking, Skirt linings, Pull strips, Bedding construction sheeting, Quilt backing, Dust covers, Flanging, Spring wrap, Quilt backings, Blankets, Wall covering backings, Acoustical wall coverings, Upholstery backings, Pillows, pillow cases, Window treatments, Drapery components, Carpet backings, carpets, Pads, Mattress pad components
Health Care	Caps, gowns, masks, Shoe covers, Sponges, dressings, wipes, Orthopedic padding, Bandages, tapes, Dental bibs, Drapes, wraps, packs, Sterile packaging, Bed linen, under pads, Contamination control gowns, Electrodes, Examination gowns, Filters for IV solutions and blood, Oxygenators and kidney, Dialyzers, Transdermal drug delivery
Household	Wipes, wet or dry polishing, Aprons, Scouring pads, Fabric softener sheets, Dust cloths, mops, Tea and coffee bags, Placemats, napkins Ironing board pads, Washcloths, Tablecloths

Industrial/Military	Coated fabrics, Filters, Semiconductor polishing pads, Wipers, Clean room apparel, Air conditioning filters, Military clothing, Abrasives, Cable insulation, Reinforced plastics, Tapes, Protective clothing, lab coats, Sorbents, Lubricating pads, Flame barriers, Packaging, Conveyor belts, Display felts, Papermaker felts, Noise absorbent felt
Leisure, Travel	Sleeping bags, Tarpaulins, tents, Artificial leather, luggage, Airline headrests, pillow cases,
Personal Care and Hygiene	Diapers, Sanitary napkins, tampons, Training pants, Incontinence products, Dry and wet wipes, Cosmetic applicators, removers, Lens tissue, Hand warmers, Vacuum cleaner bags, Tea and coffee bags, Buff pads
School, Office	Book covers, Mailing envelopes, labels, Maps, signs, pennants, Floppy disk liners, Towels, Promotional items, Pen nibs

10.1.3 TECHNOLOGIES FOR NONWOVEN MANUFACTURING [6]

All nonwovens are manufactured by two general steps, which sometimes can even be combined into one. The first is actual web formation, and the second is some method of bonding the web fibers together. Sometimes, but not always a finishing process is required to provide the specific needs of a particular end use, in a similar way to woven or knitted fabrics. Nonwoven materials are produced by a wide variety of processes ranging from hydraulic formation, air assisted processing, direct polymer to web systems, and combinations of these. These materials find application in diverse fields that include filtration, automotive, hygiene products, battery separators, medical, and home furnishings. In the wet-laid process, a mixture of fibers is suspended in a fluid, the fibers are deposited onto a screen or porous surface to remove the fluid and the web is then consolidated mechanically, chemically, or thermally. Wet-laid materials have outstanding uniformity when compared with other nonwovens. Meltblown materials are produced by melting and extruding molten polymer resin directly into a web. Fine fibers with a narrow fiber distribution can be produced using the meltblown process. Composites of wet-laid and meltblown materials can be used to obtain improved performance and strength characteristics in the final application. To make finer fibers, typically in the nanometer range, an electrospinning process is used. During the electrospinning process, electrical charge is applied to polymer solution to form nanometer size fibers on a collector. The structures thus produced have a high surface area to volume ratio, high length to diameter ratio, high porosity, and interconnected pores. In fiber processing it is common to make first a thin layer of fiber called a web and then to lay several webs on top of each other to form a batt, which goes directly to bonding. The words web and batt are cases where it is difficult to decide if a fiber layer is a web or a batt. Nevertheless the first stage of nonwoven processing is normally called batt production.

10.1.3.1   CARDING METHOD

The machines in the nonwoven industry use identical principles and are quite similar but there are some differences. In particular in the process of yarn manufacture there are opportunities for further opening and for improving the levelness of the product after the carding stage, but in nonwoven manufacture there is no further opening at all and very limited improvements in levelness are possible. It therefore follows that the opening and blending stages before carding must be carried out more intensively in a nonwoven plant and the card should be designed to achieve more opening, for instance by including one more cylinder, though it must be admitted that many nonwoven manufacturers do not follow this maxim. Theoretically either short-staple revolving flat cards or long-staple roller cards could be used, the short-staple cards having the advantages of high production and high opening power, especially if this is expressed per unit of floor space occupied. However, the short-staple cards are very narrow, whereas long-staple cards can be many times wider, making them much more suitable for nonwoven manufacture, particularly since nonwoven fabrics are required to be wider and wider for many end-uses. Hence a nonwoven installation of this type will usually consist of automatic fiber blending and opening feeding automatically to one or more wide long-staple cards. The cards will usually have some form of auto-leveler to control the mass per unit area of the output web.

10.1.3.2   AIR LAYING METHOD

The air-laying method produces the final batt in one stage without first making a lighter weight web. It is also capable of running at high production speeds but is similar to the parallel-lay method in that the width of the final batt is the same as the width of the air-laying machine, usually in the range of 3–4 m. The degree of fiber opening available in an air-lay machine varies from one manufacturer to another, but in all cases it is very much lower than in a card (Fig. 10.1). As a consequence of this more fiber opening should be used prior to air laying and the fibers used should be capable of being more easily opened, otherwise the final batt would show clumps of inadequately opened fiber. In the past the desire for really good fiber opening led to a process consisting of carding, cross laying, then feeding the cross-laid batt to an air-laying machine.

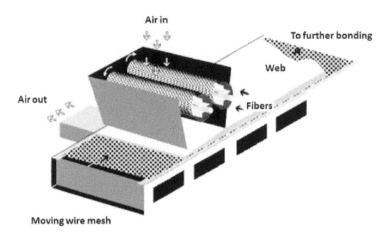

FIGURE 10.1   A schematic of Air-Laid system.

10.1.3.3   WET LAYING METHOD

The wet-laid process is a development from papermaking that was undertaken because the production speeds of papermaking are very high compared with textile production. Textile fibers are cut very short by textile standards (6–20 mm), but at the same time these are very long in comparison with wood pulp, the usual raw material for paper. The fibers are dispersed into water; the rate of dilution has to be great enough to prevent the fibers aggregating (Fig. 10.2). The required dilution rate turns out to be roughly ten times that required for paper, which means that only specialized forms of paper machines can be used, known as inclined-wire machines. In fact most frequently a blend of textile fibers together with wood pulp is used as the raw material, not only reducing the necessary dilution rate but also leading to a big reduction in the cost of the raw material. It is now possible to appreciate one of the problems of defining nonwoven. It has been agreed that a material containing 50% textile fiber and 50% wood pulp is a nonwoven, but any further increase in the wood pulp content results in a fiber-reinforced paper. A great many products use exactly 50% wood pulp. Wet-laid nonwovens represent about 10% of the total market, but this percentage is tending to decline. They are used widely in disposable products, for example in hospitals as drapes, gowns, sometimes as sheets, as one-use filters, and as cover stock in disposable nappies.

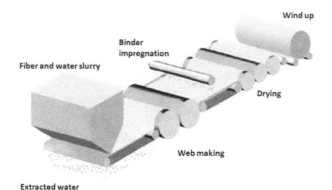

FIGURE 10.2   A schematic of Wet-Laid system.

10.1.3.4 DRY LAYING METHOD

The paper industry has attempted for many years to develop a dry paper process because of the problems associated with the normal wet process, that is, the removal of very large volumes of water by suction, by squeezing and finally by evaporation (Fig. 10.3). Now a dry process has been developed using wood pulp and either latex binders or thermoplastic fibers or powders to bond the wood pulp together to replace hydrogen bonding which is no longer possible. Owing to the similarity of both the bonding methods and some of the properties to those of nonwovens, these products are being referred to as nonwovens in some areas, although it is clear from the definition above they do not pass the percentage wood pulp criterion. Hence although the drylaid paper process cannot be regarded as a nonwoven process at present, it is very likely that the process will be modified to accept textile fibers and will become very important in nonwovens in the future.

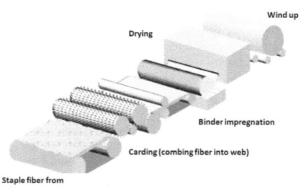

FIGURE 10.3   A schematic of Dry-Laid system.

10.1.3.5 SPUN LAYING METHOD

Spun laying includes extrusion of the filaments from the polymer raw material, drawing the filaments and laying them into a batt. As laying and bonding are normally continuous, this process represents the shortest possible textile route from polymer to fabric in one stage (Fig. 10.4). In addition to this the spun-laid process has been made more versatile. When first introduced only large, very expensive machines with large production capabilities were available, but much smaller and relatively inexpensive machines have been developed, permitting the smaller nonwoven producers to use the spun-laid route. Further developments have made it possible to produce microfibers on spun-laid machines giving the advantages of better filament distribution, smaller pores between the fibers for better filtration, softer feel and also the possibility of making lighter-weight fabrics. For these reasons spun-laid production is increasing more rapidly than any other nonwoven process. Spun laying starts with extrusion. Virtually all commercial machines use thermoplastic polymers and melt extrusion. Polyester and polypropylene are by far the most common, but polyamide and polyethylene can also be used. The polymer chips are fed continuously to a screw extruder which delivers the liquid polymer to a metering pump and thence to a bank of spinnerets, or alternatively to a rectangular spinneret plate, each containing a very large number of holes. The liquid polymer pumped through each hole is cooled rapidly to the solid state but some stretching or drawing in the liquid state will inevitably take place.

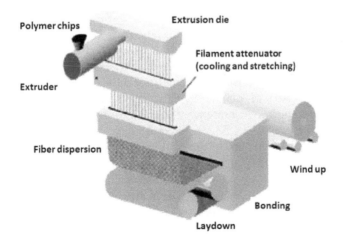

FIGURE 10.4 A schematic of Spun-Laid system.

In spun laying the most common form of drawing the filaments to obtain the correct modulus is air drawing, in which a high velocity air stream is blown past the filaments moving down a long tube, the conditions of air velocity and tube length being chosen so that sufficient tension is developed in the filaments to cause draw-ing to take place. In some cases air drawing is not adequate and roller drawing has to be used as in normal textile extrusion, but roller drawing is more complex and tends to slow the process, so that air drawing is preferred. The laying of the drawn filaments must satisfy two criteria; the batt must be as even as possible in mass per unit area at all levels of area, and the distribution of filament orientations must be as desired, which may not be isotropic. Taking the regularity criterion first, the air tubes must direct the filaments onto the conveyor belt in such a way that an even distribution is possible. However, this in itself is not sufficient because the fila-ments can form agglomerations that make 'strings' or 'ropes' which can be clearly seen in the final fabric. A number of methods have been suggested to prevent this, for instance, charging the spinneret so that the filaments become charged and repel one another or blowing the filaments from the air tubes against a baffle plate, which tends to break up any agglomerations. With regard to the filament orientation, in the absence of any positive traversing of the filaments in the cross direction, only very small random movements would take place. However, the movement of the con-veyor makes a very strong machine direction orientation; thus the fabric would have a very strong machine-direction orientation. Cross-direction movement can most easily be applied by oscillating the air tubes backwards and forwards. By control-ling the speed and amplitude of this oscillation it is possible to control the degree of cross-direction orientation.

10.1.3.6   FLASH SPINNING METHOD

Flash spinning is a specialized technique for producing very fine fibers without the need to make very fine spinneret holes. In flash spinning the polymer is dissolved in a solvent and is extruded as a sheet at a suitable temperature so that when the pres-sure falls on leaving the extruder the solvent boils suddenly (Fig. 10.5). This blows the polymer sheet into a mass of bubbles with a large surface area and consequently with very low wall thickness. Subsequent drawing of this sheet, followed by me-chanical fibrillation, results in a fiber network of very fine fibers joined together at intervals according to the method of production. This material is then laid to obtain the desired mass per unit area and directional strength. Flash-spun material is only bonded in two ways, so that it seems sensible to discuss both the bonding and the products at this juncture. One method involves melting the fibers under high pres-sure so that virtually all fibers adhere along the whole of their length and the fabric is almost solid with very little air space. This method of construction makes a very stiff material with high tensile and tear strengths. It is mainly used in competition with paper for making tough waterproof envelopes for mailing important documents and for making banknotes. The fact that the original fibers were very fine means that the material is very smooth and can be used for handwriting or printing. The alter-native method of bonding is the same, that is, heat and pressure but is only applied to small areas, say 1 mm square, leaving larger areas, say 4 mm square, completely unbonded. The bonded areas normally form a square or diagonal pattern. This material, known as Tyvek, has a lower tensile strength than the fully bonded fabric, but has good strength and is flexible enough to be used for clothing. Because the fine fibers leave very small pores in the fabric, it is not only waterproof but is also resistant to many other liquids with surface tensions lower than water. The presence of the pores means that the fabric is permeable to water vapor and so is comfortable to wear. Tyvek is used principally for protective clothing in the chemical, nuclear and oil industries, probably as protection for the armed forces and certainly in many industries not requiring such good protection but where it is found to be convenient. The garments can be produced so cheaply that they are usually regarded as disposable.

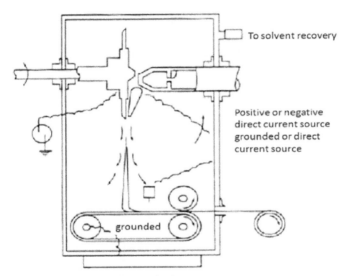

FIGURE 10.5   A schematic of Flash-spinning system.

10.1.3.7   MELT BLOWING METHOD

The process of melt blowing is another method of producing very fine fibers at high production rates without the use of fine spinnerets (Fig. 10.6). As the polymer leaves the extrusion holes it is hit by a high-speed stream of hot air at or even above its melting point, which breaks up the flow and stretches the many filaments until they are very fine. At a later stage, cold air mixes with the hot and the polymer solidifies. Depending on the air velocity after this point a certain amount of aerodynamic drawing may take place but this is by no means as satisfactory as in spun laying and the fibers do not develop much tenacity. At some point the filaments break into staple fibers, but it seems likely that this happens while the polymer is still liquid because if it happened later this would imply that a high tension had been applied to the solid fiber, which would have caused drawing before breaking. The fine staple fibers produced in this way are collected into a batt on a permeable conveyor as in air laying and spun laying. The big difference is that in melt blowing the fibers are extremely fine so that there are many more fiber-to-fiber contacts and the batt has greater integrity. For many end-uses no form of bonding is used and the material is not nonwoven, but simply a batt of loose fibers. Such uses include ultrafine filters for air conditioning and personal face masks, oil-spill absorbents and personal hygiene products. In other cases the melt-blown batt may be laminated to another nonwoven, especially a spun-laid one or the melt-blown batt itself may be bonded but the method must be chosen carefully to avoid spoiling the openness of the very fine fibers. In the bonded or laminated form the fabric can be used for breathable protective clothing in hospitals, agriculture and industry, as battery separators, industrial wipes and clothing interlinings with good insulation properties. Melt blowing started to develop rapidly in about 1975, although the process was known before 1950. It is continuing to grow at about 10% per annum.

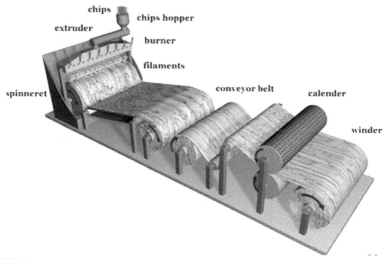

FIGURE 10.6   A schematic of Melt-Blowing system.

10.1.3.8   BONDING METHOD

The various methods for bonding are:
1.   Adding an adhesive.
2.   Thermally fusing the fibers or filaments to each other or to the other meltable fibers or powders.

3. Fusing fibers by first dissolving, and then resolidifying their surfaces.
4. Creating physical tangles or tuft among the fibers.
5. Stitching the fibers or filaments in place.

10.1.3.8.1   CHEMICAL BONDING

Chemical bonding involves treating either the complete batt or alternatively isolated portions of the batt with a bonding agent with the intention of sticking the fibers together. Although many different bonding agents could be used, the modern industry uses only synthetic lattices, of which acrylic latex represents at least half and styrene–butadiene latex and vinyl acetate latex roughly a quarter each. When the bonding agent is applied it is essential that it wets the fibers, otherwise poor adhesion will be achieved. Most lattices already contain a surfactant to disperse the polymer particles, but in some cases additional surfactant may be needed to aid wetting. The next stage is to dry the latex by evaporating the aqueous component and leaving the polymer particles together with any additives on and between the fibers. During this stage the surface tension of the water pulls the binder particles together forming a film over the fibers and a rather thicker film over the fiber intersections. Smaller binder particles will form a more effective film than larger particles, other things being equal. The final stage is curing and in this phase the batt is brought up to a higher temperature than for the drying. The purpose of curing is to develop crosslinks both inside and between the polymer particles and so to develop good cohesive strength in the binder film. Typical curing conditions are 120–140°C for 2–4 min.

10.1.3.8.2   THERMAL BONDING

Thermal bonding is increasingly used at the expense of chemical bonding for a number of reasons. Thermal bonding can be run at high speed, whereas the speed of chemical bonding is limited by the drying and curing stage. Thermal bonding takes up little space compared with drying and curing ovens. Also thermal bonding requires less heat compared with the heat required to evaporate water from the binder, so it is more energy efficient. Thermal bonding can use three types of fibrous raw material, each of which may be suitable in some applications but not in others. First, the fibers may be all of the same type, with the same melting point. This is satisfactory if the heat is applied at localized spots, but if overall bonding is used it is possible that all the fibers will melt into a plastic sheet with little or no value. Second, a blend of fusible fiber with either a fiber with a higher melting point or a nonthermoplastic fiber can be used. This is satisfactory in most conditions except where the fusible fiber melts completely, losing its fibrous nature and causing the batt to collapse in thickness. Finally, the fusible fiber may be a bicomponent fiber, that is, a fiber extruded with a core of high melting point polymer surrounded by a sheath of lower melting point polymer (Fig. 10.7). This is an ideal material to process because the core of the fiber does not melt but supports the sheath in its fibrous state. Thermal bonding is used with all the methods of batt production except the wet-laid method, but it is worth pointing out that the spun-laid process and point bonding complement each other so well that they are often thought of as the standard process.

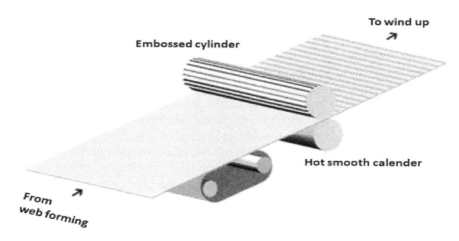

FIGURE 10.7   Thermal Bonding method.

10.1.3.8.3   SOLVENT BONDING

This form of bonding is only rarely used but it is interesting from two points of view; first, the solvent can be recycled, so the process is ecologically sound, although whether or not recycling is practical depends on the economics of recovering the solvent. Second, some of the concepts in solvent bonding are both interesting and unique. In one application of the method, a spun-laid polyamide batt is carried through an enclosure containing the solvent gas, NO_2, which softens the skin of the filaments. On leaving the enclosure bonding is completed by cold calendar rolls and the solvent is washed from the fabric with water using traditional textile equipment. This is a suitable method of bonding to follow a spun-laid line because the speeds of production can be matched. The other application uses a so-called latent solvent, by which is meant one that is not a solvent at room temperature but becomes a solvent at higher temperatures. This latent solvent is used in conjunction with carding and cross laying and is applied as a liquid before carding. The action of carding spreads the solvent and at the same time the solvent lubricates the fibers during carding. The batt is passed to a hot air oven, which first activates the solvent and later evaporates it. The product will normally be high loft, but if fewer lofts are required a compression roller could be used.

10.1.3.8.4   MECHANICAL BONDING

10.1.3.8.4.1   NEEDLE FELTING

All the methods of bonding discussed so far have involved adhesion between the fibers; hence they can be referred to collectively as adhesive bonding. The final three methods, needle felting, hydro entanglement and stitch bonding rely on frictional forces and fiber entanglements, and are known collectively as mechanical bonding. The basic concept of needle felting is apparently simple; the batt is led between two stationary plates, the bed and stripper plates as shown in Fig. 10.8. While between the plates the batt is penetrated by a large number of needles. The needles are usually made triangular and have barbs cut into the three edges. When the needles descend into the batt the barbs catch some fibers and pull them through the other fibers. When the needles return upwards, the loops of fiber formed on the down stroke tend to remain in position, because they are released by the barbs. This downward pressure repeated many times makes the batt much denser, that is, into a needle felt. The above description illustrates how simple the concept seems to be. Without going into too much detail it may be interesting to look at some of the complications. First, the needles can only form vertical loops or 'pegs' of fiber and increase the density of the batt. This alone does not form a strong fabric unless the vertical pegs pass through loops already present in the horizontal plane of the batt. It follows from this that parallel-laid fabric is not very suitable for needling since there are few fiber loops present, so most needling processes are carried out with cross laid, air-laid and spun-laid batts. Second, the amount of needling is determined partly by the distance the drawing rollers move between each movement of the needle board, the 'advance,' and partly by the number of needles per meter across the loom. If the chosen advance happens to be equal to, or even near the distance between needle rows, then the next row of needles will come down in exactly the same position as the previous row, and so on for all the rows of needles. The result will be a severe needle patterning; to avoid this distance between each row of needles must be different. Computer programs have been written to calculate the best set of row spacings. Third, if it is necessary to obtain a higher production from a needleloom, is it better to increase the number of needles in the board or to increase the speed of the needleboard. Finally, in trying to decide how to make a particular felt it is necessary to choose how many needle penetrations there will be per unit area, how deep the needles will penetrate and what type of needles should be used from a total of roughly 5000 different types. The variations possible seem to be infinite, making an optimization process very difficult.

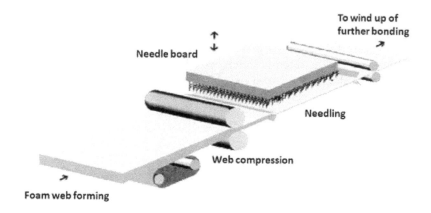

FIGURE 10.8 A schematic of Needle-Felting system.

10.1.3.8.4.2 STITCH BONDING

The idea of stitch bonding was developed almost exclusively in Czechoslovakia, in the former East Germany and to some extent in Russia, though there was a brief development in Britain. The machines have a number of variants, which are best discussed separately; many possible variants have been produced but only a limited number are discussed here for simplicity.

10.1.3.8.4.3 BATT BONDING BY THREADS

Stitch bonding uses mainly cross-laid and air-laid batts. The batt is taken into a modification of a warp-knitting machine and passes between the needles and the guide bar(s). The needles are strengthened and are specially designed to penetrate the batt on each cycle of the machine. The needles are of the compound type having a tongue controlled by a separate bar. After the needles pass through the batt the needle hooks open and the guide bar laps thread into the hooks of the needles. As the needles withdraw again the needle hooks are closed by the tongues, the old loops knock over the needles and new loops are formed. In this way a form of warp knitting action is carried out with the overlaps on one side of the batt and the underlaps on the other. Generally, as in most warp knitting, continuous filament yarns are used to avoid yarn breakages and stoppages on the machine. Two structures are normally knitted on these machines, pillar (or chain) stitch, or tricot. When knitting chain stitch the same guide laps round the same needle continuously, producing a large number of isolated chains of loops. When knitting tricot structure, the guide bar shogs one needle space to the left, then one to the right. Single-guide bar structure is called tricot, whereas the two-guide bar structure is often referred to as full tricot. The nature of this fabric is very textile-like, soft and flexible. At one time it was widely used for curtaining but is now used as a backing fabric for lamination, as covering material for mattresses and beds and as the fabric in training shoes. In deciding whether to use pillar or tricot stitch, both have a similar strength in the machine direction, but in the cross direction the tricot stitch fabric is stronger, owing to the underlaps lying in that direction. A cross-laid web is already stronger in that direction so the advantage is relatively small. The abrasion resistance is the same on the loop or overlap side, but on the underlap side the tricot fabric has significantly better resistance owing to the longer underlaps. However, continuous filament yarn is very expensive relative to the price of the batt, so tricot fabric costs significantly more.

10.1.3.8.4.4   STITCH BONDING WITHOUT THREADS

In this case the machine is basically the same as in the previous section, but the guide bar(s) are not used. The needle bar moves backwards and forwards as before, pushing the needles through the batt. The main difference is that the timing of the hook closing by the tongues is somewhat delayed, so that the hook of the needle picks up some of the fiber from the batt. These fibers are formed into a loop on the first cycle and on subsequent cycles the newly formed loops are pulled through the previous loops, just as in normal knitting. The final structure is felt-like on one side and like a knitted fabric on the other. The fabric can be used for insulation and as a decorative fabric.

10.1.3.8.4.5   STITCH BONDING IN PILE FABRIC

To form a pile fabric two guide bars are usually used, two types of warp yarns (pile yarn and sewing yarn) and also a set of pile sinkers, which are narrow strips of metal over which the pile yarn is passed and whose height determines the height of the pile. The pile yarn is not fed into the needle's hook so does not form a loop; it is held in place between the underlap of the sewing yarn and the batt itself (Fig. 10.9). It is clear that this is the most efficient way to treat the pile yarn, since any pile yarn in a loop is effectively wasted. This structure has been used for making toweling with single-sided pile and also for making loop-pile carpeting in Eastern Europe. The structure has not been popular in the West owing to competition with double-sided terry toweling and tufted carpets. Equally it has not been used in technical textiles, but it could be a solution in waiting for the correct problem. Strangely a suitable problem has been proposed. Car seating usually has a polyester face with a foam backing, but this material cannot be recycled, because of the laminated foam. It has been suggested that a polyester nonwoven pile fabric could replace the foam and would be 100% recyclable.

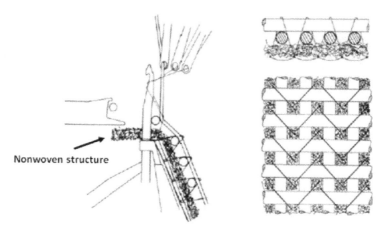

FIGURE 10.9   Stitch-Bonding method (A) and product (B).

10.1.3.8.4.6   BATT LOOPING

In this technique the needles pass through the supporting fabric and pick up as much fiber from the batt as possible. Special sinkers are used to push fiber into the needle's hook to increase this pick-up. The fiber pulled through the fabric forms a chain of loops, with loose fiber from the batt on the other surface of the fabric. The fabric is finished by raising, not as one might expect on the loose side of the fabric but instead the loops are raised because this gives a thicker pile. This structure was widely used in Eastern Europe, particularly for artificial fur, but in the West it never broke the competition from silver knitting, which gives a fabric with similar properties. The method could be used for making good quality insulating fabrics.

10.1.3.8.4.7   SWING LAID YARNS

Two distinct types of fabric can be made using the same principle. The first is a simulated woven fabric in which the cross-direction yarns are laid many at a time in a process a bit like cross laying. The machine direction yarns, if any are used, are simply unwound into the machine. These two sets of yarns are sewn together using chain stitch if there are only cross-direction threads and tricot stitch if machine direction threads are present, the under laps holding the threads down. Although fabric can be made rapidly by this system this turns out to be a situation in which speed is not everything and in fact the system is not usually economically competitive with normal weaving. However, it has one great technical advantage; the machine and cross threads do not interlace but lie straight in the fabric. Consequently the initial modulus of this fabric is very high compared with a woven fabric, which can first extend by straightening out the crimp in the yarn. These fabrics are in demand for making fiber-reinforced plastic using continuous filament glass and similar high modulus fibers or filaments. The alternative system makes a multidirectional fabric. Again sets of yarns are cross-laid but in this case not in the cross direction but at, say, 45° or 60° to the cross direction. Two sets of yarns at, say, +45° and −45° to the cross direction plus another layer of yarns in the machine direction can be sewn together in the usual way. Again high modulus yarns are used, with the advantage that the directional properties of the fabric can be designed to satisfy the stresses in the component being made.

10.1.3.8.4.8   HYDRO-ENTANGLEMENT

The process of hydro entanglement was invented as a means of producing an entanglement similar to that made by a needleloom, but using a lighter weight batt. A successful process was developed during the 1960s by Du Pont and was patented. However, Du Pont decided in the mid-1970s to dedicate the patents to the public domain, which resulted in a rush of new development work in the major industrial countries, Japan, USA, France, Germany and Britain. As the name implies the process depends on jets of water working at very high pressures through jet orifices with very small diameters (Fig. 10.10). A very fine jet of this sort is liable to break up into droplets, particularly if there is any turbulence in the water passing through the orifice. If droplets are formed the energy in the jet stream will still be roughly the same, but it will spread over a much larger area of batt so that the energy per unit area will be much less. Consequently the design of the jet to avoid turbulence and to produce a needle-like stream of water is critical. The jets are arranged in banks and the batt is passed continuously under the jets held up by a perforated screen, which removes most of the water. Exactly what happens to the batt underneath the jets is not known, but it is clear that fiber ends become twisted together or entangled by the turbulence in the water after it has hit the batt. It is also known that the supporting screen is vital to the process; changing the screen with all other variables remaining constant will profoundly alter the fabric produced. Although the machines have higher throughputs compared with most bonding systems, and particularly compared with a needleloom, they are still very expensive and require a lot of power, which is also expensive. The other considerable problem lies in supplying clean water to the jets at the correct pH and temperature. Large quantities of water are needed, so recycling is necessary, but the water picks up air bubbles, bits of fiber and fiber lubricant/fiber finish in passing through the process and it is necessary to remove everything before recycling.

It is said that this filtration process is more difficult than running the rest of the machine. Fabric uses include wipes, surgeons' gowns, disposable protective clothing and backing fabrics for coating. The wipes produced by hydro entanglement are guaranteed lint free, because it is argued that if a fiber is loose it will be washed away by the jetting process. It is interesting to note that the hydro entanglement process came into being as a process for entangling batts too light for a needleloom, but that the most recent developments are to use higher water pressures (400 bar) and to process heavier fabrics at the lower end of the needleloom range.

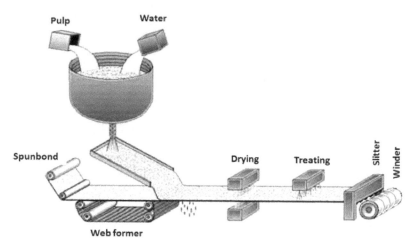

FIGURE 10.10   A schematic of hydro-entanglement process.

10.1.4. PRODUCTION AND SHIPMENTS OF NONWOVENS

The Nonwoven industry is one of the fastest growing industries in the world. It is rapidly developing a sophisticated and diverse market. It has been exhibiting an average growth of about 10% over the past 20 years and should continue this rate of growth in the next ten years. The technology in Nonwoven industry has been improved significantly in nearly all available major manufacturing processes, including those of spun bond, meltblown, needle punched, spunlaced, wet laid and dry laid fabrication. The most important point in rapid development and commercial acceptance of nonwovens is the ability to produce materials of special properties in less time and at reasonable prices. The relative production speeds of various textile technologies are compared in Table 10.2.

A large number of fibers are available in the market, but the Nonwovens market is mainly dominated by three fibers, namely polyolefin's, polyester, and rayon. These three fiber types make up a substantial part of the overall Nonwovens markets for fibers. The North American Nonwovens industry is the largest in the world and accounts for almost one third of the worldwide sales of roll goods, according to estimates from INDA. A major portion of the polyolefin and polyester fiber in the U.S. market is consumed by the Nonwovens industry [7, 8].

TABLE 10.2 Relative Production Rates of Different Textile Technologies

Technology	Relative Production Rate
Weaving	1–6
Knitting	3–16
Nonwovens – web forming:	
➢ Carding	120–400
➢ Spunbond	200–2000
➢ Wet-laid	2300
Nonwovens – bonding	
➢ Stitch bonding	40
➢ Needling	30–500
➢ Calendaring	2000
➢ Hot air bonding	5000

10.2 STRUCTURAL PROPERTIES OF NONWOVEN FABRICS

10.2.1 CLASSIFICATION

The structure and properties of a nonwoven fabric are determined by fiber properties, the type of bonding elements, the bonding interfaces between the fibers and binder elements and the fabric structural architecture. Examples of dimensional and structural parameters may be listed as follows:

1. Fiber dimensions and properties: fiber diameter, diameter variation (e.g., in meltblown microfiber and electrospun nanofiber webs), cross-sectional shape, crimp wave frequency and amplitude, length, density; fiber properties (Young's modulus, elasticity, tenacity, bending and torsion rigidity, compression, friction coefficient), fibrillation propensity, surface chemistry and wetting angle.
2. Fiber alignment: fiber orientation distribution.
3. Fabric dimensions and variation: dimensions (length, width, thickness, and weight per unit area), dimensional stability, density and thickness uniformity.
4. Structural properties of bond points: bonding type, shape, size, bonding area, bonding density, bond strength, bond point distribution, geometrical arrangement, the degree of liberty of fiber movement within the bonding

points, interface properties between binder and fiber; surface properties of bond points.

5. Porous structural parameters: fabric porosity, pore size, pore size distribution, pore shape.

Examples of important nonwoven fabric properties are:

• Mechanical properties: tensile properties (Young's modulus, tenacity, strength and elasticity, elastic recovery, work of rupture), compression and compression recovery, bending and shear rigidity, tear resistance, burst strength, crease resistance, abrasion, frictional properties (smoothness, roughness, friction coefficient), energy absorption.

• Fluid handling properties: permeability, liquid absorption (liquid absorbency, penetration time, wicking rate, rewet, bacteria/particle collection, repellency and barrier properties, run-off, strike time), water vapor transport and breathability.

• Physical properties: thermal and acoustic insulation and conductivity, electrostatic properties, dielectric constant and electrical conductivity, opacity and others.

• Chemical properties: surface wetting angle, oleophobicity and hydrophobicity, interface compatibility with binders and resins, chemical resistance and durability to wet treatments, flame resistance, dyeing capability, flammability, soiling resistance.

• Application specific performance: linting (particle generation), esthetics and handle, filtration efficiency, biocompatibility, sterilization compatibility, biodegradability and health and safety status.

10.2.2   DIMENSIONAL PARAMETERS

The structure and dimensions of nonwoven fabrics are frequently characterized in terms of fabric weight per unit area, thickness, density, fabric uniformity, fabric porosity, pore size and pore size distribution, fiber orientation distribution, bonding segment structure and the distribution. Nonwoven fabric weight (or fabric mass) is defined as the mass per unit area of the fabric and is usually measured in g/m^2 (or gsm). Fabric thickness is defined as the distance between the two fabric surfaces under a specified applied pressure, which varies if the fabric is high-loft (or compressible). The fabric weight and thickness determine the fabric packing density, which influences the freedom of movement of the fibers and determines the porosity (the proportion of voids) in a nonwoven structure. The freedom of movement of the fibers plays an important role in nonwoven mechanical properties and the proportion of voids determines the fabric porosity, pore sizes and permeability in a nonwoven structure. Fabric density, or bulk density, is the weight per unit volume of the nonwoven fabric (kg/m^3). It equals the measured weight per unit area (kg/m^2) divided by the measured thickness of the fabric (m). Fabric bulk density together with fabric porosity is important because they influence how easily fluids, heat and sound transport through a fabric.

10.2.3   WEIGHT UNIFORMITY OF NONWOVEN FABRICS

The fabric weight and thickness usually varies in different locations along and across a nonwoven fabric. The variations are frequently of a periodic nature with a recurring wavelength due to the mechanics of the web formation and/or bonding process. Persistent cross-machine variation in weight is commonly encountered, which is one reason for edge trimming. Variations in either thickness and/or weight per unit area determine variations of local fabric packing density, local fabric porosity and pore size distribution, and therefore influence the appearance, tensile properties, permeability, thermal insulation, sound insulation, filtration, liquid barrier and penetration properties, energy absorption, light opacity and conversion behavior of nonwoven products. Fabric uniformity can be defined in terms of the fabric weight (or fabric density) variation measured directly by sampling different regions of the fabric. The magnitude of the variation depends on the specimen size, for example the variation in fabric weight between smaller fabric samples (e.g., consecutive fabric samples of 1 m² or 10 mm²) will usually be much greater than the variation between bigger fabric samples (e.g., rolls of fabric of hundreds of meters). Commercially, to enable on-line determination of fabric weight variation, the fabric uniformity is measured in terms of the variation in the optical density of fabric images [9], the gray level intensity of fabric images [10] or the amount of electromagnetic rays absorbed by the fabric depending on the measurement techniques used [11, 12]. The basic statistical terms for expressing weight uniformity in the industry are the standard deviation(s) and the coefficient of variation (CV) of measured parameters as follows:

$$\text{Standard deviation: } \sigma^2 = \frac{\sum_{i=1}^{n}(w_i - \bar{w})^2}{n} \quad (1)$$

$$\text{Coefficient of variation: } CV = \frac{\sigma}{w} \quad (2)$$

$$\text{Index of dispersion: } I_{\text{dispersion}} = \frac{\sigma^2}{w} \quad (3)$$

where n is the number of test samples, w is the average of the measured parameter and w_i is the local value of the measured parameter. Usually, the fabric uniformity is referred to as the percentage coefficient of variation (CV%). The fabric uniformity in a nonwoven is normally anisotropic, that is, the uniformity is different in different directions (MD and CD) in the fabric structure. The ratio of the index of dispersion has been used to represent the anisotropy of uniformity [13]. The local anisotropy of mass uniformity in a nonwoven has also been defined by Scharcanski and Dodson in terms of the 'local dominant orientations of fabric weight' [14].

10.2.4 FIBER ORIENTATION IN NONWOVEN

The fibers in a nonwoven fabric are rarely completely randomly orientated, rather, individual fibers are aligned in various directions mostly in-plane. These fiber alignments are inherited from the web formation and bonding processes. The fiber segment orientations in a nonwoven fabric are in two and three dimensions and the orientation angle can be determined (Fig. 10.11). Although the fiber segment orientation in a nonwoven is potentially in any three-dimensional direction, the measurement of fiber alignment in three dimensions is complex and expensive [15]. In certain nonwoven structures, the fibers can be aligned in the fabric plane and nearly vertical to the fabric plane. The structure of a needle-punched fabric is frequently simplified in this way. In this case, the structure of a three-dimensional nonwoven may be simplified as a combination of two-dimensional layers connected by fibers orientated perpendicular to the plane (Fig. 10.12). The fiber orientation in such a three-dimensional fabric can be described by measuring the fiber orientation in two dimensions in the fabric plane [16].

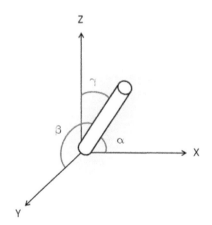

FIGURE 10.11   Fiber orientation angle in 3-D nonwoven fabric.

In the two-dimensional fabric plane, fiber orientation is measured by the fiber orientation angle, which is defined as the relative directional position of individual fibers in the structure relative to the machine direction as shown in Fig. 10.13. The orientation angles of individual fibers or fiber segments can be determined by evaluating photomicrographs of the fabric or directly by means of microscopy and image analysis.

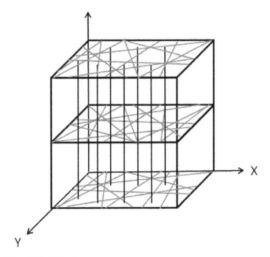

FIGURE 10.12   Simplified 3-D nonwoven structure.

The frequency distribution (or statistical function) of the fiber orientation angles in a nonwoven fabric is called fiber orientation distribution (FOD) or ODF (orientation distribution function). Frequency distributions are obtained by determining the fraction of the total number of fibers (fiber segments) falling within a series of pre-defined ranges of orientation angle. Discrete frequency distributions are used to estimate continuous probability density functions. The following general relationship is proposed for the fiber orientation distribution in a two-dimensional web or fabric:

$$\int_0^{\pi} \Omega(\alpha)d\alpha = 1 \left(\Omega(\alpha) \geq 0 \right) \tag{4}$$

or

$$\sum_{\alpha=0}^{\pi} \Omega(\alpha)\Delta\alpha = 1 \left(\Omega(\alpha) \geq 0 \right) \tag{5}$$

where α is the fiber orientation angle, and $\Omega(\alpha)$ is the fiber orientation distribution function in the examined area. The numerical value of the orientation distribution indicates the number of observations that fall in the direction α, which is the angle relative to the examined area. Attempts have been made to fit the fiber orientation distribution frequency with mathematical functions including uniform, normal and exponential distribution density functions. The following two functions in combination with the constrain in the equation $\int_0^{\pi} \Omega(\alpha)d\alpha = 1$ have been suggested by Petterson [17] and Hansen [18], respectively.

Petterson: $\Omega(\alpha) = A + B\cos\alpha + C\cos^3\alpha + D\cos^8\alpha + E\cos^{16}\alpha$ (6)

Hansen: $\Omega(\alpha) = A + B\cos^2(2\alpha)$ (7)

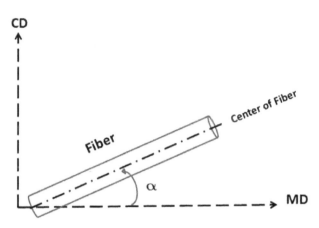

FIGURE 10.13   Fiber orientation and orientation angle.

Fiber alignments in nonwoven fabrics are usually anisotropic, that is, the number of fibers in each direction in a nonwoven fabric is not equal. The differences between the fiber orientation in the fabric plane and in the direction perpendicular to the fabric plane (i.e., transverse direction or fabric thickness direction) are particularly important. In most nonwovens except some air- laid structures, most of the fibers are preferentially aligned in the fabric plane rather than in the fabric thickness. Significant in-plane differences in fiber orientation are also found in the machine direction and in the fabric cross direction in nonwovens. Preferential fiber (either staple fiber or continuous filament) orientation in one or multiple directions is introduced during web formation and to some extent during mechanical bonding processes. A simplified example of an anisotropic nonwoven structure is a unidirectional fibrous bundle in which fibers are aligned in one direction only. Parallel-laid or cross-laid carded webs are usually anisotropic with a highly preferential direction of fiber orientation. Fiber orientation in airlaid structures is usually more isotropic than in other dry-laid fabrics both in two and three dimensions. In perpendicular-laid webs, fibers are orientated in the direction of the fabric thickness. Spunlaid nonwovens composed of filaments are less anisotropic in the fabric plane than layered carded webs [19], however, the anisotropy of continuous filament webs depends on the way in which the webs are collected and tensioned. This structural anisotropy can be characterized in terms of the fiber orientation distribution functions. This anisotropy is important because of its influence on the anisotropy of fabric mechanical and physical properties including tensile, bending, thermal insulation, acoustic absorption, dielectric behavior and permeability. The ratio of physical properties obtained in different directions in the fabric, usually the MD/CD, is a well established means of expressing the anisotropy. The MD/CD ratio of tensile strength is most commonly encountered, although the same approach may be used to express directional in-plane differences in elongation, liquid wicking distance, liquid transport rate, dielectric constant and permeability. However, these anisotropy terms use indirect experimental methods to characterize the nonwoven structure, and they are just ratios in two specific directions in the fabric plane, which can misrepresent the true anisotropy of a nonwoven structure.

10.2.5   FABRIC POROSITY

The pore structure in a nonwoven may be characterized in terms of the total pore volume (or porosity), the pore size, pore size distribution and the pore connectivity. Porosity provides information on the overall pore volume of a porous material and is defined as the ratio of the nonsolid volume (voids) to the total volume of the nonwoven fabric. The volume fraction of solid material is defined as the ratio of solid fiber material to the total volume of the fabric. While the fiber density is the weight of a given volume of the solid component only (i.e., not containing other materials), the porosity can be calculated as follows using the fabric bulk density and the fiber density:

$$\varnothing(\%) = \frac{\rho_{fabric}}{\rho_{fibre}} \times 100\% \tag{8}$$

$$\varepsilon(\%) = (1 - \varnothing) \times 100\% \tag{9}$$

where ε is the fabric porosity (%), ϕ is the volume fraction of solid material (%), ρ_{fabric} (kg/m^3) is the fabric bulk density and ρ_{fiber} (kg/m^3) is the fiber density.

In resin coated, impregnated or laminated nonwoven composites, a small proportion of the pores in the fabric is not accessible (i.e., they are not connected to the fabric surface). The definition of porosity as shown above refers to the so-called total porosity of the fabric. Thus, the open porosity (or effective porosity) is defined as the ratio of accessible pore volume to total fabric volume, which is a component part of the total fabric porosity. The majority of nonwoven fabrics have porosities >50% and usually above 80%. A fabric with a porosity of 100% is a totally open fabric and there is no such fabric, while a fabric with a porosity of 0% is a solid polymer without any pore volume; there is no such fabric either. High-loft nonwoven fabrics usually have a low bulk density because they have more pore space than a heavily compacted nonwoven fabric; the porosity of high-loft nonwovens can reach

>98%. Pore connectivity, which gives the geometric pathway between pores cannot be readily quantified and described. If the total pore area responsible for liquid transport across any distance along the direction of liquid transport is known, its magnitude and change in magnitude are believed to indicate the combined characteristics of the pore structure and connectivity.

10.3   CHARACTERIZATION OF NONWOVEN FABRICS

Fabric thickness Testing [20, 21] of nonwoven fabric thickness and fabric weight is similar to other textile fabrics but due to the greater compressibility and unevenness a different sampling procedure is adopted. The thickness of a nonwoven fabric is defined as the distance between the face and back of the fabric and is measured as the distance between a reference plate on which the nonwoven rests and a parallel presser-foot that applies a pressure to the fabric (*see* BS EN ISO 9703–2:1995, ITS 10.1).

Nonwoven fabrics with a high specific volume, that is, bulky fabrics, require a special procedure. In this context, bulky fabrics are defined as those that are compressible by 20% or more when the pressure applied changes from 0.1 kPa to 0.5 kPa. Three procedures are defined in the test standard (BS EN ISO 9703–2:1995). Three test methods, [i.e., ASTM D5729–97 (ITS 120.1), ASTM D5736–01 (ITS 120.2), ASTM D6571–01 (ITS 120.3)] are defined for the measurement of the thickness, compression and recovery of conventional nonwovens and high-loft nonwovens (it is defined, in the ASTM, as a low density fiber network structure characterized by a high ratio of thickness to mass per unit area. High-loft batts have no more than a 10% solid volume and are greater than 3 mm in thickness). Two more test standards (ITS 120.4, ITS 120.5) are defined for rapid measurement of the compression and recovery of high-loft nonwovens.

10.3.1   FABRIC MASS PER UNIT AREA

The measurement of a nonwoven weight per unit area requires a specific sampling procedure, specific dimensions for the test samples, and a greater balance accuracy than for conventional textiles. According to the ISO standards (BS EN 29073–1:1992, ISO 9073–1:1989, ITS10.1), the measurement of nonwoven fabric mass per unit area of nonwovens requires each piece of fabric sample to be at least 50,000 mm^2. The mean value of fabric weight is calculated in grams per square meter and the coefficient of variation is expressed as a percentage.

10.3.2   FABRIC WEIGHT UNIFORMITY

Nonwoven fabric uniformity refers to the variations in local fabric structures, which include thickness and density, but is usually expressed as the variation of the weight per unit area. Both subjective and objective techniques are used to evaluate the fabric uniformity. In subjective assessment, visual inspection can distinguish non-uniform areas as small as about 10 mm² from a distance of about 30 cm. Qualitative assessments of this type can be used to produce ratings of nonwoven fabric samples by a group of experts against benchmark standards. The consensual benchmark standards are usually established by an observer panel using paired comparison, graduated scales or similar voting techniques; these standard samples are then used to grade future samples. Indirect objective measurements of the web weight uniformity have been developed based on variations in other properties that vary with fabric weight including the transmission and reflection of beta rays, gamma rays (CO60), lasers, optical and infra-red light [22], and variation in tensile strength. With optical light scanning methods, the fabrics are evaluated for uniformity using an optical electronic method, which screens the nonwoven to register 32 different shades of gray [23]. The intensity of the points in the different shades of gray provides a measure of the uniformity. A statistical analysis of the optical transparency and the fabric uniformity is then produced. This method is suitable for lightweight nonwovens of 10–50 g/m². Optical light measurements are commonly coupled with image analysis to determine the coefficient of variation of gray level intensities from scanned images of nonwoven fabrics [24]. In practice, nonwoven fabric uniformity depends on fiber properties, fabric weight and manufacturing conditions It is usually true that the variation in fabric thickness and fabric weight decreases as the mean fabric weight increases. Wet-laid nonwovens are usually more uniform in terms of thickness than dry-laid fabrics. Short fiber airlaid fabrics are commonly more uniform than carded and cross laid and parallel-laid fabrics, and spunbond and meltblown fabrics are often more uniform than fabrics produced from staple fibers.

10.3.3   MEASURING FIBER ORIENTATION DISTRIBUTION

In modeling the properties of nonwoven fabrics and particularly in any quantitative analysis of the anisotropic properties of nonwoven fabrics, it is important to obtain an accurate measurement of FOD. A number of measuring techniques have been developed. A direct visual and manual method of measurement was first described by Petterson [25]. Hearle and co-workers [26, 27] found that visual methods produce accurate measurements and it is the most reliable way to evaluate the fiber orientation. Manual measurements of fiber segment angles relative to a given direction were conducted and the lengths of segment curves were obtained within a given range. Chuleigh [28] developed an optical processing method in which an opaque mask was used in a light microscope to highlight fiber segments that are orientated in a known direction. However, the application of this method is limited by the tedious and time-consuming work required in visual examinations.

To increase the speed of assessment, various indirect-measuring techniques have been introduced including both the zero span [29, 30] and short span [31] tensile analysis for predicting the fiber orientation distribution. Stenemur [32] devised a computer system to monitor fiber orientation on running webs based on the light diffraction phenomenon. Methods that employ X-ray diffraction analysis and X-ray diffraction patterns of fiber webs have also been studied [33, 34]. In this method the distribution of the diffraction peak of the fiber to X-ray is directly related to the distribution of the fiber orientation. Other methods include the use of microwaves [35], ultrasound [36], light diffraction methods [37], light reflection and light refraction [38], electrical measurements [39, 40] and liquid-migration-pattern analysis [41, 42] In the last few decades, image analysis has been employed to identify fibers and their orientation [43–46] and computer simulation techniques have come into use for the creation of computer models of various nonwoven fabrics [47–50]. Huang and Bressee89 developed a random sampling algorithm and software to analyze fiber orientation in thin webs. In this method, fibers are randomly selected and traced to estimate the orientation angles; test results showed excellent agreement with results from visual measurements. Pourdeyhimi et al. [51–54] completed a series of studies on the fiber orientation of nonwovens by using an image analyzer to determine the fiber orientation in which image processing techniques such as computer simulation, fiber tracking, Fourier transforms and flow field techniques were employed. In contrast to two-dimensional imaging techniques suitable only for thin nonwoven fabrics, the theory of Hilliard-Komori-Makishima [55] and the visualizations made by Gilmore et al. [56] using X-ray tomographic techniques have provided a means of analyzing the three-dimensional orientation.

Image analysis is a computer-based means of converting the visual qualitative features of a particular image into quantitative data. The measurement of the fiber orientation distribution in nonwoven fabrics using image analysis is based on the assumption that in thin materials a two-dimensional structure can be assumed, although in reality the fibers in a nonwoven are arranged in three dimensions. However, there is currently no generally accepted way of characterizing the fabric structure in terms of the three-dimensional geometry. The fabric geometry is reduced to two dimensions by evaluation of the planar projections of the fibers within the fabric. The assumption of a two-dimensional fabric structure is adequate to describe thin fabrics. The image analysis system in the measurement of the fiber orientation distribution is based on a computerized image capture system operating with an integrated image analysis software package in which numerous functions can be performed [57]. A series of sequential operations is required to perform image analysis and, in a simple system, the following procedures are carried out: production of a gray image of the sample fabric, processing the gray image, detection of the gray image and conversion into binary form, storage and processing of the binary image, measurement of the fiber orientation and output of results.

10.3.4 MEASURING POROSITY

Porosity can be obtained from the ratio of the fabric density and the fiber density. In addition to the direct method of determination for resin impregnated dense nonwoven composites, the fabric porosity can be determined by measuring densities using liquid buoyancy or gas expansion porosimetry [58]. Other methods include small angle neutron, small angle X-ray scattering and quantitative image analysis for total porosity. Open porosity may be obtained from xylene and water impregnation techniques [59], liquid metal (mercury) impregnation, nitrogen adsorption and air or helium penetration. Existing definitions of pore geometry and the size of pores in a nonwoven are based on various physical models of fabrics for specific applications. In general, cylindrical, spherical or convex shaped pores are assumed with a distribution of pore diameters. Three groups of pore size are defined: (i) the near-largest pore size (known as apparent opening pore size, or opening pore size); (ii) the constriction pore size (known as the pore-throat size); and (iii) the pore volume size. Pore size and the pore size distribution of nonwoven fabrics can be measured using optical methods, density methods, gas expansion and adsorption, electrical resistance, image analysis, porosimetry and porometry. The apparent pore opening (or opening pore) size is determined by the passage of spherical solid glass beads of different sizes (50 μm to 500 μm) through the largest pore size of the fabric under specified conditions. The pore size can be measured using sieving test methods (dry sieving, wet sieving and hydrodynamic sieving). The opening pore sizes are important for determining the filtration and clogging performance of nonwoven geotextiles and it enables the determination of the absolute rating of filter fabrics. The constriction pore size, or porethroat size, is different from the apparent pore opening size. The constriction pore size is the dimension of the smallest part of the flow channel in a pore and it is important for fluid flow transport in nonwoven fabrics. The largest pore-throat size is called the bubble point pore size, which is related to the degree of clogging of geotextiles and the performance of filter fabrics. The pore-throat size distribution and the bubble point pore size can be obtained by liquid expulsion methods. However, it is found that wetting fluid, air pressure and equipment type affects the measured constriction pore size [60, 61].

10.3.4.1 LIQUID POROSIMETRY

Liquid porosimetry, also referred to as liquid porometry, is a general term to describe procedures for the evaluation of the distribution of pore dimensions in a porous material based on the use of liquids. Both pore volumes and pore throat dimensions are important quantities in connection with the use of fiber networks as absorption and barrier media. Pore volumes determine the capacity of a network to absorb liquid, that is, the total liquid uptake. Pore throat dimensions, on the other hand, are related to the rate of liquid uptake and to the barrier characteristics of a network. Liquid porosimetry evaluates pore volume distributions (PVD) by measuring the volume of liquid located in different size pores of a porous structure. Each pore is sized according to its effective radius, and the contribution of each pore size to the total free volume of the porous network is determined. The effective radius R of any pore is defined by the Laplace equation:

$$R = \frac{2\gamma \cos\theta}{\Delta p} \tag{10}$$

where γ = liquid surface tension, θ = advancing or receding contact angle of the liquid, ΔP = pressure difference across the liquid/air meniscus For liquid to enter or drain from a pore, an external gas pressure must be applied that is just enough to overcome the Laplace pressure ΔP.

In the case of a dry heteroporous network, as the external gas pressure is decreased, either continuously or in steps, pores that have capillary pressures lower than the given gas pressure ΔP will fill with liquid. This is referred to as liquid intrusion porosimetry and requires knowledge of the advancing liquid contact angle. In the case of a liquid-saturated heteroporous network, as the external gas pressure is increased, liquid will drain from those pores whose capillary pressure corresponds to the given gas pressure ΔP. This is referred to as liquid extrusion porosimetry and requires knowledge of the receding liquid contact angle. In both cases, the distribution of pore volumes is based on measuring the incremental volume of liquid that either enters a dry network or drains from a saturated network at each increment of pressure.

10.3.4.1.1   INSTRUMENTATION

Until recently, the only version of this type of analysis to evaluate PVDs in general use was mercury porosimetry [62]. Mercury was chosen as the liquid because of its very high surface tension so that it would not be able to penetrate any pore without the imposition of considerable external pressure. For example, to force mercury into a pore 5 μm in radius requires a pressure increase of about 2 atm. While this might not be a problem with hard and rigid networks, such as stone, sand structures, and ceramics, it makes the procedure unsuitable for use with fiber materials that would be distorted by such compressive loading. Furthermore, mercury intrusion porosimetry is best suited for pore dimensions less than 5 μm, while important pores in typical textile structures may be as large as 1000 μm. Some of the other limitations of mercury porosimetry have been discussed by Winslow [63] and by Good [64].

A more general version of liquid porosimetry for PVD analysis, particularly well suited for textiles and other compressible planar materials, has been developed by Miller and Tyomkin [65]. The underlying concept was earlier demonstrated for low-density webs and pads by Burgeni and Kapur [66]. Any stable liquid of relatively low viscosity that has a known $\cos^\theta > 0$ can be used. In the extrusion mode, the receding contact angle is the appropriate term in the Laplace equation, while in the intrusion mode the advancing contact angle must be used. There are many advantages to using different liquids with a given material, not the least of which is the fact that liquids can be chosen that relate to a particular end use of a material. The basic arrangement for liquid extrusion porosimetry is shown in Fig. 11.14. In the case of liquid extrusion, a pre-saturated specimen is placed on a microporous membrane, which is itself supported by a rigid porous plate. The gas pressure within the closed chamber is increased in steps, causing liquid to flow out of some of the pores, largest ones first. The amount of liquid removed at each pressure level is monitored by the top-loading recording balance. In this way, each incremental change in pressure (corresponding to a pore size according to the Laplace equation) is related to an increment of liquid mass. To induce stepwise drainage from large pores requires very small increases in pressure over a narrow range that are only slightly above atmospheric pressure, whereas to analyze for small pores the pressure changes must be quite large. In early versions of instrumentation for liquid extrusion porosimetry, pressurization of the specimen chamber was accomplished either by hydrostatic head changes or by means of a single-stroke pump that injected discrete drops of liquid into a free volume space that included the chamber [65]. In the most recent instrumentation developed by Miller and Tyomkin [67], the chamber is pressurized by means of a computer-controlled, reversible, motor-driven piston/cylinder arrangement that can produce the required changes in pressure to cover a pore radius range from 1 to 1000 μm.

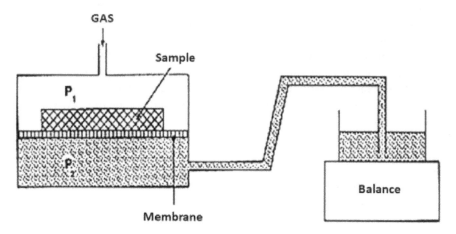

FIGURE 10.14 Basic arrangement for liquid porosimetry to quantify pore volume distribution.

10.3.4.1.2   DATA ANALYSIS AND APPLICATIONS

Prototype data output for a single cycle incremental liquid extrusion run is shown in Fig. 10.15. The experiment starts from the right as the pressure is increased, draining liquid first from the largest pores. The cumulative curve represents the amount of liquid remaining in the pores of the material at any given level of pressure. The first derivative of this cumulative curve as a function of pore size becomes the pore volume distribution, showing the fraction of the free volume of the material made up of pores of each indicated size. PVD curves for two typical fabrics woven from spun yarns are shown in Fig. 10.15. PVD curves for several nonwoven materials are shown in Fig. 10.16. These materials normally have unimodal PVD curves, but generally the pores are larger than those associated with typical woven fabrics. Several interesting points can be noted. First, the pore volumes become smaller with increasing compression (decreasing mat thickness), and at the same time the structures appear to have become less heteroporous; that is, the breadth of the PVD curves decreases with increasing compression. Also, the total pore volume (area under each curve), and therefore the sorptive capacity, decreases with compression. Since in many end-use applications fibrous materials are used under some level of compression, it is particularly important to evaluate pore structure under appropriate compression conditions. The PVD instrumentation described here, referred to as the TRI Autoporosimeter, is extremely versatile and can be used with just about any porous material, including textiles, paper products, membrane filters, particulates, and rigid foams. It also allows quantification of interlayer pores, surface pores, absorption/desorption hysteresis, uptake and retention capillary pressures, and effective contact angles in porous networks [67]. The technique has also been used to quantify pore volume dimensions and sorptive capacity of artificial skin in relation to processing conditions [68].

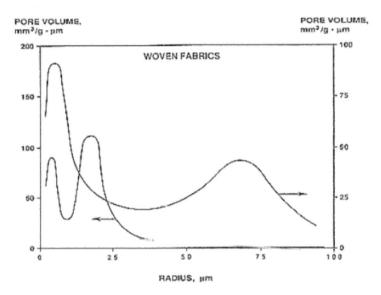

FIGURE 10.15   PVD curves for two typical spun yarn woven fabrics [67].

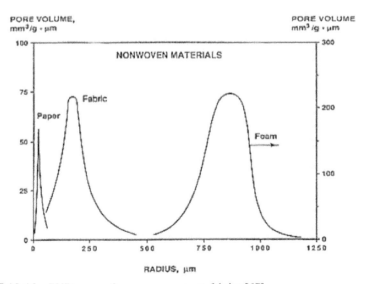

FIGURE 10.16   PVD curves for some nonwoven fabrics [67].

10.3.5   IMAGE ANALYSIS [69]

The dynamic development of computer techniques creates broad possibilities for their application, including identifying and measuring the geometrical dimensions of very small objects including textile objects. Using digital image analysis permits a more detailed analysis of such basic structural parameters of nonwoven products. The digital analysis of two-dimensional images is based on processing the image acquirement, with the use of a computer. The image is described by a two-dimensional matrix of real or imaginary numbers presented by a definite number of bytes. Image modeling is based on digitizing the real image. This process consists of sampling and quantifying the image. The digital image can be described in the form of a two-dimensional matrix, whose elements include quantified values of the intensity function, referred to as gray levels. The digital image is defined by the spatial image resolution and the gray level resolution. The smallest element of the digital image is called the pixel. The number of pixels and the number of brightness levels may be unlimited, although while presenting computer technique data it is customary to use values which are multiplications of the number 2, for example, 512×512 pixels and 256 gray levels.

10.3.5.1   IMAGE QUALITY IMPROVEMENT

Image quality improvement and highlighting its distinguished features are the most often used application techniques for image processing. The process of image quality improvement does not increase the essential information represented by the image data, but increases the dynamic range of selected features of the acquired object, which facilitates their detection.

The following operations are carried out during image quality improvement:
- changes of the gray level system and contrast improvement
- edge exposition
- pseudo-colorization
- improvement of sharpness, decreasing the noise level
- space filtration
- interpolation and magnification
- compensation of the influence of interference factors, for example, possible under-exposure.

10.3.5.1.1   REINSTATING DESIRED IMAGE FEATURES

Reinstating desired image features is connected with eliminating and minimizing any image features which lower its quality. Acquiring images by optical, opto-electronic or electronic methods involves the unavoidable degradation of some image features during the detection process. Aberrations, internal noise in the image sensor, image blurring caused by camera defocusing, as well as turbulence and air

pollution in the surrounding atmosphere may cause a worsening of quality. Reinstating the desired image features differs from image improvement, whose procedure is related to highlighting or bringing to light the distinctive features of the existing image. Reinstating the desired image features mainly includes the following corrections:
- reinstating the sharpness lowered as the result of disadvantageous features of the image
- sensor or its surrounding
- noise filtration
- distortion correction
- correction of sensors' nonlinearity

10.3.5.1.2   IMAGE DATA COMPRESSION

Image data compression is based on minimizing the number of bytes demanded for image representation. The compression effect is achieved by transforming the given digital image to a different number table in such a manner that the preliminary information amount is packed into a smaller number of samples.

10.3.5.2. MEASURING FIBER ORIENTATION IN NONWOVEN USING IMAGE PROCESSING

Fiber orientation is an important characteristic of nonwoven fabrics, since it directly influences their mechanical properties and performances. Since visual assessments of fiber orientation are not reliable, image-processing techniques are used to assess them automatically. The image of a fabric is taken and digitized. The digitized image of the surface of a fabric is progressively simplified by a line operator and an edge-smoothing and thinning operation to produce an image in which fibers are represented by curves of one pixel in thickness. The fiber-orientation distribution of a fabric is then obtained by tracing and measuring the length and orientation of the curves. The image-analysis method for automatically measuring the fiber-orientation distributions is validated by manual measurements made by experts. Spatial uniformity of fibrous structures can been described statistically using an index of dispersion. The spatial uniformity of web mass is described by dispersion of its surface relief distribution. Surface relief is representative of mass density gradients in local regions. In principle, this concept is similar to topographical surface relief that represents the variations in elevation of the earth's surface: the higher is the elevation, the greater the mass present on a surface. The surface relief in a local region is representative of the average mass density gradient in that local region. Surface relief of a web can be measured from its gray-scale images taken in transmitted or reflected light or any other radiation. Each pixel of a gray-scale image (with 256 gray-levels or 8-bit) is considered a column of cubes with each cube having area

(1 pixel) × (1 pixel) and height equal to one intensity level. Figure 11.17 shows the translation of square pixels of Fig. 11.18 into columns of cubes. Surface relief of a pixel is the difference in height (or gray level intensity, G) of two adjacent pixels sharing a side. The surface relief area was calculated as the total number of exposed faces of cube columns in a "quadrat," which is defined as a rectangular region of interest. Only lateral surfaces were considered and flat tops of columns were ignored in the calculation of relief area.

FIGURE 10.17   Pixels represented as columns of cubes.

FIGURE 10.18   Detailed image pixels.

Various techniques are used to estimate fiber orientation distribution in non-wovens, such as direct tracking, flow field analysis and the Fourier transform (FT) and Hough transform (HT) methods. In direct tracking, the actual pixels of the lines are tracked. The algorithm used is time consuming and therefore is mostly used for research and development. In denser structures, as the number of crossover points increases, this method becomes less efficient. Flow field analysis is based on the assumption that edges in an image are representative of orientation fields in the image. This method is mainly used to obtain the mean orientation distribution angle, but not the orientation distribution function. The FT method is widely used in many image processing and measurement operations. It transforms the gray scale intensity domain to a frequency spectrum. Compression of the image, leveling of the image brightness and contrast and enhancement can more conveniently be performed on the frequency spectrum. In most cases the spatial domain needs to be recovered from the frequency spectrum. Measuring the orientation is one of the applications of the FT method. The algorithm is fast and does not need high computational power for the calculations. The HT method is used in estimating fiber orientation distribution directly. The advantage of this method compared with other indirect methods is that the actual orientation of the lines is included directly in the computation of the transform. The capability of the HT method in object recognition is used to measure the length of the straight lines in the image. However, more computational time and resources are needed to run the algorithm for larger-scale images with higher accuracy. The images acquired using backlighting and optical microscopes were reported to be of low quality having poor contrast. Consequently, they limited the image acquisition technique to very thin webs for which they could obtain images that could be successfully processed. When images are digitized at high magnification, the depth of field is poor, resulting in fuzzy fiber segments and inferior image quality. To overcome this problem, in the present work a scanning electron microscope (SEM) has been used to produce high quality images. The SEM provides high depth of field even at high magnifications. This technique is not restricted to thin webs and can also be used for fabrics with very high density.

10.3.6 PORE VOLUME DISTRIBUTION AND MERCURY POROSIMETRY

Unlike porometry where the measurement of the pore-throat size distribution is based on measurement of the airflow rate through a fabric sample, the pore volume distribution is determined by liquid porosimetry, which is based on the liquid uptake concept proposed by Haines [70]. A fabric sample (either dry or saturated) is placed on a perforated plate and connected to a liquid reservoir. The liquid having a known surface tension and contact angle is gradually forced into or out of the pores in the fabric by an external applied pressure. Porosimetry is grouped into two categories based on the liquid used, which is either nonwetting (e.g., mercury) or wetting (e.g., water). Each is used for intrusion porosimetry and extrusion porosimetry where the advancing contact angle and receding contact angle are applied in liquid intrusion and extrusion porosimetry, respectively.

Mercury has a high surface tension and is strongly nonwetting on most fabrics at room temperature. In a typical mercury porosimetry measurement, a nonwoven fabric is evacuated to remove moisture and impurities and then immersed in mercury. A gradually increasing pressure is applied to the sample forcing mercury into increasingly smaller 'pores' in the fabric. The pressure P required to force a nonwetting fluid into a circular cross-section capillary of diameter d is given by:

$$P = \frac{4\sigma_{Hg} \cos \gamma_{Hg}}{d} \tag{11}$$

where σ_{Hg} is the surface tension of the mercury (0.47 N/m), 130 and gHg γ_{Hg} is the contact angle of the mercury on the material being intruded (the contact angle ranges from 135°~180°), and d is the diameter of a cylindrical pore.

The incremental volume of mercury is recorded as a function of the applied pressure to obtain a mercury intrusion curve. The pore size distribution of the sample can be estimated in terms of the volume of the pores intruded for a given cylindrical pore diameter d. The pressure can be increased incrementally or continuously (scanning porosimetry). The process is reversed by lowering the pressure to allow the mercury to extrude from the pores in the fabric to generate a mercury extrusion curve. Analysis of the data is based on a model that assumes the pores in the fabric are a series of parallel nonintersecting cylindrical capillaries of random diameters [71]. However, as a consequence of the nonwetting behavior of mercury in mercury intrusion porosimetry, relatively high pressure is needed to force mercury into the smaller pores therefore compressible nonwoven fabrics are not suitable for testing using the mercury porosimetry method.

Liquids other than mercury find use in porosimetry [72–74] and have been commercialized [75]. Test procedure is similar to that of mercury porosimetry but any liquid that wets the sample, such as water, organic liquids, or solutions may be utilized. The cumulative and differential pore volume distribution, total pore volume, porosity, average, main, effective and equivalent pore size can be obtained.

10.3.7 MEASUREMENT OF SPECIFIC SURFACE AREA BY USING GAS ADSORPTION

The number of gas molecules adsorbed on the surface of nonwoven materials depends on both the gas pressure and the temperature. An experimental adsorption isotherm plot of the incremental increases in weight of the fabric due to absorption against the gas pressure can be obtained in isothermal conditions. Prior to measurement, the sample needs to be pretreated at an elevated temperature in a vacuum or flowing gas to remove contaminants. In physical gas adsorption, when an inert gas (such as nitrogen or argon) is used as an absorbent gas, the adsorption isotherm indicates the surface area and/or the pore size distribution of the objective material by applying experimental data to the theoretical adsorption isotherm for gas adsorption on the polymer surface. In chemical gas adsorption, the chemical properties of a polymeric surface are revealed if the absorbent is acidic or basic. In some experiments, a liquid absorbent such as water is used in the same manner [76–78].

10.3.8   MEASURING TENSILE PROPERTIES

Some of the most important fabric properties governing the functionality of nonwoven materials include mechanical properties (tensile, compression, bending and stiffness), gaseous and liquid permeability, water vapor transmission, liquid barrier properties, sound absorption properties and dielectric properties.

Mechanical properties of nonwoven fabrics are usually tested in both machine direction (MD) and cross-direction (CD), and may be tested in other bias directions if required. Several test methods are available for tensile testing of nonwovens, chief among these are the strip and grab test methods. In the grab test, the central section across the fabric width is clamped by jaws a fixed distance apart. The edges of the sample therefore extend beyond the width of the jaws. In the standard grab tests for nonwoven fabrics, the width of the nonwoven fabric strip is 100 mm, and the clamping width in the central section of the fabric is 25 mm. The fabric is stretched at a rate of 100 mm/min (according to the ISO standards) or 300 mm/min (according to the ASTM standards) and the separation distance of the two clamps is 200 mm (ISO standards) or 75 mm (ASTM standards). Nonwoven fabrics usually give a maximum force before rupture. In the strip test, the full width of the fabric specimen is gripped between the two clamps. The width of the fabric strip is 50 mm (ISO standard) or either 25 mm or 50 mm (ASTM standards). Both the stretch rate and the separation distance of the two clamps in a strip test are the same as they are in the grab test. The separation distance of the two clamps is 200 mm (ISO standards) or 75 mm (ASTM standards). The observed force for a 50 mm specimen is not necessarily double the observed force for a 25 mm specimen [78, 79].

10.3.9   MEASURING WATER VAPOR TRANSMISSION

The water vapor transmission rate through a nonwoven refers to the mass of the water vapor (or moisture) at a steady state flow through a thickness of unit area per unit time. This is taken at a unit differential pressure across the fabric thickness under specific conditions of temperature and humidity ($g/Pa.s.m^2$). It can be tested by two standard methods, the desiccant method and water methods. In the desiccant method, the specimen is sealed to an open mouth of a test dish containing a desiccant, and the assembly is placed in a controlled atmosphere. Periodic weighing's determining the rate of water vapor movement through the specimen into the desiccant. In the water method, the dish contains distilled water, and the weighing's determining the rate of water vapor movement through the specimen to the controlled atmosphere. The vapor pressure difference is nominally the same in both methods except when testing conditions are with the extremes of humidity on opposite sides [80].

10.3.10   MEASURING WETTING AND LIQUID ABSORPTION

There are two main types of liquid transport in nonwovens. One is the liquid absorption, which is driven by the capillary pressure in a porous fabric and the liquid is taken up by a fabric through a negative capillary pressure gradient. The other type of liquid transport is forced flow in which liquid is driven through the fabric by an external pressure gradient. The liquid absorption that takes place when one edge of a fabric is dipped in a liquid so that it is absorbed primarily in the fabric plane is referred to as wicking. When the liquid front enters into the fabric from one face to the other face of the fabric, it is referred to as demand absorbency or spontaneous uptake [81, 82].

10.3.11   MEASURING THERMAL CONDUCTIVITY AND INSULATION [83, 84]

The thermal resistance and the thermal conductivity of flat nonwoven fabrics, fibrous slabs and mats can be measured with a guarded hot plate apparatus according to BS 4745: 2005, ISO 5085-1:1989, ISO 5085-2:1990. For testing the thermal resistance of quilt, the testing standard is defined in BS 5335 Part 1:1991. The heat transfer in the measurement of thermal resistance and thermal conductivity in current standard methods is the overall heat transfer by conduction, radiation, and by convection where applicable.

The core components of the guarded hot plate apparatus consist of one cold plate and a guarded hot plate. A sample of the fabric or insulating wadding to be tested, 330 mm in diameter and disc shaped, is placed over the heated hot metal plate. The sample is heated by the hot plate and the temperature on both sides of the sample is recorded using thermocouples. The apparatus is encased in a fan-assisted cabinet and the fan ensures enough air movement to prevent heat build up around the sample and also isolates the test sample from external influences. The test takes approximately eight hours including warm-up time. The thermal resistance is calculated based on the surface area of the plate and the difference in temperature between the inside and outside surfaces. When the hot and cold plates of the apparatus are in contact and a steady state has been established, the contact resistance, R_c ($m^2 \, K \, W^{-1}$), is given by the equation:

$$\frac{R_c}{R_s} = \frac{\theta_2 - \theta_3}{\theta_1 - \theta_2} \tag{12}$$

where R_s is the thermal resistance of the 'standard,' θ_1 is the temperature registered by thermocouple T_1, θ_2 is the temperature registered by T_2 and θ_3 is the temperature registered by T_3. Thus, the thermal resistance of the test specimen, $R_f\,(m^{2\,K}W^{-1})$, is given by the equation:

$$\frac{R_f}{R_s} = \frac{\theta_2' - \theta_3'}{\theta_1' - \theta_2'} - \frac{\theta_2 - \theta_3}{\theta_1 - \theta_2} \tag{13}$$

where θ_1' is the temperature registered by T_1, θ_2' is the temperature registered by T_2 and θ_3' is the temperature registered by T_3. Since $R_s\,(m^2KW^{-1})$ is a known constant and can be calibrated for each specific apparatus, R_f (m^2KW^{-1}) can thus be calculated. Then the thermal conductivity of the specimen, $k\,(Wm^{-1}K^{-1})$ can be calculated from the equation:

$$k = \frac{d\,(mm)^*10^{-3}}{R_f\,(m^2KW^{-1})} \tag{14}$$

The conditioning and testing atmosphere shall be one of the standard atmospheres for testing textiles defined in ISO139, that is, a relative humidity of 65%+/–2% R.H. and a temperature of 20°C +/– 2°C.

10.4 MODELING

Several theoretical models have been proposed to predict nonwoven fabric properties that can be used or adapted for spunbonded filament networks:

Backer and Petterson [85]	– The filaments are assumed to be straight and oriented in the machine direction
	– Filament properties and orientation are assumed to be uniform from point to point in the fabric
Hearle et al. [86, 87]	– The model accounts for the local fiber curvature (curl)
	– The fiber orientation distribution, fiber stress–strain relationships and the fabric's Poisson ratio must be determined in advance
Komori and Makishima [88]	– Estimation of fiber orientation and length
	– Assumed that the fibers are straight-line segments of the same length and are uniformly suspended in a unit volume of the assembly
Britton et al. [89]	– Demonstrated the feasibility of computer simulation of nonwovens
	– The model is not based on real fabrics and is designed for mathematical convenience

Grindstaff and Hansen [90]	– Stress–strain curve simulation of point-bonded fabrics – Fiber orientation is not considered
Mi and Batra [91]	– A model to predict the stress–strain behavior of certain point-bonded geometries – Incorporated fiber stress–strain properties and the bond geometry into the model
Kim and Pourdeyhimi [92]	– Image simulation and data acquisition – Prediction of stress–strain curves from fiber stress–strain properties, network orientation, and bond geometry – Simulated fibers are represented as straight lines

10.4.1 MODELING BENDING RIGIDITY

The bending rigidity (or flexural rigidity) of adhesive bonded nonwovens was evaluated by Freeston and Platt [93]. A nonwoven fabric is assumed to be composed of unit cells and the bending rigidity of the fabric is the sum of the bending rigidities of all the unit cells in the fabric, defined as the bending moment times the radius of curvature of a unit cell. The analytical equations for bending rigidity were established in the two cases of 'no freedom' and 'complete freedom' of relative motion of the fibers inside a fabric. The following assumptions about the nonwoven structure are made for modeling the bending rigidity.

1. The fiber cross-section is cylindrical and constant along the fiber length.
2. The shear stresses in the fiber are negligible.
3. The fibers are initially straight and the axes of the fibers in the bent cell follow a cylindrical helical path.
4. The fiber diameter and fabric thickness are small compared to the radius of curvature; the neutral axis of bending is in the geometric centerline of the fiber.
5. The fabric density is high enough that the fiber orientation distribution density function is continuous.
6. The fabric is homogeneous in the fabric plane and in the fabric thickness.

The general unit cell bending rigidity, $(EI)_{cell}$, is therefore as follows:

$$(EI)_{cell} = N_f \int_{-\pi/2}^{\pi/2} [Ef\ If\ cos\ 4\theta + GIp\ sin\ 2\theta\ cos\ 2\theta]\ \Omega\ (\theta)\ d\theta \qquad (15)$$

where N_f = number of fibers in the unit cell, $E_f\ I_f$ = fiber bending rigidity around the fiber axis, G = shear modulus of the fiber, I_p = polar moment of the inertia of the fiber cross section, which is a torsion term and $\Omega(\theta)$ = the fiber orientation distribution in the direction, θ.

The bending rigidities of a nonwoven fabric in two specific cases of fiber mobility are as follows:

1. 'Complete freedom' of relative fiber motion. If the fibers are free to twist during fabric bending (e.g., in a needle-punched fabric), the torsion term ($GI_p \sin^2 \theta \cos^2 \theta$) will be zero. Therefore,

$$(EI)_{cell} = \pi \, d^4_f N_f E_f / 64 \int_{-\pi/2}^{\pi/2} \Omega (\theta) \cos^4\theta \, d\theta \tag{16}$$

where d_f = fiber diameter, E_f = Young's modulus of the fiber.

2. 'No freedom' of relative fiber motion. In chemically bonded nonwovens, the freedom of relative fiber motion is severely restricted. It is assumed in this case that there is no freedom of relative fiber motion and the unit cell bending rigidity $(EI)_{cell}$, is therefore:

$$(EI)_{cell} = \pi \, N_f E_f d^2_f h / 48 \int_{-\pi/2}^{\pi/2} \Omega (\theta) \cos^4 \theta \, d\theta \tag{17}$$

where h = fabric thickness and d_f = fiber diameter.

10.4.2 MODELING SPECIFIC PERMEABILITY

The specific permeability of a nonwoven fabric is solely determined by nonwoven fabric structure and is defined based on D'Arcy's Law [94] which may be written as follows:

$$Q = - \frac{k}{\eta} \frac{\Delta p}{h} \tag{18}$$

where Q is the volumetric flow rate of the fluid flow through a unit cross-sectional area in the porous structure (m/s), h is the viscosity of the fluid ($Pa.s$), Δp is the pressure drop (Pa) along the conduit length of the fluid flow h (m) and k is the specific permeability of the porous material (m^2). Numerous theoretical models describing laminar flow through porous media have been proposed to predict permeability. The existing theoretical models of permeability applied in nonwoven fabrics can be grouped into two main categories based on:

1. Capillary channel theory, for example Kozeny [106], Carman [95], Davies [96], Piekaar and Clarenburg [97] and Dent [98].
2. Drag force theory, for example Emersleben [99], Brinkman [100], Iberall [101], Happel [102], Kuwabara [103], Cox [104], and Sangani and Acrivos [105].

Many permeability models established for textile fabrics are based on capillary channel theory or the hydraulic radius model, which is based on the work of Kozeny [106] and Carman [95]. The flow through a nonwoven fabric is treated as a conduit flow between cylindrical capillary tubes. The Hagen–Poiseuille equation for fluid flow through such a cylindrical capillary tube structure is as follows:

$$q = \frac{\pi r^4}{8} \frac{\Delta p}{h} \tag{19}$$

where r is the radius of the hydraulic cylindrical tube. However, it has been argued that models based on capillary channel theory are suitable only for materials having a low porosity and are unsuitable for highly porous media where the porosity is greater than 0.8, see, for example, Carman [95].

10.4.3 MODELING PORE SIZE AND PORE SIZE DISTRIBUTION

In the design and engineering of nonwoven fabrics to meet the performance requirements of industrial applications, it is desirable to make predictions based on the fabric components and the structural parameters of the fabric. Although work has been conducted to simulate isotropic nonwoven structures in terms of, for example, the fiber contact point numbers [107] and intercross distances [108], only the models concerned with predicting the pore size are summarized in this section.

10.4.3.1 MODELS OF PORE SIZE

Although it is arguable if the term 'pore' accurately describes the voids in a highly connective, low density nonwoven fabric, it is still helpful to use this term in quantifying a porous nonwoven structure. The pore size in simplified,

$$r = \left(0.075737 \sqrt{\frac{Tex}{\rho_{fibric}}} \right) - \frac{d_f}{2} \tag{20}$$

where Tex = fiber linear density (tex), ρ_{fabric} is the fabric density (g/cm³) and d_f is the fiber diameter (m). Several other models relating pore size and fiber size by earlier researchers can also be used in nonwoven materials. For example, both the largest pore size and the mean pore size can be predicted as follows, by using, Goeminne's equation. The porosity is defined as nonwoven structures can be approximately estimated by Wrotnowski's model [109, 111] although the assumptions that are made for the fabric structure are based on fibers that are circular in cross-section, straight, parallel, equidistant and arranged in a square pattern (Fig. 11.19). The radius of a pore in Wrotnowski's model is shown as follows,

$$\text{largest pore size } (2r_{max}): r_{max} = \frac{d_f}{2(1-\varepsilon)} \tag{21}$$

mean pore size ($2r$) (porosity < 0.9):$r = \dfrac{d_f}{4\,(1-\varepsilon)}$ (22)

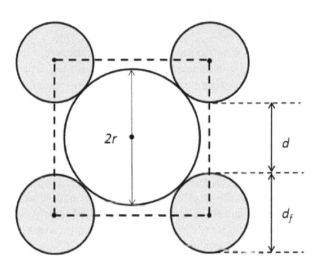

FIGURE 10.19   Wrotnowski's model for pore size in a bundle of parallel cylindrical fibers arranged in a square pattern.

In addition, pore size ($2r$) can also be obtained based on Hagen-Poiseuille's law in a cylindrical tube,

$$r = \sqrt[4]{\dfrac{8k}{\pi}}$$ (23)

where k is the specific permeability (m²) in Darcy's law.

10.4.3.2 MODELS OF PORE SIZE DISTRIBUTION

If it is assumed that the fibers are randomly aligned in a nonwoven fabric following Poisson's law, then the probability, $P(r)$, of a circular pore of known radius, r, is distributed as follows [112],

$$P(r) = -\left(2\pi v'\right)exp\,(-\pi r^2 v')$$ (24)

where $v = \dfrac{0.36}{r^2}$ and is defined as the number of fibers per unit area.

Giroud [113] proposed a theoretical equation for calculating the filtration pore size of nonwoven geotextiles. The equation is based on the fabric porosity, fabric thickness, and fiber diameter in a nonwoven geotextile fabric.

$$O_f = \left[\frac{1}{\sqrt{1-\varepsilon}} - 1 + \frac{\xi \varepsilon d_f}{(1-\varepsilon)^h} \right] d_f \tag{25}$$

where d_f = fiber diameter, ε = porosity, h = fabric thickness, ζ = an unknown dimensionless parameter to be obtained by calibration with test data to account for the further influence of geotextile porosity and $\zeta = 10$ for particular experimental results, and O_f = filtration opening size, usually given by the nearly largest constriction size of a geotextile.

Lambard [114] and Faure [115] applied Poissonian line network theory to establish a theoretical model of the 'opening sizing' of nonwoven fabrics. In this model, the fabric thickness is assumed to consist of randomly stacked elementary layers, each layer has a thickness T_e and is simulated by two- dimensional straight lines, (a Poissonian line network). Faure et al. [116] and Gourc and Faure [117] also presented a theoretical technique for determining constriction size based on the Poissonian polyhedra model. In Faure's approach, epoxy-impregnated nonwoven geotextile specimens were sliced and the nonwoven geotextile was modeled as a pile of elementary layers, in which fibers were randomly distributed in planar images of the fabric. The cross-sectional images were obtained by slicing at a thickness of fiber diameter d_f and the statistical distribution of pores was modeled by inscribing a circle into each polygon defined by the fibers (Fig. 11.20). The pore size distribution, which is obtained from the probability of passage of different spherical particles through the layers forming the geotextile, can thus be determined theoretically using the following equation:

$$Q(d) = (1-\phi)\left(\frac{2+\lambda(d+d_f)}{2+\lambda d_f}\right)^{2N} e^{-\lambda N d} \tag{26}$$

$$\lambda = \frac{4(1-\phi)}{\pi} \frac{1}{d_f}, and\ N = \frac{T}{d_f} \tag{27}$$

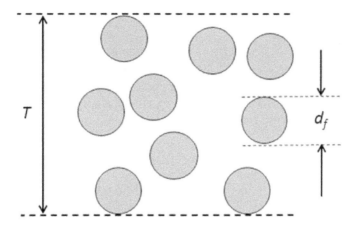

FIGURE 10.20 Model for constriction pore size in a nonwoven fabric consisting of randomly stacked elementary layers of fibers.

where $Q(d)$ = probability of a particle with a diameter d passing through a pore channel in the geotextile, ϕ = fraction of solid fiber materials in the fabric, λ = total length of straight lines per unit area in a planar surface (also termed specific length) and N = number of slices in a cross-sectional image. Because of the assumption in Faure's approach that the constriction size in geotextiles of relatively great thicknesses tends to approach zero, Faure's model generally produces lower values [113]. The use of this method is thus not recommended for geotextiles with a porosity of 50% or less [118].

10.4.3.3 MODELING OF THE PORE NETWORK BY IMAGE PROCESSING

Industry to extract the influence of the structural properties on the functional properties. It is also widely used in theoretical modeling or specific applications to infer nonwoven hydraulic and other properties of nonwoven behavior. Specific devices, like porometer, have been developed to provide indirect measurements of the PSD in small textile structure. Other devices have been developed to provide direct measurements: X-ray, Electronic Scan Microscopy, optical profiling system. The use of porometer is based on a variation of the pressure of the liquid through the sample. It saves instantly the quantity of liquid passed through the pores beginning by the largest pore and finishing by the smallest ones. It gives so a PSD based on the equivalent radius. However, this distribution is not a real distribution. In fact, the number of smallest pores is vague because it requires a big air pressure to be measured. Moreover the geometry and the shape of pores are neglected. So, it is important, to take into account the morphological aspect of the pores. In this approach, we want to study the PSD without neglecting the geometry and the shape of pores. Moreover, all of those devices are not able to measure the dynamic aspect of nonwovens and so it is difficult to understand hydraulic properties with such kind of devices. For all those reasons, we set up a test bed that allows us to study more easily the PSD and the hydraulic properties of pores through time.

10.4.3.4   THE PORES SIZE AND GEOMETRY DISTRIBUTION FUNCTION

The nonwoven materials are made up of pores in which gas effluents or liquids can run out. Fatt introduced, in 1956, the model of porous media [119] in the rock science. A lot of works are done also in the same field [120–128]. For nonwoven research, some papers have been written [129–137]. A good description of this structure permits to model the fluid transport in the medium. It permits also to extract relevant parameters characterizing the pore network. It allows also a classification of porous materials according to their hydraulic properties [138–140]. We take a black and white image of the sample. To determine the PSD, our approach is based on the detection of the 8-connected pixels belonging to the background of the image [141]. The 8-connected pixels are connected if their edges or corners touch. This means that if two adjoining pixels are on, they are part of the same object, regardless of whether they are connected along the horizontal, vertical, or diagonal direction. The algorithm consists on many steps. First of all, we convert the gray image into a binary image by using Otsu threshold. This method consists in calculating the white pixels [142]. In the second step of the algorithm, we label the resulting image by using the following general procedure:

- Scan all image pixels, assigning preliminary labels to nonzero pixels and recording label equivalences in a union-find table.
- Solve the equivalence classes using the union-find algorithm [143].
- Relabel the pixels based on the resolved equivalence classes.

The resulting matrix contains positive integer elements that correspond to different regions. For example, the set of elements equal to 1 corresponds to region 1; the set of elements equal to 2 corresponds to region 2; and so on. We can extract from the labeled matrix all the geometric parameters concerning every pore (equivalent radius, orientation, area, eccentricity etc.). We can model a pore with an ellipse or a rectangle or a convex hull. In this chapter, we choose to characterize a pore with his equivalent radius r, which represents the size of the pore, and his eccentricity e, which represents the geometry of the pore. The equivalent radius corresponds to the radius of a circle with the same area as the region:

$$r = (Area/\pi)^{0.5} \tag{28}$$

where *Area* is the actual number of pixels in the region. Another approach is based on the extraction of the biggest disk contained in a region [144], but some pixels will be ignored. A such PSD is a histogram describing the:

$$\text{Frequency (\%)} = \frac{\text{Number of pores having radius } r}{\text{Total number of pores}} \tag{29}$$

By using the same algorithm, we extract the pores geometry distribution (PGD). More exactly; we determine the eccentricity of each pore. The eccentricity is the ratio of the distance between the foci of the ellipse and its major axis length. The value varies between 0 and 1 (an ellipse whose eccentricity is 0 is actually a circle, while an ellipse whose eccentricity is 1 is a line segment). The resulting PGD is a histogram describing the:

$$\text{Frequency (\%)} = \frac{\text{Number of pores having eccentricity } e}{\text{Total number of pores}} \tag{30}$$

10.4.4 MODELING ABSORBENCY AND LIQUID RETENTION

Liquid absorbency (or liquid absorption capacity), C, is defined as the weight of the liquid absorbed at equilibrium by a unit weight of nonwoven fabric. Thus, liquid absorbency is based on determining the total interstitial space available for holding fluid per unit dry mass of fiber. The equation is shown as follows [145]:

$$C = A\frac{T}{W_f} - \frac{1}{\rho_f} + (1-\alpha)\frac{V_d}{W_f} \tag{31}$$

where, A is the area of the fabric, T is the thickness of the fabric, W_f is the mass of the dry fabric, ρ_f is the density of the dry fiber, V_d is the amount of fluid diffused into the structure of the fibers and a is the ratio of increase in volume of a fiber upon wetting to the volume of fluid diffused into the fiber. In the above equation, the second term is negligible compared to the first term, and the third term is nearly zero if a fiber is assumed to swell strictly by replacement of fiber volume with fluid volume [146]. Thus, the dominant factor that controls the fabric absorbent capacity is the fabric thickness per unit mass on a dry basis (T/W_f).

10.5 CONCLUSION

Nonwovens are porous fiber assemblies that are engineered to meet the technical requirements of numerous industrial, medical and consumer products. In this review, we briefly summarized the properties of nonwoven fabrics that govern their suitability for use in various applications. Several techniques were employed to characterize the structure and properties of nonwoven fabrics. Also, we reviewed models of nonwoven fabric structure and introduce examples of analytical models that link nonwoven fabric structure and properties. We concluded that analytical models are helpful in providing insights in to proposed mechanisms or interactions; they can show whether a mechanism is at least theoretically feasible and help to suggest experiments that might further elucidate and discriminate the influence of individual variables on fabric properties. Therefore, understanding and estimation of nonwoven behavior are very important for manufacturers where the reduction of production costs is of strategic imperative to stay in business.

KEYWORDS

- **characterization**
- **modeling**
- **nonwoven structure**
- **paper**
- **polymer**
- **textiles**

REFERENCES

1. Arthur, Drelich (1998). A Nonwoven Classification, a Simple System, Nonwovens Industry, October, 54–55.
2. http://www.inda.org/about/nonwoven.html.
3. EDANA (2004) Nonwoven Statistics.
4. ISO 9092:1988. BS EN 29092: (1992).
5. The Nonwoven Fabrics Handbook, Association of the Nonwoven Fabrics Industry, Cary, NC.
6. Handbook of Nonwovens, Russell, S. J., CRC Press LLC, 6000 Broken Sound Parkway, NW, Suite 300, Boca Raton, USA (2007).
7. Lunenscholss, J., Albrecht, J., & Vliesstoffe, W. (1982). Georg Thieme Verlag, Stuttgart, NY.
8. Principles of Nonwovens (1993). INDA, Cary, NC.
9. Pound, W. H. (2001). Real World Uniformity Measurement in Nonwoven Cover stock, Interval Nonwovens Journal, *10(1)*, 35–39.
10. Huang, X., & Bresee, R. R. (1993). Characterizing Nonwoven web Structure Using Image Analysis Techniques, Part III Web Uniformity Analysis, Interval Nonwovens J, *5(3)*, 28–38.
11. Aggarwal, R. K., Kennon, W. R., & Porat, I. (1992). A Scanned laser Technique for Monitoring Fibrous Webs and Nonwoven Fabrics, J Text Inst, *83(3)*, 386–398.
12. Boeckerman, P. A. (1992). Meeting the Special Requirements for On-line Basis Weight Measurement of Lightweight Nonwoven Fabrics, Tappi J, *75(12)*, 166–172.
13. Chhabra, R. (2003). Nonwoven Uniformity Measurements using Image Analysis, Intl, Nonwovens J, *12(1)*, 43–50.

14. Scharcanski, J., & Dodson, C. T. (1996). Texture Analysis for Estimating Spatial Variability and Anisotropy in Planar Stochastic Structures, Optical Engineering, *35(08)*, 2302–2309.
15. Gilmore, T., Davis, H., & Mi, Z. (1993). Tomographic Approaches to Nonwovens Structure Definition, National Textile Center Annual Report, USA, September.
16. Mao, N., & Russell, S. J. (2003). Modeling of Permeability in Homogeneous Three Dimensional Non Woven Fabrics, Text Resistance J, *91*, 243–258.
17. Petterson, D. R. (1958). The Mechanics of Nonwoven Fabrics, Science D, Thesis, MIT, Cambridge, MA.
18. Hansen, S. M. (1993). Non woven Engineering Principles, (Turbak, A. F. Ed.) Nonwoven Theory, Process, Performance & Testing, Tappi Press, Atlanta.
19. Groitzsch, D. Ultrafine Microfiber Spun bond for Hygiene and Medical Application, http://www.technica.net/NT/NT2/eedana.htm.
20. ASTM D5729–5797. Standard Test Method for Thickness of Nonwoven Fabrics.
21. ASTM D5736–5795. Standard Test Method for Thickness of High loft Nonwoven Fabrics.
22. Chen, H. J., & Huang, D. K. (1999). Online Measurement of Non Woven weight Evenness Using Optical Methods, ACT Paper.
23. Hunter Lab Color Scale, http://www.hunterlab.com/appnotes/an08_96 a.pdf.
24. Chhabra, R. (2003). Nonwoven Uniformity Measurements using Image Analysis, Interval, Nonwovens J, *12(1)*, 43–50.
25. Petterson, D. R. (1958). The Mechanics of Non woven Fabrics, Science D, Thesis, MIT, Cambridge, MA.
26. Hearle, J. W. S., & Stevenson, P. J. (1963). Nonwoven Fabric Studies, Part 3 The Anisotropy of Nonwoven Fabrics, Text Resistance Journal, *33*, 877–888.
27. Hearle, J. W. S., & Ozsanlav, V. (1979). Nonwoven Fabric Studies, Part 5 Studies of Adhesive Bonded Nonwoven Fabrics, Journal Text Institute, *70*, 487–497.
28. Chuleigh, P. W. (1983). Image Formations by Fibers and Fibers Assemblies, Text Resistance Journal, *54*, 813.
29. Kallmes, O. J. (1969). Techniques for Determining the Fiber Orientations Distributions Throughout the Thickness of a Sheet, TAPPI, No *52*, 482–485.
30. Votava, A. (1982). Practical Method-Measuring Paper A Symmetry Regarding Fiber Orientation Tappi J, *65(67)*.
31. Cowan, W. F., & Cowdrey, E. J. K. (1973). Evaluations of Paper Strength Components by Short Span Tensile Analysis, Tappi J, *57(2)*, 90.
32. Stenemur, B. (1992). Method and Device for Monitoring Fiber Orientation Distributions based On Light Diffraction Phenomenon, Interval Nonwovens J, *4*, 42–45.
33. Comparative Degree of Preferred Orientation in Nineteen Wood Pulps as Evaluated From X-Ray Diffraction Patterns (1950). Tappi J, *33*, 384.
34. Prud'homme, B. et al. (1975). Determinations of Fiber Orientation of Cellulosic Samples by X-Ray Diffraction, Journal Polymer Science, *19*, 2609.
35. Osaki, S. (1989). Dielectric Anisotropy of Nonwoven Fabrics by using the Microwave Method, Tappi J, *72*, 171.
36. Lee, S. (1989). Effect of Fiber Orientations on Thermal Radiations in Fibrous Media, Journal heat Mass Transfer, *32(2)*, 311.
37. McGee, S. H., & McCullough, R. L. (1983). Characterizations of Fiber Orientation in short Fiber, Composites, Journal Applied Physics, *55(1)*, 1394.
38. Orchard, G. A. (1953). The Measurements of Fiber Orientation in card Webs, Journal of Textile Institute, *44*, T380.
39. Tsai, P. P., & Bresse, R. R. (1991). Fiber orientation Distribution from Electrical Measurements, Part 1, Theory, Interval Nonwovens J, *3(3)*, 36.

40. Tsai, P. P., & Bresse, R. R. (1991). Fiber Orientation Distribution from Electrical Measurements, Part 2, Instrument and Experimental Measurements, Int Nonwovens J, *3(4)*, 32.

41. Chaudhray, M. M. (1972). MSC Dissertations University of Manchester.

42. Judge, S. M. (1973). MSC Dissertations University of Manchester.

43. Huang, X. C., & Bressee, R. R. (1993). Characterizing Nonwoven Web Structure Using Image Analyzing Techniques, Part 2: Fiber Orientation Analysis in Thin Webs, Interval Nonwovens J, *2*, 14–21.

44. Pourdeyhimi, B., & Nayernouri, A. (1993). Assessing Fiber Orientation in Non woven Fabrics, INDA J, Nonwoven Resistance, *5*, 29–36.

45. Pouredyhimi, B., & Xu, B. (1993). Characterizing Pore Size in Nonwoven Fabrics, Shape Considerations, Int Nonwoven J, *6(1)*, 26–30.

46. Gong, R. H., & Newton, A. (1996). Image Analysis Techniques Part II the Measurement of Fiber Orientation in Nonwoven Fabrics, Text Resistance Journal, *87*, 371.

47. Britton, P. N., Sampson, A. J., Jr Elliot, C. F., Grabben, H. W., & Gettys, W. E. (1983). Computer Simulation of the Technical Properties of Nonwoven Fabrics, Part 1 the Method, Text Resistance J, *53*, 363–368.

48. Grindstaff, T. H., & Hansen, S. M. (1986). Computer Model for Predicting Point-Bonded Nonwoven Fabric Strength, Part 1: Text Res J, *56*, 383–388.

49. Jirsak, O., Lukas, D., & Charrat, R. (1993). A Two-Dimensional Model of Mechanical Properties of Textiles, J. Text. Inst., *84*, 1–14.

50. Xu, B., & Ting, Y. (1995). Measuring Structural Characteristics of Fiber Segments in Non Woven Fabrics, Textile Res. J., *65*, 41–48.

51. Pourdeyhimi, B., Dent, R., & Davis, H. (1997). Measuring Fiber Orientation in Nonwovens, Part 3: Fourier Transform, Text Res. J., *67*, 143–151.

52. Pourdeyhimi, B., Ramanathan, R., & Dent, R., (1996). Measuring Fiber Orientation in Non Woven, Part 2: Direct Tracking, Text Resistance J., *66*, 747–753.

53. Pourdeyhimi, B., Ramanathan, R., & Dent, R., (1996). Measuring Fiber Orientation in Nonwovens, Part 1: Simulation, Text Resistance J, *66*, 713–722.

54. Pourdeyhimi, B., & Dent, R. (1997). Measuring Fiber Orientation in Nonwovens, Part 4 Flow Field Analysis, Text Resistance J, *67*, 181–187,

55. Komori, T., & Makishima, K. (1977). Number of Fiber-to-Fiber Contacts in General Fiber Assemblies, Text Res J, *47*, 13–17.

56. Gilmore, T., Davis, H., & Mi, Z. (1993). Tomo Graphic Approaches to Nonwovens Structure Definition, National Textile Center Annual Report, USA, September.

57. Manual of Quant met 570, Leica Microsystems Imaging Solutions (1993) Cambridge, UK.

58. BS 1902–1903.8, Determination of Bulk Density, True Porosity and Apparent Porosity of Dense Shaped Products (Method 1902–2308).

59. BS EN 993–1001: (1995), BS 1902–1903: 8: (1995) Methods of Test for Dense Shaped Refractory Products, Determination of Bulk Density, Apparent Porosity and True Porosity.

60. Bhatia, S. K., & Smith, J. L. (1995). Application of the Bubble Point Method to the Characterization of the Pore Size Distribution of Geo Textile, Geo Tech Test J, *18(1)*, 94–105.

61. Bhatia, S. K., & Smith, J. L. (1996). Geo Textile Characterization and Pore Size Distribution, Part II A Review of Test Methods and Results, Geo synthetically Interval, *3(2)*, 155–180.

62. Good, R. J. (1992). Contact Angle, Wetting, and Adhesion, a Critical Review, J Adhesion Science Technology, *6(12)*, 1269–1302.

63. Olderman, G. M. (1984). Liquid Repellency and Surgical Fabric Barrier Properties, Engineering Medicine, *13(1)*, 35–43.

64. Block S. S. (1996). Disinfection, Sterilization, and Preservation, 4th ed, Lea & Febiger, Philadelphia, PA.

65. Weast, R. C., Astle, M. J., & Beyer, W. H. (Eds.) (1988). CRC Handbook of Chemistry and Physics, 69th ed., CRC Press, Boca Raton, FL.
66. Lentner, C. Ed. (1981). Geigy Scientific Tables, Volume 1, Units of Measurement, Body Fluids, Composition of the Body, Nutrition, Medical Education Division, Ciba-Geigy Corporation, West Caldwell, NJ.
67. Lentner, C. Ed. (1984). Geigy Scientific Tables, Volume 3, Physical Chemistry, Composition of Blood, Hematology, Somato metric Data, Medical Education Division, Ciba-Geigy Corporation, West Caldwell, NJ.
68. Internal Research, Gore, W. L., & Associates, Inclusive, Elkton, MD.
69. Behera, B. K. "Image Processing in Textiles" Textile Institute.
70. Haines, W. B. (1930). J Agriculture Science 20, 97–116.
71. Washburn, E. (1921). The Dynamics of Capillary Flow, Physics Review, 17(3), 273–283.
72. ASTM E 1294–1389 Standard Test Method for Pore Size Characteristics of Membrane Filters Using Automated Liquid Porosimetry.
73. Miller, B., Tyomkin, I., & Wehner, J. A. (1986). Quantifying the Porous Structure of Fabrics for Filtration Applications, Fluid Filtration Gas 1, Raber, R. R. (Editor) ASTM Special Technical Publication 975, Proceedings of a Symposium Held in Philadelphia, Pennsylvania, USA, 97–109.
74. Miller, B., & Tyomkin, I. (1994). An Extended Range Liquid Extrusion Method for Determining Pore Size Distributions, Textile Research Journal, 56(1), 35–40.
75. http://www.triprinceton.org/instrument_sales/autoporosimeter.html
76. ISO/DIS 15901–16002, Pore size distribution and porosimetry of Materials Evaluation by Mercury Posimetry and Gas Adsorption, Part 2: Analysis of Meso-Pores and Macro-Pores by Gas Adsorption.
77. ISO/DIS 15901–15903, Pore Size Distribution and Porosity of Solid Materials by Mercury Porosimetry and Gas Adsorption, Part 3: Analyses of Micro-Pores by Gas Adsorption.
78. BS 7591–7592 Porosity and Pore Size Distribution of Materials Method of Evaluation by Gas Adsorption (1992).
78. Hearle, J. W. S., & Stevenson, P. J. (1964). Studies in Nonwoven Fabrics, Prediction of Tensile Properties, Text Resistance Journal, 34, 181–191,
79. Hearle, J. W. S., & Ozsanlav, V. (1979). Nonwoven Fabric Studies, Part 1: A Theoretical Model of Tensile Response Incorporating Binder Deformation, J Text Inst, 70, 19–28.
80. WSP 70.4: WSP70.5; WSP70.6.
81. Kissa, E. (1996). Wetting and Wicking, Text Res J 66, 660.
82. Harnett, P. R., & Mehta, P. N. (1984). A Survey and Comparison of Laboratory test Methods for Measuring Wicking, Text Resistance Journal, 54, 471–478.
83. Grewal, R. S., & Banks-Lee, P. (1999). Development of Thermal Insulation for Textile Wet Processing Machinery Using Needle Punched Nonwoven Fabrics, Interval Nonwovens J, 2, 121–129.
84. Baxter, S. (1946). The Thermal Conductivity of Textiles, Proceedings of the Physical Society, 58, 105–118,
85. Backer, S., & Patterson, D. R. (1960). Textile Research Journal, 30, 704–711.
86. Hearle, J. W. S., & Stevenson, P. J. (1968). Textile Research Journal, 38, 343–351.
87. Hearle, J. W. S., & Stevenson, P. J. (1964). Textile Research Journal, 34, 181–191.
88. Komori, T., & Makishima, K. (1977). Textile Research Journal, 47, 13–17.
89. Britton, P., & Simpson, A. J. (1983). Textile Research Journal, 53, 1–5 and 363–368.
90. Grindstaff, T. H., & Hansen, S. M. (1986). Textiles Researches Journal, 56(6) 383.
91. Gilmore, T. F., Mi, Z., & Batra, S. K. May (1993). Proceedings of the TAPPI Conference.
92. Kim, H. S., & Pourdeyhimi, B. (2001). Journal of Textile and Apparel Technology and Management, 1(4) 1–7.

93. Freeston, W. D., & Platt, M. M. (1965). Mechanics of Elastic Performance of Textile Materials Part XVI Bending Rigidity of Nonwoven Fabrics, Text Res J, *35(1)*, 48–57.
94. Darcy, H. (1856) Les Fontaines Publiques de la Ville de Dijon (Paris Victor Valmont).
95. Carman, P. C. (1956) Flow of Gases through Porous Media, Academic Press, New York.
96. Davies, C. N. (1952). The Separation of Airborne Dust and Particles Proc Instn Mech. Engrs, I. B., 185–213.
97. Piekaar, H. W., & Clarenburg, L. A. (1967). Aerosol Filters Pore Size Distribution in Fibrous filters, Chemistry Eng Sci., *22*, 1399.
98. Dent, R. W. (1976). The Air Permeability of Nonwoven Fabrics, J Text Inst, *67*, 220–223.
99. Emersleben, V. O. (1925). Das Darcysche Filtergesetz, Physikalische Zeitschrift, 26p 601.
100. Brinkman, H. C. (1948). On the Permeability of Media Consisting of Closely Packed Porous Particles, Applied Scientific Research, *A1*, 81.
101. Iberall, A. S. (1950). Permeability of Glass Wool and other Highly Porous Media, Journal Resistance National Bureau Standards, *45*, 398.
102. Happel, J. (1959). Viscous Flow Relative to Arrays of Cylinders, AIChE J., *5*, 174–177.
103. Kuwabara, S. J. (1959). The Forces Experienced by Randomly Distributed Parallel Circular Cylinder or Spheres in a Viscous Flow at Small Reynolds Numbers, Journal Physics Society Japan, 14, 527.
104. Cox, R. G. (1970). The Motion of Long Slender Bodies in a Viscous Fluid, Part 1, J Fluid Mechanics, 44, 791–810.
105. Sangani, A. S., & Acrivos, A. (1982). Slow Flow Past Periodic Arrays of Cylinders with Applications to Heat Transfer, Interval Journal Multiphase Flow, *8*, 193–206.
106. Kozeny, J. (1927) Royal Academy of Science, Vienna, Process Class 1, *136*, 271.
197. Dodson, C. T. J. (1996). Fiber Crowding, Fiber Contacts and Fiber Flocculation, Tappi J, *79(9)*, 211–216.
108. Dodson, C. T. J., & Sampson, W. W. (1999). Spatial Statistics of Stochastic Fiber Networks, J Stat Phys, *96(1–2)*, 447–458.
109. Wrotnowski, A. C. (1962). Nonwoven Filter Media, Chemical Engineering Progress, *58(12)*, 61–67.
110. Wrotnowski, A. C. (1968). Felt Filter Media, Filtration and Separation, September/October, 426–431.
111. Goeminne, H. (1974). The Geometrical and Filtration Characteristics of Metal-Fiber Filters Filtration and Separation (August), 350–355.
112. Rollin, A. L., Denis, R., Estaque, L., & Masounave, J. (1982). Hydraulic Behaviour of Synthetic Nonwoven Filter Fabrics, the Canadian Journal of Chemical Engineering, 60, 226–234.
113. Giroud, J. P. (1996). Granular Filters and Geo textile Filters, Process, Geo-filters'96, Montréal, 565–680.
114. Lambard, G., et al. (1988). Theoreticals and Experimental Opening Size of Heat-Bonded Geotextiles, Text Resistance Journal, April, 208–217.
115. Faure, Y. H., et al., (1989). Theoretical and Experimental determination of the filtration opening size of geotextiles, 3rd International Conference on Geotextiles, Vienna, Austria, 1275–1280.
116. Faure, Y. H., Gourc, J. P., & Gendrin, P. (1990). Structural Study of Porometry and Filtration Opening Size of Geo textiles, Geo synthetics Microstructure and Performance, ASTM STP 1076, Peggs, I. D. (ed.), Philadelphia, 102–119.
117. Gourc, J. P., & Faure, Y. H. (1990). Soil Particle, Water and Fiber, a Fruitful Interaction Non Controlled, Proc., 4th Int. Conference on Geo textiles, Geo membranes and Related Products, the Hague, The Netherlands, 949–971.
118. Aydilek, A. H., Oguz, S. H., & Edil, T. B. (2005). Constriction Size of Geo textile Filters, Journal of Geotechnical and Geo environmental Engineering, *131(1)*, 28–38.

119. Fatt I. (1956). "The Network model of Porous Media I Capillary Pressure Characteristics," AIME Petroleum Transactions, *207,* 144.
120. Bakke, S., & Oren, P. E. (1996). 3d Pore-Scale Modeling of Heterogeneous Sandstone Reservoir Rocks and Quantitative Analysis of the Architecture, Geometry and Spatial Continuity of the Pore Network, in European 3D Reservoir Modeling Conference, Stavanger, Norway, SPE 35–45.
121. Blunt, M. J., & King, P. (1991). Relative Permeability's from Two and Three-Dimensional Porescale Network Modeling, Transport in Porous Media, *6,* 407–433.
122. Blunt, M. J., & Sher, H. (December 1995). Pore Network Modeling of Wetting, Physical Review E, 5263–5287.
123. Feyen, & Wiyo, K. (Ed) (1999). Modeling of Transport Processes in Soils, Wageningen Pers, the Netherlands, 153–163.
124. Jean-Fran, Ois Delerue, & Edith Perrier (2002) DX Soil, a Library for 3D Image Analysis in Soil Science, Computer & Geosciences, *28,* 1041–1050.
125. Denesuk, M., Smith, G. L., Zelinski, B. J. J., Kreidl, N. J., & Uhlmann, D. R. (1993). "Capillary Penetration of Liquid Droplets into a Porous Materials" Journal of Colloid and Interface Science, *158,* 114–120.
126. Hidajat Rastogi, A., Singh, M., & Nohanty, K. (March 2002). Transport Properties of Porous Media Reconstructed from Thin Sections, SPE Journal, 40–48.
127. Lindquist, W. B., &. Venkatarangan A. (1999). Investigating 3D Geometry of Porous Media From High Resolution Images, Physics Chemistry Earth A, *25(7),* 593–599.
128. Oren, P. E., Bakke, S., & Arntzen, O. J. (2001). Extending Predictive Capabilities to Network Models, SPE Journal (December 1998) 324–336 Yarn Torsion", Polymer Testing, *20,* 553–561.
129. Pourdeyhimi, B., Dent, R., & Davis, H. (1997). "Measurings Fiber Orientation in Nonwovens Part III Fourier Transform", Textile Research Journal, *2,* 143–151.
130. Pourdeyhimi, B., & Ramanathan, R. (1996). "Measuring Fiber Orientation in Nonwovens Part II Direct Tracking", Textile Research Journal, *12,* 747–753.
131. Pourdeyhimi, B. (1998). Reply to "Comments on Measuring Fiber Orientation in Nonwovens", Textile Research Journal, *4, 307–308,* 593–599.
132. Marmur, A. (1992). "Penetration and Displacement in Capillary Systems" Advances in Colloid and Interface Science, *39,* 13–33.
133. Marmur, A., Cohen, R. D. (1997). "Characterization of Porous Media by the Kinetics of Liquid Penetration, the Vertical Capillaries Model", Journal and Colloid and Interface Science, *189,* 299–304.
134. MeBratney, A. B., & Moran, C. J. (1994). Soil Pore Structure Modeling Using Fuzzy Random Pseudo Fractal Sets, International Working Meeting on Soil Micro Morphology, 495–506.
135. Rebenfeld, L., & Miller, B. (1995). "Using Liquid Flow to Quantify the Pore Structure of Fibrous Materials", Journal Text Inst, *2,* 241–251.
136. Sedgewick, R. (1998). Algorithms in C, 3rd ed, Addison-Wesley, 11–20.
137. Zeng, X., Vasseur, C., & Fayala, F. (2000). Modeling Micro Geometric Structures of Porous Media with a Predominant axis for Predicting Diffusive Flow in Capillaries, Applied Mathematical Modelling, 24pp 969–986.
138. Anderson, A. N., McBratney, A. B., & FitzPatrick, E. A. (1996). Soil Mass, Surface, and Spectral Fractal Dimensions Estimated from Thin Section Photographs, Soil Science Society American Journal, *60,* 962–969.
139. Perwelz, A., Mondon, P., & Caze, C. (2000). "Experimental Study of Capillary Flow in Yarns", Textile Research Journal, *70(4),* 333–339.
140. Perwelz, A., Cassetta, M., & Caze, C. (2001). "Liquid Organization during Capillary Rise in Yarns-Influence of Yarn Torsion", Polymer Testing, *20,* 553–561.

141. J Serra Image Analysis and Mathematical Morphology (1982). Academic Press, New York.
142. Otsu, N. (1979). "A Threshold Selection Method from Grey-Level Histograms", IEEE Transactions Systems, Man, and Cybernetics, *9(1)*, 62–66.
143. Sedgewick, R. (1998). Algorithms in C, 3rd ed, Addison-Wesley, 11–20.
144. Delerue, J. F., Perrier, E., Timmerman, A., Rieu, M., & Leuven, K. U. (1999). New Computer Tools to quantify 3D Porous Structures in Relation with Hydraulic Properties, In Journal Feyen & Wiyo, K. (Ed), Modeling of Transport Processes in Soils, Wageningen Pers, The Netherlands, 153–163.
145. Gupta, B. S., & Smith, D. K. (2002). Nonwovens in Absorbent Materials, Textile Science and Technol, *13*, 349–388.
146. Gupta, B. S. (1988). The Effect of Structural Factors on Absorbent Characteristics of Nonwovens, Tappi J, *71*, 147–152.

INDEX

A

Aarony-Stauffer rule, 183
Acrylonitrile–butadiene–styrene, 102
Activation energy, 128, 134
Activation-relaxation technique, 61
Aerogel, 2, 35, 36, 38–43, 45–49, 51,
    53–57, 59, 60, 65, 67, 70–73, 75, 76
    first introduced by Kistler, 35
Agricultural sciences, 2
Airglass (Sweden), 54
Air-laying method, 276
Algorithm types, 21
    Beeman's algorithm, 21
    Leap-frog algorithm, 21
    Velocity Verlet, 21, 22
    Verlet algorithm, 21, 22
Aliphatic polyisocyanate, 257
Ambient pressure drying, 41, 44
Amorphous polymer matrix, 156
Amorphous silica, 62, 72, 76
Amorphous state, 162, 174, 193, 195, 197,
    198, 203, 210, 222, 232, 236, 248
An overheated liquid to solid body
    transition, 224, 225
Analysis of variance, 265, 266
Ångströms, 151
Annealing, 51, 112, 196–198, 200, 202
Arc xenon lamp DKSSh, 129
Association of the Nonwovens Fabrics
    Industry, 273
ASTM standards, 311
Atom movement types, 10
    bending between bonds, 10
        harmonic potential, 10, 11
    rotating around bonds, 10
        torsion angle potential, 11
    stretching along the bond, 10
        harmonic potential, 10, 11
Attractive force, 7, 28

arises from fluctuations in charge
    distribution, 7
    in the electron clouds, 7
Avogadro number, 153, 166
Aza-substitution, 126, 127, 131–133, 142,
    144, 145, 147

B

Backbone characteristics, 46
Backbone elements, 37
Baily Hirsh relationship, 215
Bathochromic shift, 132, 133
Batt bonding by threads, 286
Batt looping, 288
Beam-bending method, 49
Beeman's algorithm, 21
Beest–Kramer-van Santeen potential, 66
Bend angles, 12
Beta rays, 299
Bhattacharya and Kieffer, 64, 65
Bhavnagar-Gross-Krook collision model, 33
Billiard balls, 68
Biodegradable polymer films, 84, 88
Biodestruction process, 87
Biomolecules, 27
    docking of molecules, 27
    membranes, 27
    micelles, 27
    protein folding, 27
    structure and dynamics of proteins, 27
Boltzmann equation, 5, 33, 34
Born–Mayer empirical potential, 65
Born–Mayer form, 66
Born–Meyer–Huggins potential, 66
Brittle-ductile transition, 239, 240
Bulk polymer, 99
Burgers vector, 215, 243, 244

C

C.I. disperse blue 56, 264
Capillary channel theory, 315, 316
Carbon nanotubes, 96, 108
Carding method, 276
Cartesian coordinates, 6
Cell's effect, 225
Central composite design, 264, 265
Cerenkov detector applications and research
    projects, 54
Characterization of aerogels, 2
CHARMM force field, 11
    improper dihedral term, 11
    Urey-Bradley term, 11
CHARMM potential function, 7
Chemical bonding, 283
Chemical modification, 84, 262
Cherenkov counters, 36
Cherenkov radiators, 36
Chitosan, 84, 86, 87, 88
Classical trajectories, 20
    ergodic properties, 20
    mixing properties, 20
Clay nanocomposites, 98, 102, 103
Cluster model, 151, 153, 155, 162, 163,
    167, 170, 174, 186, 193, 195, 197–199,
    203, 210, 217, 220, 222, 223, 225, 228,
    232, 236, 248
Coarse graining potential model, 13
Collective behavior, 27
    coupling of translational and rotational
       motion, 27
    decay of space and time correlation
       functions, 27
    dielectric properties, 27
    orientational order, 27
    spectroscopic measurements, 27
    vibration, 13, 27, 47
Complex fluids, 8
    films, 27
    fluid interfaces, 27
    ionic liquids, 27
    liquid crystals, 27
    molecular liquids, 27
    monolayers, 27
    pure water and aqueous solutions, 27

    structure and dynamics of glasses, 27
Computational methods, 2
Computer budget, 3
Computer experiment, 4, 5
Computer simulations, 2, 3, 5, 56
Conventional data, by Polymer Laboratories
    software, 116
    heat release rate, 116
    peak of heat release, 116
    time to ignition, 116
    total heat release, 116
    weight loss, 107, 116
Coulomb interactions, 63
Coulomb potential, 9
β-cristobalite, 66, 72, 76
Cross-direction, 280, 288, 311
Cross-linked polymer, 220, 258, 262
Cryogenic drying, 41
Crystalline lattices, 210
Crystalline solids, 210, 248
    stretched chains, 153, 156, 166, 198
Crystallization process, 223
Cubic system, 71

D

D'Arcy's Law, 315
D2Q9 model,
    see, lattice Boltzmann method
D3Q19 model, 35
Darcy's law, 317
Debye regime, 48
Debye-Huckel theory, 5
Density functional theory, 57, 58, 128
Design of Experiment, 266
Differential scanning calorimetry, 106
Diffusion limited cluster aggregation
    network, 50
Diffusion-limited cluster–cluster
    aggregation, 49
Dilute gases, 5
    transport properties, 5, 27, 46, 48
Dipole–dipole interaction energy, 191
Dislocation theory, 211
Dispersion force, 7.
    See also attractive force
Dry heteroporous network, 302

Dry solid skeleton, 35
Drying process, 41, 43, 61
  ambient pressure drying, 41, 44
  cryogenic drying, 41
  supercritical drying, 35, 36, 41–43
Dynamic Monte Carlo, 60, 61

E

E-bonded, 6, 10
E-nonbonded, 7
Eigen modes, 48
Elastic modulus of silica, 75
Elastomechanical behavior, 46, 48
Electron density, 10, 126–128, 132, 134,
  135, 146, 147
Electron-electron interaction, 7
Electron microscope jem-100b, 133
Electrostatic charges, 9
Electrostatic Potential, 7
Electrostatic multipoles, distribution of, 10
Engineering, 2, 91, 181, 316
Entropic effects, 14
Epoxy coatings, 261
Epoxy-isocyanate catalyst systems, 258
Epoxy materials, 262
Epoxy oligomers, 256, 262
Epoxy resin, 256, 257, 259, 260, 261, 262
  compositions, 88, 256, 257, 259–261
Epoxy resin ED-20 as a modifier, 256
Epoxy units, 256
Epoxy-urethane oligomers, 256
EPR-spectroscopy, 172, 191
Ethylene-co-vinyl acetate, 90, 100, 101, 123
Euclidean geometry, 163
Euclidean space, 151, 202, 230, 231, 234,
  236
European Disposables and Nonwovens
  Association, 273
EVA copolymers, 91
EVA nanocomposites, 90, 102, 105, 108,
  114, 116, 117
External load (mechanical energy)
  application, 236
External mechanical stress field, 218

F

Fabric plane, 294, 296, 297, 312, 314
Fabric porosity, 292, 293, 297, 301, 318
Fabric thickness testing, 298
Fabric weight uniformity, 298
Faure's approach, 318, 319
Fermi level, 134
Feuston and Garofalini model, 64
Fiber-like clays particles, 102
  sepiolite, 90, 92, 95–97, 102, 108,
    116–118, 121
Fiber orientation distribution, 291, 292, 295,
  296, 300, 309, 313, 314
Film-forming composites, 256, 262
  composition, 46, 51, 88, 94, 117, 177,
    256, 259, 261, 262
Flash spinning method, 280
  handwriting or printing, 280
  Tyvek, 281
  very fine fibers, 280, 281, 282
Flash-spun material, 280
Flight simulation explosion simulation, 3
Fluid dynamics, 27
  boundary layers, 27
  laminar flow, 27
  rheology of non-Newtonian fluids, 27
  unstable flow, 27
Fluid transport, 320
Fluorohectorite, 111
Force-field development and refinement, 13
Fourier transform, 72, 309
Fractal analysis, 150, 152, 158, 162, 185,
  203, 229, 231, 233, 234
Fractal dimension of, 69, 71
  backbone, 37, 41–49, 51–53
  silica aerogels, 2, 35, 36, 43, 46, 47, 50,
    54, 55, 65, 68, 70–74, 76
Fractal (hausdorff) dimension, 236
Fractals, 47, 48, 156, 230
Fracton, 48
Free volume microvoids number, 229, 238
Frenkel model, 212–214
  conclusion, 155, 162, 170, 176, 213, 217,
    225, 226, 228, 231, 233, 241, 242, 246

mechanism, 47, 94, 100, 117, 134, 163, 166, 184, 204, 211–213, 215, 220, 221, 248, 322
Fundamental studies, 27
   diffusion, 3, 27, 28, 47, 49, 50, 60, 66, 109, 112, 113, 115, 117
   equilibration, 4, 27
   kinetic theory, 27, 228
   potential functions, 7, 14, 27, 28
   size dependence, 27
   tests of models, 27
   tests of molecular chaos, 27
   transport properties, 5, 27, 46, 48

G

Gamma rays, 299
GAUSSIAN 03 program, 135
Gaussian distribution, 19
Gaussian random fields, 59
Gel aging, 61
Gel backbone, 37, 43, 44, 48, 52, 53
Gel drying process, 61
Gelation step, 61
GelSil 200 material, 60
Gibbs Ensemble Monte Carlo technique, 62
   binary mixtures, 62
Gibbs free energy, 14
Gibbs function, 150, 151, 223
Gibbs specific function, 157, 225
β-glycoside bond break, 87
Goeminne's equation, 316
Gourc and Faure technique, 318
Ground-state spin, 145, 146

H

Hagen–Poiseuille equation, 316
Hamiltonian system, 17, 20
Harmonic crystalline solids, 5
Harmonic potential, 10, 11
Hartree-Fock method, 128, 136
Hashin–Shtrikman principle, 45
Hausdorff dimension, 157
Heat barriers, 45
Heat release rate, 116
Hectorite, 111

Helmholtz free energy, 29
Hexamethyldisilazane, 43
Hierarchical systems behavior, 222
High pressure simulation wind channel simulation, 3
High elasticity theory, 153, 154, 220, 237
High resolution crystal structures, 11
Highest occupied state, 140
Homogeneous fractal, 156
Hilliard-Komori-Makishima theory, 300
Hough transform, 309
Hopping charge-transfer mechanism, 134
Hybrid polymer systems, 162, 204
Hydraulic radius model, 315
Hydro-entanglement, 290
Hydrolysis of the alcoxysilane, 38
Hyperbranched polymer, 264, 265, 269
Hypochromic shift, 126

I

Ideal gas, 5
Improper dihedral term, 11
Inelastic deformation process, 218, 231
Infra-red light, 299
Inorganic material, 100
Institutional Research Development Program, 118
Integration algorithms, 19, 21
Interaction energy, 7, 8, 28–31, 191, 192, 221
Interatomic potential, 2, 16, 56, 62, 66
Intercommunication, 150, 151, 157, 158, 191, 222, 241
Interfacial polymer, 99
Ion exchange media, 45
Ionization potential, 128, 133
IP (clusters), 218, 219
IR spectroscopy
   gas phase, 11, 42, 44
Isocyanates, 256, 257
Isoelectrical point, 38
   point of zero charge, 38
ISO standards, 298, 311
Isothermal-isobaric conditions, 14
   constant pressure, 14, 66
   constant system size, 14

constant temperature, 14, 61
IUPAC, 77, 102

J

Jetting process, 289

K

Kantor's set (dust), 230
Kerner, 176–180, 185, 188, 194
Kerner equation, 177–180, 188, 194

L

Laboratory, 3, 63, 74
LAMMPS software, 68
Langevin dynamics, 67, 74
Langevin equation, 57
Laplace equation, 302, 303
Lattice Boltzmann method, 32, 33
    D2Q9 model, 33, 34
    quiescent state, 32, 33
    two-dimensional lattice model, 32
Lattice models, 5, 238
    two-dimensional Ising model, 5
Lattice nodes number, 238
Law of Inertia, 16
    see, Newton's law
Layered double hydroxides, 102
Leap-frog algorithm, 21
Lennard-Jones potential, 9
    24–6 potential, 68
    6–12 potential, 8, 9
Ligand bond ionicity, 126, 127
Liquid 1 to liquid 2 transition, 165
Liquid porosimetry, 301
    liquid porometry, 301
Liquid–vapor interface, 53
Liraza agent, 85
Loading cycles, 221
    first, 221
    second, 221
    third, 221
Local order regions (clusters), 154, 162, 229
Local order thermofluctuational effect, 156

Loosely packed matrix, 154, 155, 162,
    163, 167, 172–174, 177, 178, 180–182,
    184–191, 195, 198, 216–221, 223, 224,
    232, 236, 237, 239, 244
Low density polyethylene, 84, 88
    Powder, 51, 84, 86, 87
Low-density silica aerogel granules, 43
Lowest free state, 140

M

Machine direction, 280, 287–289, 294, 296,
    311, 313
Macromolecular entanglements cluster
    network density, 154, 200
Macroscopic, 2, 3, 5, 14, 15, 23, 32, 33, 35,
    44–47, 53, 167, 169, 171, 213, 218, 219,
    240, 246
    density, 33
    velocities, 32
    world, 2, 3
    yield stress, 218
Magadiite, 111
Mars Exploration rovers (2003), 36
Matching method, 238
Materials, 2, 4–6, 15, 16, 36, 38, 45, 47, 51,
    54, 55, 59, 61, 63, 77, 84, 86, 87, 90–93,
    95–98, 100–102, 104, 107, 109, 117, 162,
    177, 195, 200, 203, 210, 215, 222, 229,
    256, 262, 272, 273, 275, 290, 297, 300,
    302–304, 310, 311, 316, 319, 320
Material science, 150
Maxwell-Boltzmann, 19
MD method, 6, 23, 28, 32
Measuring instrument, 4
    nometer, 4
    thermometer, 4
    viscosimeter, 4
Mechanical and chemical properties, 256,
    262
Mechanical bonding, 285, 296
    batt bonding by threads, 286
        warp-knitting machine, 286
    batt looping, 288
    hydro-entanglement, 290
    needle felting, 285
        needleboard, 285

needleloom, 285, 289, 290
stitch bonding in pile fabric, 287
stitch bonding without threads, 287
stitch bonding, 272, 285, 286
  Czechoslovakia, 286
swing laid yarns, 288
  jetting process, 289
Mechanical film properties, 85
elongation, 85, 88, 100, 108, 297
tensile strength, 85, 88, 107, 115, 116,
  281, 297, 299
Medical, 2, 275, 321
Melt blended polymer composite, 100
enthalpic loss, 100
  unfavorable polymer clay interactions,
  100
entropic gain, 100
  increased polymer mobility over
  confinement, 100
Meltblown process, 275
Melt flow index, 112
Melt peak, 105, 106
Melt processing, 102
Meniscus, 53, 302
Metropolis method, 29, 30
Microcomposite models, 179
Micropores in carbon aerogels, 53
Microporous membrane, 303
Microscopic, 2, 3, 6, 14, 15, 23, 28, 29–31,
  60, 75
scale, 2
states, 15, 28–30
Microwave spectroscopy data, 11
Mie equation, 221
Miller and Tyomkin instrumentation, 303
Mimetic simulations, 2, 57
Miscibility diagram, 36
ethanol, 36, 38, 42, 44
tetraethoxysilane system, 36
water, 27, 36, 38, 43, 44, 58, 61–64, 76,
  84–87, 260–262, 277, 278, 281, 283,
  284, 289, 290, 292, 301, 310–312
Mittag-Lefelvre function, 231, 232
Model systems, 4
essential physics, 4, 8
explore consequences of, 4
simple model, test theory using, 4

Modeling structure, 58
atomistics, 58
  quantum-mechanical potential, 57, 58
coarse-grained descriptions, 57
Mmolecular dynamics method, 2, 65
Molecular mechanics force fields, 13
Molecular science, 158
Molecular simulations, 14
Molecular system, 9
Monsanto, 54
Monte Carlo, 2, 3, 28, 60–62, 75, 80
Montmorillonite, 92, 96, 98, 100, 101, 111,
  116, 185, 186, 193–195
Mozambican Research Foundation, 118
Multiphase (multicomponent) systems, 180

N

N-body system, 14
Nanocoatings, 97
Nanolayers, 97
Nanopore Inc, 54
Nanoscale fillers, 98
Nanostructured materials, 90–92, 95, 117
one-dimensional nanostructured
  materials, 90
two-dimensional nanostructured
  materials, 90
zero-dimensional nanostructured
  materials, 90
National Research Foundation, 118
Natural composites, 162, 204
Natural polysaccharide, 84
Needle felting, 285
Needle-punched fabric, 294, 315
Neighbor list, 25
Newton's equations, 4, 6, 19, 57, 60
Newton's law, 16
first law, 16
  law of Inertia, 16
motions, 48
second law, 16
third law, 16
Nitrogen atoms, 126, 127, 132
NMR spectroscopy, 19
Nometer, 4
Nonbonded interactions, 7, 8, 68

in CHARMM potential function, 7
  electrostatic interaction energy, 7
  Van der Waals interaction energy, 7
  energy, 7
Nonwoven, 271–275, 290, 292–294, 298,
  307, 311, 322–326
Nonwoven applications, 274
  industry, 100, 104, 260, 261, 264, 272,
    276, 278, 282, 283, 290, 293
  materials, 2–6, 15, 16, 36, 38, 45, 47,
    51, 54, 55, 59, 61, 63, 77, 84, 86, 87,
    90–93, 95–98, 100–102, 104, 107, 109,
    117, 162, 177, 195, 200, 203, 210, 215,
    222, 229, 256, 262, 272, 273, 275, 290,
    297, 300, 302–304, 310, 311, 316, 319,
    320
Nonwoven products, 274
  agriculture and landscaping, 274
  automotive, 91, 275
  clothing, 275, 281, 282, 289
  construction, 56, 135, 175, 212, 274, 280
  geotextiles, 301, 318, 319, 326
  health care, 275
  home furnishings, 275
  household, 274
  industrial/military, 275
  leisure, travel, 275
  personal care and hygiene, 275
  school and office, 275
Nonwoven steps (manufacturing), 275
  actual web formation, 275
  bonding the web fibers together, 275
Nonwoven structure, 273, 292, 295–297,
  314, 316, 322
Nuclear magnetic resonance, 73

O

Oligomeric molecules, 256
Oligomerization, 59, 61–64, 76
One-dimensional euclidean space, 230
One-dimensional nanostructured materials,
  90
One-step process, 39
Organic polymer, 100
Organic semiconductors, 126, 147
Orientation distribution function, 295, 309

Overheated liquid to solid body transition,
  224, 225

P

Pair distribution function, 68, 76
Partial melting, 227–229, 248
Particle traps, 45
Pathfinder, 55
Peak of heat release, 116
PEC properties, 128
Pendulum device, 258
Performance windows, 45
Periodic boundary conditions, 24
Pharmaceutical, 2
Phase diagram of co_2, 42
Phase transition, 27
  critical phenomena, 27
  order parameters, 27
  phase coexistence, 13, 27
Photoactive maxima, 128
Photoactivity of pigments, 132
Photoelectrochemical characteristics, 126,
  127
Phthalocyanine films, 132
Phthalocyanines, 127
Physics-chemistry of polymers, 150
Pigments, 126–130, 132, 133, 147
Plastics, 91, 104, 275
Poisson's ratio, 177, 178, 200, 214, 216
Poissonian polyhedra model, 318
Polyamide 6 (PA6), 102
Polyarylate, 150, 151, 154, 165, 193, 195
Polycarbonate, 150, 151, 154, 163, 193, 203
Polyethylenes, 155, 223, 246, 248
Poly (ethylene terephthalate), 224, 264
Polyhedral oligomeric silsesquioxane, 90,
  92, 93, 101
Polymer chains, 13, 101, 104, 108, 109,
  111, 112, 114–116, 163, 191, 259
Polymer laboratories software, 116
Polymer-clay nanocomposites, 103
Polymer films basing, 84
Polymeric cluster, 150, 158
Polymer-like clusters, 37
Polymer melt, 13
Polymer-surface interaction, 102

Polymers, 27
    chains, 27
    equilibrium conformation, 27
    relaxation process
    rings and branched molecules, 27
    transport process, 27
Polymers yielding, 197, 211, 212, 223, 229, 241, 248
Poly (methyl methacrylate), 173, 230
Poly oxazolidones, 256
Polypiromellithimide structure, 233
Polypropylene, 102, 107, 185, 186, 223, 242, 264, 279
Polytetrafluoroethylene, 214, 223
Polyurethane, 102, 256, 262
Pores geometry distribution, 321
Pore volume distributions, 302
Porosity, 37, 39, 40, 45, 46, 47, 51, 53, 62, 67, 73, 93, 260, 275, 292, 293, 297, 301, 310, 316–319
    measurement, 299, 300
Porphyrins, 127, 147
Positrons annihilation method, 226
POSS nanocomposites, 99, 100, 106, 107
Powder, 51, 84, 86, 87
Power law, 66, 69, 72, 74, 76, 164
Predictions, 3, 5, 316
Pt-znpc electrode, 128
Pure science, 2

Q

Quantum-chemical calculation, 10
Quantum-chemical calculation, 126, 128
Quantum-mechanical potentials, 57
Quartz resonator, 129
Quasistatic tensile tests, 242
Quaternary ammonium salts, 102

R

Rahman at Argonne, 6
Raman spectroscopy, 11
Ramsey's theorem, 157
Redox potential, 128, 132, 134
Reformation process, 163
Relative fiber motion, 315
    complete freedom, 314
    no freedom, 314, 315
Renyi dimensions, 157
Repulsive force, 7, 9, 28
    arises at short distances, 7
        electron-electron interaction, 7
Response surface methodology, 264, 269
Reverse nonequilibrium md, 72, 73
Ring currents, 126, 133, 147
RSM model, 269
Rubber-like state, 165, 237
Rubbers, 91, 231, 237, 248
Rutan approach, 135

S

Scaling behavior, 47, 49
Scanning electron microscopy, 45
Science and technology, 2, 90
Secondary hydroxyl groups of epoxy-oligomers, 258
Semicrystalline polymers, 155, 156, 162, 213, 219, 223, 224, 242, 243, 245, 246
Sepiolite, 90, 92, 95–97, 102, 108, 116–118
Sequential aza-substitution, 126, 127, 132, 133, 142
Shear model, 214
    Frenkel model, 212, 213, 214
Shear modulus, 154, 178, 179, 202, 211, 218, 242, 245, 314
Shear-thinning behavior, 115
Shear thinning exponent, 115, 116
Shear yield stress, 244
SHG1 device, 258
Short-range repulsive force, 9
Shrinking, 44
Silanol groups, 96, 108
Silica aerogels, 2, 35, 36, 43, 46, 47, 50, 54, 55, 65, 68, 70–74, 76
Silicate platelets, 108–110, 113, 115
Silsesquioxanes, 94
Sir william rowan hamilton, 18
Sliding plane, 213
Sol–gel processing, 36, 58, 62, 64
Solid structure, 40, 248
    defect formation and migration, 27

elastic and plastic mechanical properties, 27
epitaxial growth, 27
fracture, 27
friction, 27
grain boundaries, 27
molecular crystals, 27
radiation damage, 6, 27
shock waves, 27
structural transformations, 27
Solvated protein-DNA complex, 15
Soret band, 126, 147
Soules potential, 66
Spun laying method, 279
STARDUST mission, 55
Statistical mechanics, 2, 14, 24, 26, 29
Steven Kistler, 38
Stitch bonding, 272, 285, 286
pile fabric, 287
without threads, 287
Stretching along the bond, 10
Structureless (defect-free) polymers, 162, 203
Structureless liquid, 165
Structureless polymers, 162, 203
Styrene–butadiene latex, 283
Super capacitors, 45
Supercritical drying, 35, 36, 41, 42, 43
Surfactant-surface interaction, 102
Swing laid yarns, 288
Synergetics principles, 163, 166, 170
Synthetic polymer, 84, 272

T

Takayanagi model, 177
Taylor series expansion, 21
TBP films, 132
TBP molecule, 126, 127
Tersoff potential, 68, 72, 73, 76
Tetrabenzoporphyrin, 126, 127, 129, 131, 132, 140–144, 146
Tetraethoxysilane, 36
Tetraethyl orthosilane, 38
Tetramethyl ortho silane, 38
Tetrapyrrole compounds, 127, 134, 147
Textile fibers, 264, 272, 277, 278

Textile technologies, 290
The North American Nonwovens, 290
Theoretical physics, 150, 158
Theoretical shear strength of crystals, 211
Thermal bonding, 283, 284
Thermal conductivity, 36, 37, 45, 51, 53, 68, 72, 73, 76, 96, 312, 313
Thermal insulate, 77
Thermodynamic equilibrium, 2, 28, 30
Thermodynamic state, 15
heat capacity, 3
heat of adsorption, 3
pressure, 3, 4
properties, 2–6, 13–16
structure, 2, 3, 11, 19
Thermogravimetric analysis, 107
Thermometer, 4
Thermophysical properties, 13
Thermoplastic elastomers, 91
Thetorsional potential, 13
Tie chains, 155
Time dependence, 26
Time dependent (kinetic) phenomenon, 15
Time to ignition, 116
Torsional potential, 9
Torsion angle potential, 11
Torsion angle potential function, 11
Torsion angles, 12
Total heat release, 116
TPC macrocycle structure, 126, 127
Transmission electron microscopy, 109
Transport, 3, 5, 26, 27, 44, 46, 48, 51, 72, 93, 292, 293, 297, 298, 301, 312, 320
diffusion coefficient, 3
viscosity, 3
Trimethylchlorosilane, 43
Triplets of atoms (three-body component), 62
Twin-screw extruder, 110
Two-dimensional ising model for ferromagnets, 5
Two-dimensional nanostructured clay platelet, 98
Two-dimensional nanostructured materials, 90
Two-step sol–gel process, 39
condensation step, 39

hydrolysis step, 39
Tyvek, 281

U

Ultrasound probes, 45
Urethane fragments, 256
Urey-Bradley term, 11

V

Van der waals, 5, 7, 8
Van der waals interaction, 7, 8
  between two atoms arises from, 7
    a balance between repulsive and
      attractive forces, 7
Van der waals potential, 7
Varian Mini-TASK, 129
Varnish to water, 260
Velocity autocorrelation function, 3, 72
Velocity–Verlet algorithm, 67
Verlet algorithm, 20–23, 67
Vibration frequencies, 13
Vineyard at brookhaven, 6
Vinyl acetate, 90, 91, 100, 101, 111, 123, 283
Violation (interruption), 210
Viscosimeter, 4

W

WCA model, 9
weight loss, 107, 116

wet-laid process, 275, 277
wetting angle, 291, 292
Witten-Sander clusters, 182
Wrotnowski's model, 316, 317

X

Xerogels, 2, 36, 39, 40, 45, 51, 56, 61, 66, 73, 76
Xerogel structures, 39
X-ray crystal structure, 19
X-ray diffraction analysis, 109, 300
X-ray diffraction patterns, 300
  POSS, 103–107
X-raying methods, 211

Y

Yarn spinning stage, 272
Yielding process, 211, 212, 217–219, 223, 225, 226, 228, 229, 236, 241, 245, 246, 248
Yield tooth, 196, 220, 233, 234, 248
  disappearance, 196
Young's modulus, 49–53, 74, 92, 108, 114, 116, 177, 291, 292, 315

Z

Zero-dimensional nanostructured materials, 90
Zero-shear viscosity, 115, 116

Milton Keynes UK
Ingram Content Group UK Ltd.
UKHW031142141024
449569UK00024B/1139